Alexander Mehler, Reinhard Köhler

Aspects of Automatic Text Analysis

Studies in Fuzziness and Soft Computing, Volume 209

Editor-in-chief
Prof. Janusz Kacprzyk
Systems Research Institute
Polish Academy of Sciences
ul. Newelska 6
01-447 Warsaw
Poland
E-mail: kacprzyk@ibspan.waw.pl

Alexander Mehler
Reinhard Köhler

Aspects of Automatic Text Analysis

 Springer

Alexander Mehler
Universität Bielefeld
Fakultät für Linguistik und
Literaturwissenschaft
Postfach 10 01 31
33501 Bielefeld, Germany
E-mail: Alexander.Mehler@uni-bielefeld.de

Reinhard Köhler
Universität Trier, FB II
Linguistische
Datenverarbeitung, Computerlinguistik
Universitätsring 15
54286 Trier, Germany
E-mail: koehler@uni-trier.de

ISSN print edition: 1434-9922
ISSN electronic edition: 1860-0808

ISBN 978-3-642-07225-3
e-ISBN 978-3-540-37522-7

Springer is a part of Springer Science+Business Media
springer.com
© Springer-Verlag Berlin Heidelberg 2007
Softcover reprint of the hardcover 1st edition 2007

Cover design: Erich Kirchner, Heidelberg

In honor of Professor Burghard B. Rieger
on the occasion of his 65th anniversary.

Contents

Part III Quantitative Linguistic Modeling

Part IV Corpus Linguistic and Text Technological Modeling

Part V Text Categorization and Classification

Part VI Cognitive Modeling

Part VII Visual Systems Modeling

Appendix

Introduction:
Machine Learning in a Semiotic Perspective*

Alexander Mehler[1] and Reinhard Köhler[2]

[1] Bielefeld University
`Alexander.Mehler@uni-bielefeld.de`
[2] University of Trier
`koehler@uni-trier.de`

1 Introduction

In order to introduce vagueness as a proper object of formal-mathematical modeling, Max Black [5] developed the notion of *consistency profile*. Other than classical logics constrained by the *principium exclusi tertii*, consistency profiles allow mapping the transition from negative to positive predication to any degree. This enabled Black to provide a framework for the classification of predicates according to their vagueness and ambiguity. In other words: *Max Black offered a first approach to distinguishing both types of informational uncertainty in, nevertheless, precise, mathematical terms.*

It was up to Zadeh's seminal work on *fuzzy set theory, linguistic variables,* and *generalized constraints* [82, 83] to provide a comprehensive mathematical framework for modeling informational uncertainty in general. The universality of Zadeh's approach is due to his encompassing view of cognitive information processing which includes the syntagmatic combination of concepts and their recursive *analysis into* as well as *synthesis out of* fuzzy granules.[1] Zadeh describes informational uncertainty as a prerequisite of efficient information processing. He developed a formal framework for representing this uncertainty *in precise, mathematical terms without the need to exclude it for guaranteeing modeling accuracy and bivalence, respectively.* Thus, with the advent of fuzzy set theory, possibility and probability theory, the vagueness and ambiguity of information processing became proper objects of formal-mathematical modeling.

A central premise of *fuzzy computational linguistics* as introduced by Burghard Rieger [57, 61, 65] is that, beyond representing concepts as fuzzy constraints on the fuzzy values of fuzzy attributes, the question arises *how these*

---

*This volume is dedicated to Professor Burghard B. Rieger on the occasion of his 65th anniversary.

[1] This framework is explained in detail by Zadeh (in this volume).

A. Mehler and R. Köhler: *Introduction: Machine Learning in a Semiotic Perspective*, StudFuzz
209, 1–29 (2007)
`www.springerlink.com`   © Springer-Verlag Berlin Heidelberg 2007

representations can be automatically learned as a model of those cognitive processes by which cognitive systems actually acquire their concept systems. Although fuzzy set theory and related approaches allow tackling informational uncertainty in general, they do not model its constitution and evolvement according to the onto- and glossogenesis of cognitive information processing systems. Thus, the question put forward by Rieger arises: *What does an algebra of models of cognitive processes look like which prevents from abstracting from the process dynamics of information processing and thus departs from focusing solely on the informational uncertainty of its input/output relation?*

As far as *machine learning* (ML) [46] is considered *the* scientific discipline to investigate computer-based models of automatic concept learning, one has to state that this claim is not met so far. This is not due to any methodic deficit of ML, as it proves to be a very well established field of mathematical, explorative data analysis. Rather, it is due to a lack of cognitive grounding [78] of the learning mechanisms being applied. Although ML routines are completely sufficient from the point of view of successful classification, they may nevertheless be deficient from the point of view of modeling cognitive processes.

An approach which encompasses both machine learning and cognitive grounding is given by the simulation approach to language evolution [8] which makes onto-, phylo- and glossogenetic [36] processes of structure formation and meaning constitution an object of modeling in a multi-agent setting [7, 9] *including their informational uncertainty ab initio.*[2] The *iterated learning model* [35] can be referred to as an architectural simulation model in this area. It addresses the bottleneck problem according to which a language is transmitted from generation to generation via agents who evidently do not have access to the totality of knowledge characterizing the language to be learned. Consequently, language change – subject to the pressure of varying speaker and hearer needs – is inescapable in this model by analogy with natural languages. In this scenario, language learning is tackled with respect to referential semantics and symbolic grounding [6, 74, 77], the learning of lexical knowledge (regarding the articulation of content and expression plane) [24, 28, 36], the learning of syntax formation [24, 36] and the interrelation of lexico-grammar and semantics (as regards, for example, the emergence of compositionality) [36]. All these approaches apply machine learning techniques (e.g. classification, grammar induction etc.) in order to model language learning of individual agents and thus relate – from a methodological point of view – to *computational linguistics.*

---

[2]Note that this focus on informational uncertainty – not as a deficit, but as a very condition of linguistic meaning – was already proposed by Rieger [55, 56] who develops the notion of *semantic vagueness* and its empirical, corpus-based measurement – see also [59] and [60] for two comprehensive volumes on aspects of semantic uncertainty; an overview of the notion of *pragmatic information* is given by Atmanspacher (in this volume). Note also that [54] is one of the first approaches to meaning constitution from the point of view of computational linguistics.

Many of these approaches utilize a rather simplistic model as their *semantic kernel*, which in the following sections will be called *meaning-signal model*. It claims that sign vehicles (either enumerated by the model constructor in advance or intentionally defined by a procedure for the construction of vehicles out of a finite alphabet) are associated with their meanings which are likewise either enumerated in advance or intentionally constrained or procedurally generated as elements of, for example, a set of possible attribute-value structures. In the case of Hurford [27], for example, the mapping of form and meaning units is predefined by the model constructor, whereas Steels [74] endows semiotic agents with a procedure for learning the informational uncertain association of form and meaning units, thereby distinguishing between meaning and reference [28].

Obviously, the meaning-signal model has its pitfalls as it introduces strong assumptions into the simulation model and thus restricts its explanatory power. Consequently, in order to better ground simulation models and related approaches from the point of view of cognitive processing, they need alternatives to simplistic *meaning-signal mappings* which, for the time being, are applied in the majority of cases. More specifically, text representation models are needed which – comparable to the computer-based processes operating on them – are not only grounded in terms of cognitive systems, but are likewise consistent with prevailing architectures of simulation models of language evolution.

As regards the learning of the meanings of lexical units, such an approach comes from computational linguistics: Rieger [65] invents the notion of *semiotic modeling* in order to grasp the dynamics of meaning constitution. Other than formal semantics, he focuses on the constitution of meaning representations without presupposing them as elements of predefined, enumerable sets. Comparable to fuzzy set theory, this does not mean to dispense with any necessary level of modeling accuracy. Rather, this process-oriented framework models vagueness as a characteristic of the input/output relation of certain *processes* of sign-meaning constitution *which in Rieger's approach come into focus of modeling*.

This approach opens up new perspectives on text-based machine learning[3] and its integration into the framework of language simulation. More specifically, it relates to challenging *supervised function learning* (e.g. text categorization) based on pre-established category sets to be learned in training phases and to be attuned in test phases:

- Firstly, the paradigm of supervised learning is opposed by *unsupervised learning*, i.e. by automatic classification with a controlled reduction of the amount of predefined knowledge regarding the composition of the learning space. *Controlled* means that what is dispensed with on the level of predefined knowledge is replaced by procedures for acquiring this knowledge.

---

[3]That is, to the approach of automatically exploring linguistic information from natural language texts.

- Secondly, it is opposed by the paradigm of *relation learning* which allows assigning more than one class per unit and allows the learned relations to be classified as being vague, ambiguous or otherwise constrained by informational uncertainty.[4]
- Thirdly, it is opposed by approaches which ground the learning mechanism in terms of *cognitive processing* and thus serve as candidates for tackling the predominance of the meaning-signal model.[5]

Following this analogy, Rieger's approach can be seen as a reference example of how to integrate unsupervised relation learning from natural language texts into the framework of language simulation models. In the following sections, this role of Rieger's approach as a thread is outlined with respect to learning the meanings of lexical units (Section 2.1), with respect to *semantic spaces* as a uniform format for representing those meanings (Section 2.2) and with respect to the status of this format as a candidate for replacing the meaning-signal model (Section 3).

2 Computing with Linguistic Items

Latent Semantic Indexing (LSI) is one of the predominant models in Information Retrieval as regards automatic learning index terms from natural language texts [2].[6] It was developed in order to tackle the so called polysemy problem [12] which relates to deficits of the vector space model [68] to retrieve relevant documents which do not share terms with the focal query. Deerwester et al. propose a solution to this problem of measuring *indirect, content based similarity relations* in the framework of *singular value decomposition*. Because of its tremendous success, Landauer & Dumais [38] developed a theoretical underpinning of this model in terms of cognitive processes. It is based on the hypothesis that similarity relations of cognitive units result from a *two-level* process of inductive learning starting from the units' contiguity relations. In the case of lexical items, they equate these contiguity relations with co-occurrence relations. More specifically, they describe the learning of similarity relations as a process of dimensionality reduction as a result of which similarities of items can be detected even if they do not at all or only rarely co-occur. That is, similarity associations of linguistic items are described as functions of their contiguity associations. According to this model, inductive learning of similarity relations of linguistic items results from exploiting the similarities of their usage contexts. Landauer & Dumais equate this with a solution of the *knowledge acquisition problem* which they describe as follows:

---

[4]See Kacprzyk & Zadrożny as well as Klir & Sentz (in this volume).

[5]See, for example, Perlovsky (in this volume).

[6]In this chapter we will speak of *Latent Semantic Analysis* (LSA) according to Landauer & Dumais [38] instead of LSI in order to stress its relevance from the point of view of modeling cognitive processes.

"One of the deepest, most persistent mysteries of cognition is how people acquire as much knowledge as they do on the basis of as little information as they get." [38, p. 212].

Obviously, any algorithmic adequate answer to this question can be seen as a first step towards an operationalization of the *symbolic theft hypothesis* [6], that is, of the hypothesis, that semiotic agents successfully learn from each other via symbolic communication even in cases where they do not have direct, sensomotoric access to the communicated events – this is what Rieger [65] more generally calls *mediate learning*.

The development of this cognitive model can be traced back to the famous notion of *syntagmatics* and *paradigmatics* introduced by Ferdinand de Saussure [11] and Louis Hjelmslev [25]. Its reconceptualization in terms of cognitive *contiguity* and *similarity associations* stems from Roman Jakobson [29]. Beyond that, its predecessors primarily relate to computational linguistics where syntagmatic and paradigmatic relations of lexical units are represented by means of *semantic spaces* – cf. Rieger [57].[7]

Regarding automatic text analysis, this research field is of utmost relevance, since it prototypically integrates unsupervised, explorative relation learning from natural language texts with linguistic grounding of the learning model. It does not only depart from supervised learning of crisp functions, but places emphasis on grounding the learning of fuzzy, probabilistic relations and thus departs from application-oriented approaches such as information retrieval and related areas. The present section sheds light on this linguistic grounding and outlines its computational linguistic continuation in terms of semantic spaces. Because of its widespread roots, we dispense with explaining all its development directions. Among others, this relates to the notion of meaning as use – see Rieger [61] for a comprehensive usage-based theory of semantic vagueness. Rather, we begin with its structuralist heritage which nowadays comes into focus in terms of contiguity and similarity associations as well as of word priming [32] and text priming [71].

2.1 Structuralist Abstraction, Cognitive Reconstruction and Corpus Linguistic Exploration

Starting from Aristotelian associationism and his laws of association, Mikołaj Kruszewski distinguished two modes of linguistic association: (i) *association by similarity* (or contrast) and (ii) *association by* spatio-temporal *contiguity*

---

[7]There are further predecessors in the area of Information Retrieval. Cf., for example, Lewis et al. [41] and especially Lesk [40] who speaks of *first* and *second order associations* which he views to be extendable by higher order associations, although he did not observe an effect of second order associations on retrieval effectiveness. The iteration of measuring the similarities of associations is studied by Gritzmann (in this volume).

[26]. Kruszewski believes that these two modes of association form a basis of language learning:

"Wir sind überzeugt, daß die Aneignung und der Gebrauch der Sprache unmöglich wären, wenn sie eine Menge von vereinzelten Wörtern darstellte. Die Wörter sind miteinander verbunden: 1. vermittelst der Ähnlichkeitsassociationen und 2. vermittelst der Angrenzungsassociationen. Daher entstehen Familien oder Systeme und Reihen von Wörtern." (cited after Heinz Happ [20, p. 38]).

Adopting the notion of similarity association, de Saussure [11] developed his famous distinction of syntagmatic, conjunctive relations *in praesentia* and disjunctive relations *in absentia*. More specifically, de Saussure described syntagmatic relations to hold between (sub-sentential) constituents of the same utterance.[8] This concept was later specified in glossematics and syntactics, where syntagmatic relations are seen to hold between (groups of) linguistic items co-occurring in instances of the same (syntactic) context type, though not necessarily side by side. In contrast to this, associative relations are seen to hold between items which are substitutable for each other within the same contexts (of a certain type under consideration) without changing their focal syntactical, semantical or pragmatical properties (e.g. grammaticality, well-formedness or acceptability etc.).

Although de Saussure factors out processes of meaning constitution and language change from synchronic linguistics, he nevertheless clarifies the constitutive dependence of language systematic associations and syntagmatic relations as follows:

"[...] die Zusammenordnung im Raum wirkt an der Schaffung assoziativer Zuordnungen mit, und diese ihrerseits sind nötig für die Analyse der Teile der Anreihung." [11, p. 153].

In other words, de Saussure identifies syntagmatic relations as a source of the constitution of linguistic similarity associations which were later called *paradigmatic* and reconstructed in terms of substitutability under invariance of certain linguistic features [25]. Starting from this hypothesis, it is evident to claim that the similarities of the syntagmatic relations into which linguistic items enter, contribute to their paradigmatic similarity or similarity associations, respectively.

The further development of de Saussure's opposition departed from its psychological roots and focused instead on structural considerations inspired by the development of formal logics in the first half of the twentieth century. The most prominent proponent of this development is Louis Hjelmslev [25] who introduced the term *paradigmatic relation* and thus opposed *syntagmatic*

---

[8]For a critical discussion of de Saussure's notion of linearity see Wildgen (in this volume).

(i.e. text-based) by *paradigmatic* (i.e. system-based) relations in order to stress this change:

"C'est pour éviter le psychologisme adopté dans le *Cours* de F. de Saussure que je substitue le terme de 'rapport paradigmatique' à celui de 'rapport associatif'." (Hjelmslev (1938) cited after Wolfgang Raible [51, p. 35]).

This was done with the help of the glossematic concept of *function*: Starting from the concept of a functor as an argument of a *linguistic function*, Hjelmslev defined *constants* to be functors whose presence is a necessary condition of the presence of those functors, with which they enter into the same function [25]. In contrast to this, he defined *variables* to be functors whose presence is not a sufficient condition in this sense. Utilizing the notion of constant and variable, Hjelmslev distinguished three types of syntagmatic (conjunctive) and paradigmatic (disjunctive) functions, respectively, in order to define *paradigms* as classes of linguistic units entering into homogeneous functions. Moreover, Hjelmslev left the narrow limitation of syntagmatics to sub-sentential units and focused instead on whole texts – *anticipating the structuralist beginning of text linguistics*.

After Hjelmslev it was Algirdas Julien Greimas [16] who applied the syntagmatics-paradigmatics opposition to semantics.[9] Although he developed a feature semantics implying a set of atomic features as constituents of meaning representations, he places emphasis on discourse specific processes of meaning calibration based on two syntagmatic operations. First, he described *expansion* as a text-internal operation relating a specifying, explaining or otherwise elaborating text span with a specified, explained or elaborated one. Second, he viewed this operation to be reversed by *condensation* which relates summaries or denominating spans with summarized or denominated ones. Both operations transcend the sentence level and thus relate syntagmatics to text constitution. The contribution of syntagmatics to text constitution was further clarified by Roland Harweg [23]. He introduced the notion of *syntagmatic substitution* which describes text constitution as a result of chains of pronominal and initial auto-semantical text constituents used to concatenate subsequent sentences. This structuralist conception is now considered one of the starting points of text linguistics which specifies syntagmatic relations in terms of cohesion and coherence relations.

Before we continue to survey this reconceptualization, we stress Paul Ricœur's [52] notion of lexis and its relation to language change. According to his conception, the lexical system serves as a mediator between the language system and discourse events, whereby this mediation is constrained by two opposing processes: On the one hand, the process of meaning expansion forces lexical meaning to reflect the lexemes' discourse specific usages. On the other hand, this lexically distributed process is restricted by limita-

---

[9]Cf. also Coseriu's diachronic semantics as an alternative approach to structural semantics [10].

tion processes inside the language system – mediated, for example, by lexical fields. According to this conception, informational uncertainty (e.g. vagueness or ambiguity of lexical meaning) is only adequately described with respect to this process dynamics.

So far, the concept of syntagmatics and paradigmatics is developed with respect to lexical (meaning) relations, their informational uncertainty and their contribution to text constitution. As text linguistics needed to leave the structuralist stance in order to investigate criteria of textuality which cannot be reduced to intratextual relations, the cognitive reconstruction of both types of relations got into focus as described at the beginning of this section with respect to Rieger's model of semantic spaces and to latent semantic analysis. This tradition was, amongst others, initiated by Roman Jakobson [29] who related paradigmatic *and* (other than de Saussure) also syntagmatic relations to cognitive associations. More specifically, he described paradigmatic selection in terms of substitutability which, in simple cases, is based on *similarity associations*. In contrast to this, he viewed syntagmatic combinations to be based, in simple cases, on contiguity associations. This distinction allowed Jakobson to distinguish two types of interpretants, that is, two aspects of meaning:

"[...] there are two references which serve to interpret the sign – one to the code, and the other to the context, whether coded or free, and in each of these ways the sign is related to another set of linguistic signs, through an *alternation* in the former case and through an alignment in the latter. A given significative unit may be replaced by other, more explicit signs of the same code, whereby its general meaning is revealed, while its contextual meaning is determined by its connection with other signs within the same sequence." [29, p. 244].

Jakobson further developed this notion in order to distinguish Broca and Wernicke aphasia according to their effect on paradigmatic selection and syntagmatic combination. Thus, he related both types of linguistic relations not only to language systematic associations, but also to the *language faculty of single speakers/hearers*. That is, both types of relations are now seen to constitute part of the *linguistic knowledge of single cognitive information processing systems* which they learn, maintain, change and lose during their growth, maturity stage and dying.

These considerations of the genesis of the concept of syntagmatic and paradigmatic learning finally lead back to its cognitive resources, challenging its structural heritage and transcending the limits of a purely associationist stance. It was Rieger [58] who first focused on the cognitive view on syntagmatic and paradigmatic learning in computational linguistics. In his approach, both threads are integrated: *the cognitive grounding of syntagmatic/paradigmatic relations as well as their automatic, corpus-based exploration*.

Generally speaking, as far as computational linguistics is concerned with *automatically learning* this kind of linguistic knowledge, three learning scenarios can be distinguished:

- *Learning language-systematic knowledge:* From the point of view of the lexical system of a (sub-)language, learning lexical meanings relates, amongst others, to learning their paradigmatic relations which in turn are seen to be reconstructable as functions of their syntagmatic relations. This scenario is described by the *weak contextual hypothesis* of Miller and Charles [45]. It says that the similarity of the contextual representations of words contributes to their semantic similarity. In structuralist terms, contextual representations are described as representations of syntagmas. In other words – and this leads back to de Saussure's initial observation –: As far as lexical units are analyzed, their paradigmatic similarity is modeled as a function of their syntagmatic relations. In order to make this a contribution to lexical meaning, paradigmatic similarity has to be focused upon. This was done by Rieger [61] who, in computational linguistics, first implemented the two-stage process of learning paradigmatic relations from natural language texts by means of semantic spaces as a format for representing contextual similarities.[10]

- *Single agent-based learning:* Jakobson's conception encompasses a second scenario according to which a computer simulation (e.g. an artificial, computer-based agent) is endowed with the learning function just described in order to autonomously acquire parts of its linguistic knowledge based on the ontogenesis of its text processing. As an example, consider the approach of Foltz et al. [15] who utilize LSA to model lexical cohesion of single texts. Consider also Kintsch's construction-integration model [33] which utilizes LSA in order to model (lexical priming as part of) text comprehension.[11] A reference example which includes a touchstone experiment for evaluating instances of the focal class of approaches is given by Rieger [65] who describes a computer simulation of a single text processing system which – on the basis of a controlled vocabulary – automatically learns semantic relations.

- *Multi agent-based learning:* The latter two approaches abstract from the embodiment of the learning procedures being modeled or, as in the case of the second approach, concentrate on single-agent models. Thus, a third learning scenario has to be considered which relates to sociogenetic, or

---

[10]Using LSA (see above) for modeling this two-stage process, concurrent models focus, for example, on resolving lexical semantic ambiguities [70] and modeling meaning calibration in predicate-argument structures [34].

[11]An alternative approach stems from Sharkey & Sharkey [71] – in order to name only one representative of the paradigm of connectionist modeling. It focuses on the simulation of context priming as a function of the ontogenetic and actual genetic evolvement of text processing, and thus interrelates both kinds of associations with the well established field of priming in cognitive linguistics.

more specifically, to glossogenetic [9] learning in language simulation models. It is based on *multi-agent* computer simulations in which groups of artificial agents communicate with each other and thus cooperatively produce the texts or dialogues which serve as input to their acquisition of linguistic knowledge. In such a scenario, the constitution of language-systematic and of idiosyncratic knowledge (specific to single agents as abstract models of interlocutors participating in the language community under consideration) is simulated [44].

It is the latter scenario in which approaches to semantic spaces come into play in order to replace the meaning-signal model in the framework of simulating language evolution. It departs from the first two approaches in the sense that the corpus underlying language learning is no longer pre-established by the model constructor (whether as a corpus of preselected (thematically or pragmatically homogeneous) texts or as a result of a deterministic text production system as demonstrated in [65]), but autonomously constituted within the simulation experiment, that is, by the communicating agents. Thus, in order to implement one of the first two approaches, the corpus linguistic question for appropriate learner corpora has to be answered first. This question leads back to a third tradition of conceptualizing syntagmatic and paradigmatic relations, now in the framework of *corpus linguistics*.

Above, we mentioned the *weak contextual hypothesis* [45] according to which the similarity of the contextual representations of words contributes to their *semantic similarity*. This hypothesis has at least two predecessors. On the one hand, it is connected with the Firthian tradition of corpus linguistics [79] which stresses the notion of meaning as use. Firth states that *collocations* of lexical units reflect lexical affinities beyond grammatical restrictions. Collocative regularities are seen to provide an integral part of *syntagmatic* functions and in this sense of lexical meaning. He summarizes this conception by the formula "You shall know a word by the company it keeps!" [13, p. 11]. In another context he stresses:

"Meaning by collocation is an abstraction at the syntagmatic level and is not directly concerned with the conceptual or idea approach to the meaning of words. One of the meanings of *night* is its collocability with *dark* [...]." [14, p. 196].

Collocation analysis is an approach to reconstruct this collocability on the basis of corpora of natural language texts. Its goal is to discover semantically similar words based on the similarity of their collocations [cf. 73, 79], though the notion of similarity was not mathematically operationalized at this early level of development.

A second prominent predecessor is Harris' *distributionalism* which refers to the notion of correlation in order to specify the interdependence of *meaning* and *distribution*:

"[...] the formal features of the discourses can be studied by distributional methods within the text; and the fact of their correlation with a particular type of situation gives a meaning-status to the occurrence of these formal features." [22, p. 316].

This conception implies that there is no homomorphism between semantic and distributional structure as claimed, for example, by the principle of compositionality with respect to the syntactic and semantic structure of compound expressions: Neither is the meaning of a word determined by its distribution, nor do words have a kernel (literal) meaning as input of a function which aggregates meanings according to a compositional semantics. Rather, Harris claims that distributional regularities correlate with some aspects of their meaning in a way that semantic differences are reflected by dissimilar distributions and vice versa:

"The various members of a distributional class or subclass have some element of meaning in common, which is stronger the more distributional characteristics the class has." [21, p. 156].

But the question, exactly which elements of meaning are shared by the elements of the same distributional class, is left unspecified. That is, distributionalism and related approaches focus solely on co-occurrence patterns according to the famous formula:

"[...] difference in meaning correlates with difference of distribution." [21, p. 156].

Nevertheless, distributionalism, corpus linguistics and related approaches claim that syntagmatic [79] and, based on that, also paradigmatic relations can be reconstructed in terms of co-occurrence patterns by exploring corpora of natural language texts.[12] Whereas the cognitive linguistic reconstruction of these relations shed light on their cognitive status, it is corpus linguistics which shows a way to *measure* the extent to which they actually hold. This raises the question, how to represent the results of such measurements. Without giving a detailed answer to this question, we concentrate on semantic spaces as a candidate format for this task.[13] Once more, this choice is due to the fact that Rieger's approach is the first one in which the *cognitive grounding* of linguistic relations is integrated with their *automatic, corpus-based exploration*.

2.2 Semantic Spaces – Requirements Analysis

From a methodological point of view the question arises, how to represent syntagmatic and paradigmatic patterns. That is, the question has to be an-

---

[12]See Stubbs [80] for an approach to the reconstruction of phrasal micro structures in the framework of corpus linguistics. Cf. also Stubbs (in this volume) for a critical review of the notion of induction in corpus linguistics.

[13]Gritzmann as well as Leopold – both in this volume – give detailed mathematical specifications of semantic spaces.

swered which format allows representing the results of measuring contiguity and similarity associations by inducing which information loss as well as which time and space complexity. As these measurement operations include multivariate analyses of the dependencies of multiple variables and their clustering along multiple dimensions, multivariate statistics comes into play [39]. This relates especially to *semantic spaces* which were explicitly developed in order to model syntagmatic and paradigmatic relations. A reference model of semantic spaces was developed by Burghard Rieger [53, 57]. In this section, the requirements analysis underlying this format is outlined in order to shed light on its benefits and drawbacks. These requirements are as follows:

1. *Sensitivity to meaning relations:* Following the line of argumentation of the *weak contextual hypothesis* of Miller & Charles (see Section 2.1), it has to be clarified that semantic spaces are used to represent usage-based similarities of linguistic items in order to map their semantic similarity. In other words: Regarding the data basis analyzed to explore certain usage regularities, it has to be substantiated how these regularities actually contribute to the semantic similarity of the signs under consideration.

2. *Linguistic grounding:* The latter requirement can be intensified by demanding that it has to be clarified which contribution to which meaning aspect is actually represented by the semantic space model under consideration. In the case that textual signs are analyzed, it has to be clarified, for example, which type(s) of cohesion and coherence relations are mapped.

These two requirements are indispensable as they target at the possibility to evaluate semantic spaces regarding the measurement operations they are used to perform. This linguistic grounding has a procedural variant:

3. *Procedural grounding:* The computation of semantic spaces has to be grounded by a *procedural model* (cf. Marr [42]) which relates this computation to linguistic processes of meaning constitution (beyond its algorithmic formalization and computer-based implementation).

Beyond Rieger's seminal work on semantic spaces, this requirement is met, for example, by LSA [38] which models the two-stage process of inductive learning by means of a dimension reduction algorithm which amalgamates learning of contiguity and similarity associations. Rieger [61] prevents this amalgamation by using *separate* formats for mapping syntagmatic and paradigmatic regularities (by means of corpus and semantic spaces, respectively).[14] Fulfilling the third requirement is indispensable, as explained above, when looking for alternatives to the meaning-signal model in simulations of language evolution.

The following requirement emphasizes more strictly the corpus analytic stance of the semantic space model:

---

[14]See also Bordag & Heyer (in this volume) who explicitly put apart both steps of learning lexical relations.

4. *Explorative corpus analysis:* Semantic spaces have to be automatically computed by referring to an amount of pre-established (linguistic) knowledge as little as possible. Moreover, the knowledge represented in semantic spaces has to be automatically explored from corpora as samples of the sub-language (or genre, register, thematic field etc.) under consideration.

This requirement targets at the transferability of the model to corpora of different languages, genres and registers and thus at its fault tolerance, robustness and adaptability. It allows distinguishing semantic spaces from approaches which exploit thesauri and related lexical reference systems like WordNet [47]. As the majority of semantic space models is insensitive to structure (as regards syntactic structure and cohesion/coherence relations which are hardly explored automatically in the case of large input corpora), this requirement (and thus transferability) is easily met, but at the price of a loss in *linguistic grounding* according to requirement (2). Requirement (4) is closely related to the following requirement which leads back to Zadeh's work and Rieger's theory of semantic vagueness:

5. *Informational uncertainty:* Linguistic units require a representational format which allows for modeling their varying aspects of informational uncertainty. More specifically, it has to be clarified which aspect of uncertainty (e.g. vagueness, ambiguity, variability) is actually mapped by which part of the model under consideration.

The majority of approaches utilizes semantic spaces in order to model the vagueness of unsystematic sense relations [19] – only a couple of them (e.g. Schütze [69, 70]) explicitly use them to model semantic ambiguity. As semantic spaces have originally been proposed as an alternative to the vector space model in order to map indirect meaning relations, the following requirement comes into focus:

6. *Implicit relations:* Lexical items should be considered as semantically related even if they do not or only rarely co-occur, but tend to be used in similar contexts. Analogously, textual units should be considered as semantically related even if they do not or share only a couple of constituents, but deal with similar topics. Additionally, lexical and textual units should be considered as semantically related even if the former do not or rarely occur in the latter, but share usage regularities with their lexical constituents.

This requirement can be summarized as follows: In order to be judged to be paradigmatically related it is neither necessary nor sufficient that the signs co-occur or share constituents (as in the case of lexical and textual units, respectively), but that they are used or composed in a way which supports their semantic relatedness. It is the *automatic* exploration of such indirect, implicit similarity relations which separates semantic spaces from competing approaches. But the concentration on syntagmatic and paradigmatic learning also reveals a central deficit of semantic spaces which – with a couple of exceptions – disregard processes of text constitution and thus syntactic as well as cohesion and coherence relations. This becomes palpable when looking for state of the art models of semantic spaces which map words and texts each onto single meaning points (i.e. feature vectors) – *irrespective of the structuring of the latter in contrast to the former*.

Nevertheless, meeting the requirements just enumerated is a big step forward to replacing the meaning-signal model based on pre-established sets of meaning and signal representations in the framework of simulating language evolution. The following sections lead back to the question how semantic spaces can be incorporated into this framework.

3 Challenging the Meaning-Signal Model

A central assumption of the semiotic (i.e. sign-based) approach to situated cognition is that systems of *Semiotic Cognitive Information Processing* (SCIP) constitute an endo-view of their environment as a result of sign processes – cf. Rieger [65]. As far as the system of signs involved in these processes is concerned, its constitution is focused and, consequently, communities of SCIP systems whose communication is both: grounded *by* as well as constitutive *for* this system [65, 76]. According to the distinction of modeling single SCIP systems vs. communities of them, two consequences have to be put apart when challenging the meaning-signal approach:

- As far as the information processing of a *single* SCIP system is focused, this necessarily rules out identifying the exo-view of the model constructor with the endo-view of the modeled system – cf. Rieger [62]. That is, neither context nor meaning representations are to be predefined by the constructor. Rather, the SCIP model needs to include routines of unsupervised sign processing which enable it to generate meaning and context representations as constituents of its endo-view. This position is due to Rieger [62] – it can be seen as a basic principle of explorative modeling in cognitive science.
- As far as the constitution and evolution of a language system is focused, this analogously implies that a *community* of SCIP systems has to be modeled whose shared meanings cannot be predefined by the model constructor. Rather, the SCIP systems have to be endowed with communication routines whose unsupervised application allows them to acquire the meanings being shared within their community.

Any effort in modeling *sign* systems by simulating agent-based communication necessarily needs to resist endowing the simulation with a meaning function, that is, a function which maps predefined sign vehicles and context representations onto likewise predefined meaning units. Rather, this approach has to be replaced by a procedural model which allows *generating* both, the sign relation as well as its arguments, i.e. context and meaning representations. In the last consequence, this also means reconstructing sign vehicles as arguments of this relation, and thus to resist presupposing them as lexicon entries [64] (cf. Bordag & Heyer and Medina-Urrea, both in this volume). This raises the question for the kind of modeling needed to pretend this identification. In order to outline an answer, Pattee's [49] distinction of *simulation* and *realization* can be utilized:

1. *Simulations*[15] symbolically stand for the entities they model. They serve, for example, to reconstruct/predict system functions as well as input and output states which are otherwise not directly observable. Thus, simulations necessarily include a time parameter, i.e. a symbolic representation of (physical, biological or social-semiotic) time. But simulations do not realize the focal functions and states, respectively – they only symbolically model them. The modeled functions and states are ontologically apart from the symbolic repertoire the simulation is composed of. Nevertheless, simulations introduce a further level of falsification: They are falsifiable with respect to the reconstructions/predictions they produce and the accuracy of the measurement being made.

2. *Realizations* are material models implementing certain functions of certain systems. This implementation is performed in the same ontic area in which the modeled system operates. Realizations are necessarily procedural models whose execution serves to realize the functions in question.[16] They are evaluated with respect to the adequacy of this realization and not with respect to modeling accuracy. If, for example, "flying as dragonflies" is the function to be realized, a helicopter can be seen as a realization of this function, but as a bad model of dragonflies.

The *computer-based simulation* of the evolution of sign systems is a much discussed topic in cognitive science and related disciplines. Starting from the insight that an agent's capability to survive correlates with its ability to process signs, a lot of computer-based models of the evolution of sign systems have been elaborated [4, 7, 62, 74, 75, 76, 81]. According to these approaches neither rule-based [72], nor statistical models alone allow to account for the dynamics of sign systems as a result of countless events in which agents make use of signs to serve their communication needs. Rather, the evolution of sign systems – which natural agents use in order to collectively survive – is necessarily *simulated by* means of computer-based *multi-agent systems*.

This challenging approach hinges on efficient routines of unsupervised learning linguistic items and their relations as well as on formats expressive enough

---

[15]See also Eikmeyer, Kindt & Strohner (in this volume) for the role of simulations in cognitive science. See also Rickheit & Strohner (in this volume) for an approach to ground automatic text analysis in the framework of cognitive science.

[16]Whether this realization is only possible in the "real" physical world or also in artificial worlds is a central question raised in *artificial life studies* [49].

for representing them. It hinges on adequate representations of textual units and of the language system manifested by these units. *But how do these two approaches – automatic text-analysis and simulation of language evolution – interrelate, if at all? How are text representation models as developed in computational linguistics, text technology[17] and related areas to be used in order to dispense with the limits of the meaning-signal model?*

These questions can be tackled with the help of systems-theoretical linguistics, or more concretely with the notion of *text* as introduced by Halliday [17, 18].[18] Halliday describes natural languages as dynamic systems which use texts as the primary unit of interaction with their environment.[19] Thereby, frequencies of text components reflect *probabilistic, fuzzy* dispositions of the corresponding lexico-grammatical system. According to Halliday, this manifestation is mediated by social context. He assumes that patterns of textual manifestation do systematically, though non-deterministically, vary with types of social context (i.e. genres and registers [43]).[20] Regarding the persistence of a language as a result of its interaction with its environment by means of countless text events, Halliday distinguishes three moments of this interaction:

- From the point of view of *phylogeny*, the change of the language system as a whole is dealt with.
- From the point of view of *ontogeny*, the life cycle of groups and single text processing systems are described with respect to their growth, maturity stage and dying.
- Finally, the point of view of *actual genesis* relates to the generation (production/processing) of single textual interaction units.

From this point of view, natural language texts are informationally uncertain instances of a constantly changing (though at a much slower rate) language system where this change is mediated by numerous social-semiotic context types and their changes (once more according to a slower time scale). Evidently, this model allows integrating approaches of unsupervised relation learning from natural language texts:

- Firstly, the system-theoretical view adopts a notion of meaning which takes informational uncertainty in terms of *diversification* [1], *ambiguity* [31], *vagueness* [50, 61] and temporal *variability* [67] into account. Following this approach, learning cannot – because of the many-to-many relation

---

[17]Grammar oriented text technological modeling is exemplified by Metzing & Pönninghaus and by Rolshoven (in this volume).

[18]For an approach to system theoretical linguistics which utilizes synergetic, quantitative modeling see Köhler (in this volume). See also Ziegler & Altmann (in this volume) for a quantitative approach in the area of system theoretical linguistics.

[19]This concept relates to Ricœur's dynamic perspective on structuralism (see above).

[20]See Lenders (in this volume) who investigates structural indicators of the membership of texts to text types and registers.

of the lexicogrammatical system, the semantic system and their textual manifestations – be performed as (fuzzy, probabilistic) function learning alone, but needs to include (fuzzy, probabilistic) relation learning.

- Thus, secondly, the system-theoretical view demands considering the contextual stratification of language. In this sense, the *meaning relation* has to include *context parameters* [3] for which the different branches of computational linguistics offer several representation formats. *But because of the interplay of changes of linguistic and contextual units, context parameters need to be learned in an unsupervised setting, too.*

At the same time, the system-theoretical view allows integrating agent-oriented models of meaning constitution and language evolution: Viewing language as the focal system, its environment is seen as consisting of text producing/processing systems, that is, of agents interacting with language by means of the texts they produce and process, respectively. In other words, the system-theoretical view necessarily comprises agent-oriented models in order to account for environmental dynamics as far as it enslaves language internal states.

Although these considerations relate automatic text-analysis to the area of meaning constitution and language evolution, they do not determine the type of modeling (i.e. simulation or realization) which is actually performed. If we conceive, for example, text-based machine learning as the effort to automatically reconstruct and represent outcomes of processes of text comprehension, the question whether this learning is a kind of simulation, realization or something else is rather delicate. From the point of view of optimizing precision and recall within the paradigm of supervised learning, answering this question seems superfluous: As long as the input/output function focused upon is learned correctly, it is irrelevant how this learning is achieved provided that it is done efficiently in the sense of time and space complexity. This does not hold for approaches related to the simulation of language evolution where language learning becomes an object of modeling *in its own right* (cf. Section 1). Consequently, a broader approach is needed which does not solely ground machine learning in statistics and mathematical modeling, but also in *modeling linguistic, cognitive processes.*

Against this background, we can outline an answer to the question raised at the beginning of this section as follows: In a narrow sense there is a negative interpretation of simulations as being models of the interaction of system and environment, where the model constructor presupposes all measurements needed to constitute the environment under consideration. *Presupposing* means that the constructor enumerates the universe of all possible contexts of the system by referring to a compositional semantics operating on a finite set of atomic units for context representation (e.g. relations, locations, time variables, etc.) and thereby reducing the system's meaning relation to a classical meaning function. Such a function may look as follows: For a system S, the meaning $\|a_i\|$ of sign $a_i \in \mathbb{V}$ in context $C_1 \in \mathbb{C}$ is $M_{i_1} \in \mathbb{M}$, in context $C_2 \in \mathbb{C}$ it is $M_{i_2} \in \mathbb{M}$, ..., the meaning $\|a_j\|$ of sign $a_j \in \mathbb{V}$ in context C_1 is M_{j_1} ... where $\mathbb{V}$ is the lexicon of the focal language, $\mathbb{M}$ is the set of meanings and $\mathbb{C}$ the set of contexts.[21] Focusing on a single sign x, such a function may be noted as follows:

$$
\|x_i\| = \begin{cases} M_{i_1} & : & C_1 \\ M_{i_2} & : & C_2 \\ \vdots & : & \vdots \\ M_{i_n} & : & C_n \end{cases} \tag{1}
$$

This blueprint of a meaning function – used, for example, in model-theoretic semantics [48] – is obviously inadequate when looking for an approach to unsupervised learning as needed in simulations of language evolution and related areas. But actually, it is this schema which underlies the meaning-signal approach. In order to arrive at a more realistic model, this blueprint of a context sensitive meaning-signal model can and has been revised according to several "relaxation" steps:

1. A first step is to view $\|\|$ no longer as a crisp function $\|\| : \mathbb{V} \times \mathbb{C} \to \mathbb{M}$, but as a fuzzy relation $R \in \mathcal{F}(\mathbb{V} \times \mathbb{C} \times \mathbb{M})$.[22] In this case, a membership function $\mu_R : \mathbb{V} \times \mathbb{C} \times \mathbb{M} \to [0,1]$ is introduced whose projection on a specific $v \in \mathbb{V}$ allows deriving the "fuzzified" meaning representation of this lexeme. Whereas this extension focuses on semantic *vagueness* [61], *ambiguity* is dealt with by means of possibility constraints [83]. Other aspects of informational uncertainty are introduced analogously [66, 83].[23]

2. Next, the fuzzy meaning relation may be relaxed by no longer viewing $\mathbb{C}$ and $\mathbb{M}$ as finite, but as uncountable infinite sets. In this case, the original meaning-signal model is enlarged in a way which does not presuppose complete extensional knowledge about possible meanings. Moreover, this approach allows mapping newly emerging context and meaning units beyond the set of initially established ones. This strategy is already followed by several approaches in the field of machine learning and language simulation [cf. the review in 28].

3. So far, elements of $\mathbb{M}$ and $\mathbb{V}$ are considered atomic units irrespective of their internal structure. Actually, the structuring of sign vehicles and meaning units is coupled in a way which, in summary, is described by the principle of semantic compositionality [30]. Thus, the next relaxation step is to account for this interdependence by redefining $\|\|$ as a homomorphism operating on (models of) the syntactic structure of signs. There are several examples in the field of language simulation which account for compositional structure either as a predefined characteristic of a (non-context

---

[21]This criticizable view presupposes the countability of meanings and contexts.

[22]$\mathcal{F}(\mathbb{V} \times \mathbb{C} \times \mathbb{M})$ denotes the set of all fuzzy sets over $\mathbb{V} \times \mathbb{C} \times \mathbb{M}$.

[23]Kacprzyk & Zadrożny (in this volume) demonstrate this relaxation step with respect to text categorization.

sensitive) meaning function or as an epiphenomenon of simulation [36].[24]
An alternative is to focus on the emergence of sign vehicles as elements of
V, that is, on the formation of expression and not only of meaning units.[25]

4. Likewise, elements of $\mathbb{C}$ can be considered according to, for example, situation semantics [3] as relational systems of necessary and sufficient conditions describing classes of contextual configurations which have to be
   validated in order to attribute a certain meaning to a certain sign in instances of the focal class (or situation type [3]) of context units. Obviously,
   these related relaxation steps (3) and (4) which incorporate structuring
   on the side of signs, their meanings and the contexts of their application
   can be extended by taking underspecification and other types of informational uncertainty into account. In the case of complex signs this means,
   for example, utilizing probabilistic grammars. In the case of meaning units,
   this means, amongst others, dealing with semantic ambiguity of complex
   units. Finally, in the case of contextual units, this relates to dealing with
   underspecifications thereof.

5. Next, the *extensional* stance of the meaning relation (whether finite or
   infinite) may be abandoned. More specifically, in the case of atomic sign
   vehicles it may be redefined *intensionally* in terms of a constraint satisfaction framework which uses, for example, production or IF-THEN rules in
   order to constrain the mapping of signs onto their meanings or interpretations in certain contexts. That is, statements of the form "*In context C,
   sign x has meaning M.*" are replaced by statements of the form "*In context C, the meaning of sign x has the properties $P_1, \ldots, P_n$.*" This leaves
   out what the meanings actually are, but only considers necessary (and
   maybe also sufficient) conditions any meaning representation has to obey
   in order to count as a candidate meaning of x in C. It is obvious that this
   approach lends itself to the framework of fuzzy constraint satisfaction [37].

Many other alternatives can be taken into account which focus on relaxing
the extensional and functional stance of the original meaning-signal model by
including structure formation and informational uncertainty. A central idea
of Rieger's approach is that what many of these alternatives ignore is the
way the focal units *are learned* – whether by single semiotic agents or within
communities thereof. They disregard the fact that context sensitivity and informational uncertainty are both: preconditions as well as epiphenomena of
meaning constitution. The basic idea of Rieger's approach is to deal no longer
with meaning functions and relations in terms of extensional or intensional
definitions invented by the model constructor(s), but to view them as units to
be automatically learned as a whole. He replaces the intensional by a procedural view which utilizes procedures to generate meaning representations as

---

[24]See Mehler (in this volume) who focuses on the principle of compositionality in
the context of semantic spaces.
[25]See, for example, Rieger [64]. See also Bordag & Heyer and Medina-Urrea (both
in this volume).

elements of an uncountable infinite search space.[26] Moreover, since language learning is not necessarily convergent, but may be temporarily divergent, this approach tackles the problem of changes of the search space too.

The series of enhancements of the meaning-signal model discussed so far mirrors, so to speak, a "naturalization" of meaning representation: First, this relates to taking more and more aspects of informational uncertainty into account, on the level of sign structuring and sign meaning and of the contexts of sign usage. Second, this relates to making learning an object of modeling, too, so that it no longer acts as an attribute of the model original to be put apart. The central idea of Rieger's approach is that *the success of these enhancements hinges upon an adequate grounding of the resulting procedural model in terms of semiotic, cognitive processes of meaning constitution.* Rieger explains this basic principle as follows:

"Trying to understand [...] understanding of signs and meanings cannot rely on the simulative processing of (symbol) structures whose representational status is declared by drawing on a pre-established semantics (known by the modeler, made accessible to the model, but not at all compulsory for the modeled system). Instead, modeling processes of meaning constitution or understanding will have to realize that very function in an implemented and operational information-processing system which is able to render some structure – in a self-organizing way – representational of something else, and also allows to identify what that structure is to stand for." [65, p. 167].

In other words, realizations in the sense of Rieger do not need and – because of the complexity of the environments these models operate on – do not even allow the symbolic enumeration of their possible meanings *and* contexts. That is, their semantic *and* contextual universes are no longer seen to be extensionally enumerable, nor intensionally definable, but only approachable by means of procedures as models of cognitive learning processes.[27] This procedural definition necessarily reflects the way natural sign processing systems learn semantic *and* contextual representations. That is, the procedural model represents the way cognitive systems are used to grasp their *Umwelt* in which they interact *meaningfully.* Consequently, following Rieger's approach means building computer simulations in which neither meaning nor context units are pre-established, but learnt by the semiotic cognitive information processing systems by *forming* the *substance* which the model constructor pre-establishes solely besides the procedural learning model.

---

[26]This is explained in detail in [65, 66].

[27]Obviously, a procedure as an abstract representation interrelated by means of its algorithmic specifications with its computer-based implementations [42] can be seen as a complex intension mapping its input onto its output. The difference is that in the present context a procedure is necessarily an intensional definition of the learning process, but not an abbreviatory specification of the input-output relation – see Ziemke [84] for a related argumentation.

4 Conclusion: The Semiotic Approach to Machine Learning

Rieger proposes an instantiation of the *semiotic approach* according to which any static, purely symbolic representation of meaning as a function of enumerable sets, whose predefinition presupposes knowledge about all possible referents of all modeled signs in all equally presupposed possible contexts, is replaced by a simulation model in which semiotic agents are endowed with a learning procedure which allows them to autonomously find out what relevant contexts are and which signs refer to which referents subject to which meanings. Rieger stresses that the procedural approach is insufficient as long as it does not *measure* those cognitive processes for which the modeling procedures symbolically stand. The semiotic approach to meaning constitution deals with the following problem scenario:

"Recent research findings [...] give rise to expect that processes which determine regularities and assemble them to form (intermediate) representations whose properties resemble (or can even be identified with) those of observable entities may indeed be responsible for (if not identical with) the emergence and usage of sign-functional structures in language understanding systems, both natural and artificial. As more abstract (theoretical) levels of representation for these processes – other than their procedural modeling – are not (yet) to be assumed, and as any (formal) means of deriving their possible results – other than by their (operational) enactment – are (still) lacking, it has to be postulated that these processes – independent of all other explanatory paradigms – will not only relate but produce different representational levels in a way that is formally controlled or computable, that can be modeled procedurally or algorithmized, and that may empirically be tested or implemented." [63, p. 544].

An implementation of this approach allows for replacing the meaning-signal model in order to approach more adequate language simulation models. Such an implementation promises a way out of the measurement fallacy as a consequence of the limits of the meaning-signal model. This fallacy is due to the fact that it forces the model constructor to make to far reaching presuppositions and thus to restrict the explanatory power of simulations based thereon. It is one of the many scientific merits of Burghard Rieger that he has theoretically explored, formally specified and empirically shown a way out of this fallacy.

5 Survey of the Book

The book is divided into seven parts covering the thematic spectrum of this introduction. It includes contributions to informational uncertainty and semantic spaces, corpus linguistics and quantitative modeling, text technology and text classification as well as cognitive modeling and computer-based simulations. The first part deals with *information modeling*. Its starting point is the notion of informational uncertainty and its alternative formalizations. On the background of his famous concept of fuzzy sets and its numerous enhancements, LOTFI A. ZADEH introduces a constraint-satisfaction language for the formally precise and computationally tractable representation of natural language sentences. It is based on the concept of generalized constraints which also play a fundamental role in the chapter of GEORGE J. KLIR & KARI SENTZ on the notion of linguistic approximation and linguistic variables. Both contributions focus on informational uncertainty from the point of view of fuzzy set theory. In contrast to this, HARALD ATMANSPACHER starts from a philosophical position in order to distinguish syntactic, semantic, and pragmatic information with a focus on the latter and a review of its formalization, amongst others, in terms of complexity theory.

The second part of the book deals with *semantic spaces* and thus leads to the representation format developed by Burghard Rieger in order to automatically learn fuzzy sets as representations of sign meanings. The starting point of PETER GRITZMANN is the two-level process of syntagmatic and paradigmatic learning which he generalizes to an n-level process in order to derive formal-mathematical constraints of clustering in semantic space. Further, he investigates methods for the lower-dimensional representation and traversal of those spaces and mathematically develops the notion of corpus immanent, so to speak, intrinsic contextuality based on lower-dimensional spaces. EDDA LEOPOLD surveys different space models starting from Rieger's *Fuzzy Linguistics*. She extends these approaches by building a semantic space by means of a family of support vector machines where each of the machines represents a single, pre-established topic category. Based on a cognitive model of text comprehension, ALEXANDER MEHLER develops a formal apparatus for making semantic spaces sensitive to sign structure. This is done w.r.t the hierarchical structure of texts and their graph-inducing coherence relations.

The third part of the book deals with quantitative models of lexical, syntactical and textual aspects of structure formation and, thus, focuses on a wide area of linguistic structure formation. STEFAN BORDAG & GERHARD HEYER develop an algorithmic framework which combines the notion of syntagmatic and paradigmatic learning with its reconstruction in a model of nested levels of linguistic resolution. A major result of their contribution is a specification of sense relations in terms of a statistical approach to lexical semantics. Next, REINHARD KÖHLER investigates quantitative aspects of syntactic structure formation. Based on synergetic linguistics, he develops a system theoretical model of order parameters which control quantitative characteristics of syntax. Finally, ARNE ZIEGLER & GABRIEL ALTMANN explore structure formation within natural language texts based on reference chains, that is, chains of interlinked lexical items which refer to the same entity. These chains and the text spanning graphs they induce are investigated, amongst others, as a preparatory work on a quantitative model of text coherence.

The fourth part of the book investigates the corpus linguistic background of approaches to automatic text analysis and exemplifies text technological modeling as well as explorative corpus analysis. MICHAEL STUBBS critically discusses induction as the underlying method of reasoning in corpus-based lexicography as well as in efforts at automatically learning linguistic patterns from large corpora. Consequently, he argues the possibility of an entirely automatic semantic analysis. In contrast to this, he pleads for developing computer-based means which support corpus linguistic work with respect to selecting and preprocessing large amounts of data. Such computer-based means include annotation tools whose underlying document grammars are focused by DIETER METZING & JENS PÖNNINGHAUS by example of the Kilivila verb morphology. One of the research questions addressed by them concerns the limitations of document grammars w.r.t modeling linguistic structures. More specifically, Metzing & Pönninghaus consider using document grammars for controlling the annotation of linguistic, multimodal corpora. Morphological structures are also focused in the contribution of ALFONSO MEDINA-URREA. But instead of grammar-based modeling, he concentrates on the exploration of morphological units, especially of affixes. In order to exemplify his algorithm, he analyzes four different languages and, thus, deals with corpus linguistics from the point of view of explorative data analysis. Finally, JÜRGEN ROLSHOVEN utilizes object-oriented modeling in the area of natural language processing, especially with respect to machine translation. This is done by means of a two-stage modeling paradigm which starts from UML-based conceptual modeling in order to derive language descriptions in a linguistic programming language. Rolshoven exemplifies his approach by means of phrase structures.

The fifth part of the book deals with text classification from the point of view of discourse analysis and fuzzy-linguistic modeling. It is introduced by WINFRIED LENDERS who investigates the selectivity of structural indicators in text *classification*. His starting point is that there is little known about structure induced similarity measures of texts. Consequently, he focuses on a kind of feature selection which combines structure and content-oriented indicators and, thus, follows a new research field in the area of text classification. JANUSZ KACPRZYK & SLAWOMIR ZADROŻNY likewise focus on an aspect of text *categorization* which is rather disregarded in the literature. Their starting point is to abandon the view that category assignments are crisp. In accordance with what is known about human cognition, they develop an algorithm for multiple, fuzzy categorizations and evaluate it by means of a well-known reference corpus. Beyond fuzziness of category assignments, the dynamics of such category systems is a topic of semiotic approaches to machine learning which is addressed by LEONID PERLOVSKY. He tackles the problem of concept formation and, thus, goes beyond those approaches to text categorization which start from pre-established category sets, whose dynamic formation they disregard.

The dynamic turn in automatic text analysis with its focus on fuzziness and structure-formation as demonstrated in part five, leads over to the sixth part of the book which stresses the point of view of cognitive modeling. GERT RICKHEIT & HANS STROHNER claim that a cognitive-theoretical embedding of automatic text analysis is indispensable especially with respect to the notion of inference based on the interaction of the textual input and the knowledge base of the text processing system. Consequently, they plead for a kind of *situated* text analysis which transcends the textual input subject to cognitive constraints of inference processes. Based on the extended experimental-simulative method, HANS-JÜRGEN EIKMEYER, WALTHER KINDT & HANS STROHNER deal with dialogical communication manifesting purpose dependent interactions of several interlocutors. They start from an annotated corpus of dialogical communication for evaluating quantitative hypotheses which are further investigated in subsequent experiments. In this framework, simulation has the function of hypothesis evaluation by means of parameter settings which are experimentally inaccessible.

The last part of the book deals with visual systems modeling. WOLFGANG WILDGEN analyses the dimensionality of the organization of pictorial signs as a basic determinant of their structure. He asks for commonalities of textual and pictorial signs and, thereby, challenges de Saussure's linearity axiom of language. In this sense, he returns to a linguistic tradition which was identified at the beginning of this introduction as one of the roots of Burghard Rieger's approach to modeling semiotic cognitive information processing systems.

Acknowledgement

The editors of this book thank Amélie Zöllner-Weber and Carolin Kram for their help in proof-reading the contributions of this book and in formatting the printer's copy.

References

[1] G. Altmann and R. Köhler. "Language forces" and synergetic modelling of language phenomena. In *Glottometrika 15*, pages 62–76. Brockmeyer, Bochum, 1996.

[2] R. Baeza-Yates and B. Ribeiro-Neto, editors. *Modern Information Retrieval*. Addison-Wesley, Reading, Massachusetts, 1999.

[3] J. Barwise and J. Perry. *Situations and Attitudes*. MIT Press, Cambridge, 1983.

[4] J. Batali. Computational Simulations of the Emergence of Grammar. In J. R. Hurford, M. Studdert-Kennedy, and C. Knight, editors, *Approaches to the Evolution of Language*, pages 405–426. Cambridge University Press, Cambridge, 1998.

[5] M. Black. Vagueness. An exercise in Logical Analysis. *Philosophy of Science*, 4:427–455, 1937.

[6] A. Cangelosi, A. Greco, and S. Harnad. Symbol Grounding and the Symbolic Theft Hypothesis. In A. Cangelosi and D. Parisi, editors, *Simulating the Evolution of Language*, chapter 9, pages 191–210. Springer, London, 2002.

[7] A. Cangelosi and D. Parisi. Computer Simulation: A New Scientific Approach to the Study of Language Evolution. In A. Cangelosi and D. Parisi, editors, *Simulating the Evolution of Language*, chapter 1, pages 3–28. Springer, London, 2002.

[8] A. Cangelosi and D. Parisi, editors. *Simulating the Evolution of Language*. Springer, London, 2002.

[9] M. H. Christiansen and S. Kirby. Language Evolution: Consensus and Controversies. *Trends in Cognitive Sciences*, 7(7):300–307, 2003.

[10] E. Coseriu. Für eine strukturelle diachrone Semantik. In H. Geckeler, editor, *Strukturelle Bedeutungslehre*, pages 90–163. Wissenschaftliche Buchgesellschaft, Darmstadt, 1978.

[11] F. de Saussure. *Grundfragen der allgemeinen Sprachwissenschaft*. de Gruyter, Berlin/New York, 1967.

[12] S. Deerwester, S. T. Dumais, G. W. Furnas, T. K. Landauer, and R. Harshmann. Indexing by Latent Semantic Analysis. *Journal of the American Society for Information Science*, 41(6):391–407, 1990.

[13] J. R. Firth. A Synopsis of Linguistic Theory, 1933-1955. In J. R. Firth, editor, *Studies in Linguistic Analysis*, pages 1–32. Blackwell, Oxford, 1957.

[14] J. R. Firth. *Papers in Linguistics. 1934-1951*. Oxford University Press, London, 1964.

[15] P. Foltz, W. Kintsch, and T. Landauer. The measurement of Textual Coherence with Latent Semantic Analysis. *Discourse Processes*, 25(2&3):285–307, 1998.

[16] A. J. Greimas. *Strukturale Semantik. Methodologische Untersuchungen*. Viehweg, Braunschweig, 1971.

[17] M. A. K. Halliday. Lexis as a Linguistic Level. In C. E. Bazell, J. Catford, M. A. K. Halliday, and R. Robins, editors, *In Memory of J. R. Firth*, pages 148–162. Longman, London, 1966.

[18] M. A. K. Halliday. Towards Probabilistic Interpretations. In E. Ventola, editor, *Functional and Systemic Linguistics*, pages 39–61. de Gruyter, Berlin/New York, 1991.

[19] M. A. K. Halliday and R. Hasan. *Cohesion in English*. Longman, London, 1976.

[20] H. Happ. 'Paradigmatisch' – 'syntagmatisch'. Zur Bestimmung und Klärung zweier Grundbegriffe der Sprachwissenschaft. Carl Winter Universitätsverlag, Heidelberg, 1985.

[21] Z. S. Harris. Distributional Structure. *Word*, 10:146–162, 1954.

[22] Z. S. Harris. Discourse Analysis. In Z. S. Harris, editor, *Papers in Structural and Transformational Linguistics*, pages 313–348. Reidel, Dordrecht, 1970.

[23] R. Harweg. *Pronomina und Textkonstitution*. Fink, München, 1968.

[24] T. Hashimoto. The Constructive Approach to the Dynamical View of Language. In A. Cangelosi and D. Parisi, editors, *Simulating the Evolution of Language*, chapter 14, pages 307–324. Springer, London, 2002.

[25] L. Hjelmslev. *Prolegomena to a Theory of Language*. University of Wisconsin Press, Madison, 1969.

[26] E. Holenstein. *Roman Jakobsons phänomenologischer Strukturalismus*. Suhrkamp, Frankfurt a. M., 1975.

[27] J. R. Hurford. Expression/Induction Models of Language Evolution: Dimensions and Issues. In T. Briscoe, editor, *Linguistic Evolution through Language Acquisition: Formal and Computational Models*, chapter 10. Cambridge University Press, Cambridge, 2002.

[28] E. Hutchins and B. Hazlehurst. Auto-Organization and Emergence of Shared Language Structure. In A. Cangelosi and D. Parisi, editors, *Simulating the Evolution of Language*, chapter 13, pages 279–306. Springer, London, 2002.

[29] R. Jakobson. *Selected Writings II. Word and Language*. Mouton, The Hague, 1971.

[30] T. M. V. Janssen. Compositionality (with an Appendix by Barbara H. Partee). In J. van Benthem and A. ter Meulen, editors, *Handbook of Logic and Language*, pages 417–473. Elsevier, Amsterdam, 1997.

[31] H. Kamp and B. Partee. Prototype Theory and Compositionality. *Cognition*, 57(2):129–191, 1995.

[32] W. Kintsch. The Role of Knowledge in Discourse Comprehension: A Construction-Integration Model. *Psychological Review*, 95(2):163–182, 1988.

[33] W. Kintsch. *Comprehension. A Paradigm for Cognition*. Cambridge University Press, Cambridge, 1998.

[34] W. Kintsch. Predication. *Cognitive Science*, 25:173–202, 2001.

[35] S. Kirby. Natural Language from Artificial Life. *Artificial Life*, 8(2):185–215, 2002.

[36] S. Kirby and J. R. Hurford. The Emergence of Linguistic Structure: An Overview of the Iterated Learning Model. In A. Cangelosi and D. Parisi, editors, *Simulating the Evolution of Language*, chapter 6, pages 121–148. Springer, London, 2002.

[37] G. J. Klir and B. Yuan. *Fuzzy Sets, Fuzzy Logic, and Fuzzy Systems. Selected Papers by Lotfi A. Zadeh*. World Scientific, Singapore, 1996.

[38] T. K. Landauer and S. T. Dumais. A Solution to Plato's Problem: The Latent Semantic Analysis Theory of Acquisition, Induction, and Representation of Knowledge. *Psychological Review*, 104(2):211–240, 1997.

[39] L. Lebart, A. Salem, and L. Berry. *Exploring Textual Data*. Kluwer, Dordrecht, 1998.

[40] M. Lesk. Word-Word Associations in Document Retrieval Systems. *American Documentation*, 1:27–38, 1969.

[41] P. Lewis, P. Baxendale, and J. Bennett. Statistical Discrimination of the Synonymy/Antonymy Relationship Between Words. *Journal of the ACM*, 14(1):20–44, 1967.

[42] D. Marr. *Vision: A Computational Investigation into the Human Representation and Processing of Visual Information*. Freeman, New York, 1982.

[43] J. R. Martin. *English Text. System and Structure*. John Benjamins, Philadelphia, 1992.

[44] A. Mehler. Stratified Constraint Satisfaction Networks in Synergetic Multi-Agent Simulations of Language Evolution. In A. Loula, R. Gudwin, and J. Queiroz, editors, *Artificial Cognition Systems*. Idea Publisher, Hershey, 2006.

[45] G. A. Miller and W. G. Charles. Contextual Correlates of Semantic Similarity. *Language and Cognitive Processes*, 6(1):1–28, 1991.

[46] T. M. Mitchell. *Machine Learning*. McGraw-Hill, New York, 1997.

[47] J. Morris and G. Hirst. Lexical Cohesion Computed by Thesaural Relations as an Indicator of the Structure of Text. *Computational Linguistics*, 17(1):21–48, 1991.

[48] B. H. Partee. Compositionality. In F. Landman and F. Veltman, editors, *Varieties of Formal Semantics. Proceedings of the fourth Amsterdam Colloquium, September 1982*, pages 281–311, Dordrecht, 1984. Foris.

[49] H. H. Pattee. Simulations, Realizations, and Theories of Life. In C. G. Langton, editor, *Artificial Life. SFI Studies in the Sciences of Complexity*, pages 63–77. Addison-Wesley, Redwood, 1988.

[50] M. Pinkal. Semantische Vagheit: Phänomene und Theorien, Teil I. *Linguistische Berichte*, 70:1–25, 1980.

[51] W. Raible. Von der Allgegenwart des Gegensinns (und einiger anderer Relationen). Strategien zur Einordnung semantischer Informationen. *Zeitschrift für romanische Philologie*, 97(1-2):1–40, 1981.

[52] P. Ricœur. *Hermeneutik und Strukturalismus. Der Konflikt der Interpretationen I*. Kösel, München, 1973.

[53] B. B. Rieger. Eine 'tolerante' Lexikonstruktur. Zur Abbildung natürlich-sprachlicher Bedeutungen auf 'unscharfe' Mengen. *Zeitschrift für Literaturwissenschaft und Linguistik*, 4(16):31–47, 1974.

[54] B. B. Rieger. Bedeutungskonstitution. Einige Bemerkungen zur semiotischen Problematik eines linguistischen Problems. *Zeitschrift für Literaturwissenschaft und Linguistik*, 27/28:55–68, 1977.

[55] B. B. Rieger. Theorie der unscharfen Mengen und empirische Textanalyse. In W. Klein, editor, *Methoden der Textanalyse*, pages 84–99. Quelle & Meyer, Heidelberg, 1977.

[56] B. B. Rieger. Vagheit als Problem der linguistischen Semantik. In K. Sprengel, W. D. Bald, and H. W. Viethen, editors, *Semantik und Pragmatik*, pages 91–101. Niemeyer, Tübingen, 1977.

[57] B. B. Rieger. Feasible Fuzzy Semantics. In K. Heggstad, editor, *Proceedings of the 7th International Conference on Computational Linguistics (COLING '78)*, pages 41–43. ICCL, Bergen, 1978.

[58] B. B. Rieger. Fuzzy Word Meaning Analysis and Representation in Linguistic Semantics. In *Proceedings of the 8th International Conference on Computational Linguistics (COLING '80)*, pages 76–84, Tokyo, 1980.

[59] B. B. Rieger, editor. *Empirical Semantics I. A Collection of New Approaches in the Field*. Quantitative Linguistics 12. Brockmeyer, Bochum, 1981.

[60] B. B. Rieger, editor. *Empirical Semantics II. A Collection of New Approaches in the Field*. Quantitative Linguistics 13. Brockmeyer, Bochum, 1981.

[61] B. B. Rieger. *Unscharfe Semantik: Die empirische Analyse, quantitative Beschreibung, formale Repräsentation und prozedurale Modellierung vager Wortbedeutungen in Texten*. Peter Lang, Frankfurt a. M., 1989.

[62] B. B. Rieger. Situation Semantics and Computational Linguistics: Towards Informational Ecology. In K. Kornwachs and K. Jacoby, editors, *Information. New Questions to a Multidisciplinary Concept*, pages 285–315. Akademie-Verlag, Berlin, 1995.

[63] B. B. Rieger. Computational Semiotics and Fuzzy Linguistics. On Meaning Constitution and Soft Categories. In A. Meystel, editor, *A Learning Perspective. Proceedings of the 1997 International Conference on Intelligent Systems and Semiotics (ISAS-97)*, NIST Special Publication 918, pages 541–551. NIST, Washington, 1997.

[64] B. B. Rieger. Warum fuzzy Linguistik? Überlegungen und Ansätze zu einer computerlinguistischen Neuorientierung. In D. Krallmann and H. W. Schmitz, editors, *Perspektiven einer Kommunikationswissenschaft. Internationales Gerold Ungeheuer Symposium, Essen 1995*, pages 153–183. Nodus, Münster, 1998.

[65] B. B. Rieger. Computing Granular Word Meanings. A Fuzzy Linguistic Approach in Computational Semiotics. In P. Wang, editor, *Computing with Words*, pages 147–208. John Wiley & Sons, New York, 2001.

[66] B. B. Rieger. Semiotic Cognitive Information Processing: Learning to Understand Discourse. A Systemic Model of Meaning Constitution. In R. Kühn, R. Menzel, W. Menzel, U. Ratsch, M. M. Richter, and I. O. Stamatescu, editors, *Perspectives on Adaptivity and Learning*, pages 347–403. Springer, Berlin, 2002.

[67] B. B. Rieger and C. Thiopoulos. A self-organizing lexical system in hypertext. In R. Köhler and B. B. Rieger, editors, *Contributions to Quantitative Linguistics*, pages 67–78, Dordrecht, 1993. Kluwer.

[68] G. Salton. *Automatic Text Processing: The Transformation, Analysis, and Retrieval of Information by Computer*. Addison-Wesley, Reading, Massachusetts, 1989.

[69] H. Schütze. *Ambiguity Resolution in Language Learning: Computational and Cognitive Models*, volume 71 of *CSLI Lecture Notes*. CSLI Publications, Stanford, 1997.

[70] H. Schütze. Automatic Word Sense Discrimination. *Computational Linguistics*, 24(1):97–123, 1998.

[71] A. J. C. Sharkey and N. E. Sharkey. Weak Contextual Constraints in Text and Word Priming. *Journal of Memory and Language*, 31(4):543–572, 1992.

[72] H. A. Simon and A. Newell. Informationsverarbeitung in Computer und Mensch. In W. C. Zimmerli and S. Wolf, editors, *Künstliche Intelligenz. Philosophische Probleme*, pages 112–145. Reclam, Stuttgart, 1994.

[73] J. Sinclair. *Corpus, Concordance, Collocation*. Oxford University Press, Oxford, 1991.

[74] L. Steels. Self-organising Vocabularies. In C. Langton and T. Shimohara, editors, *Proceedings of Artificial Life V, Nara, Japan*, 1996.

[75] L. Steels. Synthesizing the Origins of Language and Meaning Using Coevolution, Self-Organization and Level Formation. In J. R. Huford, M. Studdert-Kennedy, and C. Knight, editors, *Approaches to the Evolution of Language*, pages 384–404. Cambridge University Press, Cambridge, 1998.

[76] L. Steels. The Puzzle of Language Evolution. *Kognitionswissenschaft*, 8:143–150, 2000.

[77] L. Steels. Grounding Symbols through Evolutionary Language Games. In A. Cangelosi and D. Parisi, editors, *Simulating the Evolution of Language*, chapter 10, pages 211–226. Springer, London, 2002.

[78] G. Strube. Cognitive Modeling: Research Logic in Cognitive Science. In N. J. Smelser and P. B. Baltes, editors, *International Encyclopedia of the Social and Behavioral Sciences*, volume 3, pages 2124–2128. Elsevier Science, Oxford, 2001.

[79] M. Stubbs. *Text and Corpus Analysis. Computer-Assisted Studies of Language and Culture*. Blackwell, Cambridge, Massachusetts, 1996.

[80] M. Stubbs. On Inference Theories and Code Theories: Corpus Evidence for Semantic Schemas. *Text*, 21(3):437–465, 2001.

[81] H. Turner. An Introduction to Methods for Simulating the Evolution of Language. In A. Cangelosi and D. Parisi, editors, *Simulating the Evolution of Language*, pages 29–50. Springer, London/Berlin, 2002.

[82] L. A. Zadeh. Fuzzy Sets. *Information and Control*, 8:338–353, 1965.

[83] L. A. Zadeh. Toward a Theory of Fuzzy Information Granulation and its Centrality in Human Reasoning and Fuzzy Logic. *Fuzzy Sets and Systems*, 90:111–127, 1997.

[84] T. Ziemke. Rethinking Grounding. In A. Riegler, M. Peschl, and A. von Stein, editors, *Understanding Representation in the Cognitive Sciences. Does Representation Need Reality?*, pages 177–190. Kluwer/Plenum, New York/Boston/Dordrecht, 1999.

Part I

Information Modeling

Precisiated Natural Language

Lotfi A. Zadeh

University of California Berkeley
zadeh@cs.berkeley.edu

Summary. This article is a sequel to an article titled "A New Direction in AI –
Toward a Computational Theory of Perceptions", which appeared in the Spring 2001
issue of *AI Magazine* (volume 22, No. 1, 73-84). The concept of precisiated natural
language (PNL) was briefly introduced in that article, and PNL was employed as
a basis for computation with perceptions. In what follows, the conceptual structure
of PNL is described in greater detail, and PNL's role in knowledge representation,
deduction, and concept definition is outlined and illustrated by examples. What
should be understood is that PNL is in its initial stages of development and that
the exposition that follows is an outline of the basic ideas that underlie PNL rather
than a definitive theory.

A natural language is basically a system for describing perceptions. Perceptions,
such as perceptions of distance, likelihood, relevance, and most other attributes of
physical and mental objects are intrinsically imprecise, reflecting the bounded ability
of sensory organs, and ultimately the brain, to resolve detail and store information.
In this perspective, the imprecision of natural languages is a direct consequence of
the imprecision of perceptions [44, 45].

How can a natural language be precisiated – precisiated in the sense of making it possible to treat propositions drawn from a natural language as objects of
computation? This is what PNL attempts to do.

In PNL, precisiation is accomplished through translation into what is termed
a *precisiation language*. In the case of PNL, the precisiation language is the
generalized-constraint language (GCL), a language whose elements are so-called
generalized constraints and their combinations. What distinguishes GCL from languages such as Prolog, LISP, SQL, and, more generally, languages associated with
various logical systems, for example, predicate logic, modal logic, and so on, is its
so much higher expressive power.

The conceptual structure of PNL mirrors two fundamental facets of human cognition, (a) partiality and (b) granularity [43]. Partiality relates to the fact that most
human concepts are not bivalent, that is, are a matter of degree. Thus, we have
partial understanding, partial truth, partial possibility, partial certainty, partial
similarity, and partial relevance, to cite a few examples. Similarly, granularity and
granulation relate to clumping of values of attributes, forming granules with words
as labels, for example, young, middle-aged, and old as labels of granules of age.

L.A. Zadeh: *Precisiated Natural Language*, StudFuzz **209**, 33–59 (2007)
www.springerlink.com

Existing approaches to natural language processing are based on bivalent logic – a logic in which shading of truth is not allowed, PNL abandons bivalence. By so doing, PNL frees itself from limitations imposed by bivalence and categoricity, and opens the door to new approaches for dealing with long-standing problems in AI and related fields [26].

At this juncture, PNL is in its initial stages of development. As it matures, PNL is likely to find a variety of applications, especially in the realms of world knowledge representation, concept definition, deduction, decision, search, and question answering.

1 Introduction

Natural languages (NLs) have occupied, and continue to occupy, a position of centrality in AI. Over the years, impressive advances have been made in our understanding of how natural languages can be dealt with on processing, logical, and computational levels. A huge literature is in existence. Among the important contributions that relate to the ideas described in this article are those of [1, 2, 3, 9, 15, 21, 22, 23, 34, 35, 36].

When a language such as precisiated natural language (PNL) is introduced, a question that arises at the outset is: What can PNL do that cannot be done through the use of existing approaches? A simple and yet important example relates to the basic role of quantifiers such as *all, some, most, many*, and *few* in human cognition and natural languages.

In classical, bivalent logic the principal quantifiers are *all* and *some*. However, there is a literature on so-called generalized quantifiers exemplified by *most, many*, and *few* [2, 29]. In this literature, such quantifiers are treated axiomatically, and logical rules are employed for deduction.

By contrast, in PNL quantifiers such as *many, most, few, about 5, close to 7, much larger than 10*, and so on are treated as fuzzy numbers and are manipulated through the use of fuzzy arithmetic [12, 14, 40]. For the most part, inference is computational rather than logical. Following are a few simple examples. First, let us consider the Brian example [40]:

> Brian is much taller than most of his close friends.
> How tall is Brian?

At first glance it may appear that such questions are unreasonable. How can one say something about Brian's height if all that is known is that he is much taller than most of his close friends? Basically, what PNL provides is a system for precisiation of propositions expressed in a natural language through translation into the generalized-constraint language (GCL). Upon translation, the generalized constraints (GCs) are propagated through the use of rules governing generalized-constraint propagation, inducing a generalized constraint on the answer to the question. More specifically, in the Brian example, the answer is a generalized constraint on the height of Brian.

Now let us look at the balls-in-box problem:

> A box contains balls of various sizes and weights.
> The premises are:
>
> Most are large.
> Many large balls are heavy.
> _____
> What fraction of balls are large and heavy?

The PNL answer is: $most \times many$, where $most$ and $many$ are fuzzy numbers defined through their membership functions, and $most \times many$ is their product in fuzzy arithmetic [14]. This answer is a consequence of the general rule

$$\frac{Q_1 \text{ As are } Bs \qquad Q_2 \ (A \text{ and } B)\text{s are } Cs}{(Q_1 \times Q_2) \text{ As are } (B \text{ and } C)\text{s}}$$

Another simple example is the tall Swedes problem (version 1):

> Swedes who are more than twenty years old range in height from 140 centimeters to 220 centimeters. Most are tall. What is the average height of Swedes over twenty?

A less simple version of the problem (version 2) is the following (a^* denotes "approximately a") – a PNL-based answer to this problem is given in the appendix:

> Swedes over twenty range in height from 140 centimeters to 220 centimeters. Over 70* percent are taller than 170* centimeters; less than 10* percent are shorter than 150* centimeters, and less than 15 percent are taller than 200* centimeters. What is the average height of Swedes over twenty?

There is a basic reason generalized quantifiers do not have an ability to deal with problems of this kind. The reason is that in the theory of generalized quantifiers there is no concept of the count of elements in a fuzzy set. How do you count the number of tall Swedes if tallness is a matter of degree? More generally, how do you define the probability measure of a fuzzy event (cf. Zadeh [39])?

What should be stressed is that the existing approaches and PNL are complementary rather than competitive. Thus, PNL is not intended to be used in applications such as text processing, summarization, syntactic analysis, discourse analysis, and related fields. The primary function of PNL is to provide a computational framework for precisiation of meaning rather than to serve as a means of meaning understanding and meaning representation. By its nature, PNL is maximally effective when the number of precisiated propositions is small rather than large and when the chains of reasoning are short rather than long. The following is intended to serve as a backdrop.

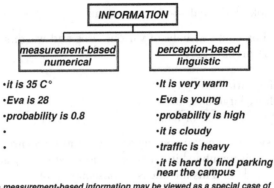

Fig. 1. Modalities of information: measurement-based and perception-based.

It is a deep-seated tradition in science to view the use of natural languages in scientific theories as a manifestation of mathematical immaturity. The rationale for this tradition is that natural languages are lacking in precision. However, what is not recognized to the extent that it should be is that adherence to this tradition carries a steep price. In particular, a direct consequence is that existing scientific theories do not have the capability to operate on perception-based information – information exemplified by "Most Swedes are tall", "Usually Robert returns from work at about 6 PM", "There is a strong correlation between diet and longevity", and "It is very unlikely that there will be a significant increase in the price of oil in the near future" (figure 1).

Such information is usually described in a natural language and is intrinsically imprecise, reflecting a fundamental limitation on the cognitive ability of humans to resolve detail and store information. Due to their imprecision, perceptions do not lend themselves to meaning representation and inference through the use of methods based on bivalent logic. To illustrate the point, consider the following simple examples.

The balls-in-box example:

A box contains balls of various sizes. My perceptions of the contents of the box are:
- There are about twenty balls.
- Most are large.
- There are several times as many large balls as small balls.

The question is: What is the number of small balls?

The Robert example (a):

My perception is:
- Usually Robert returns from work at about 6 PM.

The question is: What is the probability that Robert is home at about 6:15 PM?

The Robert example (b):

- Most tall men wear large-sized shoes.
- Robert is tall.
- What is the probability that Robert wears large-sized shoes?

An immediate problem that arises is that of meaning precisiation. How can the meaning of the perception "There are several times as many large balls as small balls" or "Usually Robert returns from work at about 6 PM" be defined in a way that lends itself to computation and deduction? Furthermore, it is plausible, on intuitive grounds, that "Most Swedes are tall" conveys some information about the average height of Swedes. But what is the nature of this information, and what is its measure? Existing bivalent-logic-based methods of natural language processing provide no answers to such questions.

The incapability of existing methods to deal with perceptions is a direct consequence of the fact that the methods are based on bivalent logic – a logic that is intolerant of imprecision and partial truth. The existing methods are categorical in the sense that a proposition, p, in a natural language, NL, is either true or not true, with no shades of truth allowed. Similarly, p is either grammatical or ungrammatical, either ambiguous or unambiguous, either meaningful or not meaningful, either relevant or not relevant, and so on. Clearly, categoricity is in fundamental conflict with reality – a reality in which partiality is the norm rather than an exception. But, what is much more important is that bivalence is a major obstacle to the solution of such basic AI problems as commonsense reasoning and knowledge representation [6, 7, 24, 34, 35, 37, 38], nonstereotypical summarization [22], unrestricted question answering, [18], and natural language computation [3].

PNL abandons bivalence. Thus, in PNL everything is, or is allowed to be, a matter of degree. It is somewhat paradoxical, and yet is true, that precisiation of a natural language cannot be achieved within the conceptual structure of bivalent logic.

By abandoning bivalence, PNL opens the door to a major revision of concepts and techniques for dealing with knowledge representation, concept definition, deduction, and question answering. A concept that plays a key role in this revision is that of a generalized constraint [42]. The basic ideas underlying this concept are discussed in the following section. It should be stressed that what follows is an outline rather than a detailed exposition.

•*standard constraint:* **X** ∈ **C**
•*generalized constraint:* **X isr R**

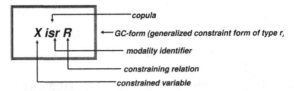

•*X= (X₁ , ..., Xₙ)*
•*X may have a structure: X=Location (Residence(Carol))*
•*X may be a function of another variable: X=f(Y)*
•*X may be conditioned: (X/Y)*
•*r := / ≤ / .../ ⊂ / ⊃ / blank / v / p / u / rs / fg / ps / ...*

Fig. 2. Schema of a generalized constraint.

2 The Concepts of Generalized Constraint and Generalized-Constraint Language

A conventional, hard constraint on a variable, X, is basically an inelastic restriction on the values that X can take. The problem is that in most realistic settings – and especially in the case of natural languages – constraints have some degree of elasticity or softness. For example, in the case of a sign in a hotel saying "Checkout time is 1 PM", it is understood that 1 PM is not a hard constraint on checkout time. The same applies to "Speed limit is 65 miles per hour" and "Monika is young". Furthermore, there are many different ways, call them modalities, in which a soft constraint restricts the values that a variable can take. These considerations suggest the following expression as the definition of generalized constraint (figure 2):

$$X \text{ isr } R,$$

where X is the constrained variable; R is the constraining relation; and r is a discrete-valued modal variable whose values identify the modality of the constraint [44]. The constrained variable may be an n-ary variable, $X = (X_1, \ldots, X_n)$; a conditional variable, $X|Y$; a structured variable, as in $Location(Residence(X))$; or a function of another variable, as in $f(X)$. The principal modalities are possibilistic ($r = blank$), probabilistic ($r = $ p), veristic ($r = $ v), usuality ($r = $ u), random set ($r = $ rs), fuzzy graph ($r = $ fg), bimodal ($r = $ bm), and Pawlak set ($r = $ ps). More specifically, in a possibilistic constraint,

$$X \text{ is } R,$$

R is a fuzzy set that plays the role of the possibility distribution of X. Thus, if $U = u$ is the universe of discourse in which X takes its values, then R is a fuzzy subset of U and the grade of membership of u in R, $\mu_R(U)$, is the possibility that $X = u$:

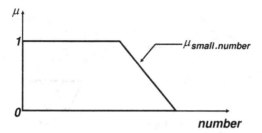

Fig. 3. Trapezoidal membership function of "Small number". "Small number" is context-dependent.

$$\mu_R(u) = Poss(X = u) \ .$$

For example, the proposition p: X *is a small number* is a possibilistic constraint in which "small number" may be represented as, say, a trapezoidal fuzzy number (figure 3), that represents the possibility distribution of X. In general, the meaning of "small number" is context-dependent.

In a probabilistic constraint:

$$X \text{ isp } R \ ,$$

X is a random variable and R is its probability distribution. For example,

$$X \text{ isp } N(m, \sigma^2)$$

means that X is a normally distributed random variable with mean m and variance σ^2.

In a veristic constraint, R is a fuzzy set that plays the role of the verity (truth) distribution of X. For example, the proposition "Alan is half German, a quarter French, and a quarter Italian", would be represented as the fuzzy set

$$Ethnicity(Alan) \text{ isv } (0.5 \,|\, German \ + \ 0.25 \,|\, French \ + \ 0.25 \,|\, Italian),$$

in which $Ethnicity(Alan)$ plays the role of the constrained variable; $0.5 \,|\, German$ means that the verity (truth) value of "Alan is German" is 0.5; and $+$ plays the role of a separator.

In a usuality constraint, X is a random variable, and R plays the role of the usual value of X. For example, X isu *small* means that usually X is small. Usuality constraints play a particularly important role in commonsense knowledge representation and perception-based reasoning.

In a random set constraint, X is a fuzzy-set-valued random variable and R is its probability distribution. For example,

$$X \text{ isrs } (0.3 \backslash small \ + \ 0.5 \backslash medium \ + \ 0.2 \backslash large),$$

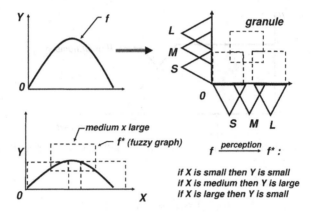

Fig. 4. Fuzzy graph of a function.

means that X is a random variable that takes the fuzzy sets *small, medium,* and *large* as its values with respective probabilities 0.3, 0.5, and 0.2. Random set constraints play a central role in the Dempster-Shafer theory of evidence and belief [30].

In a fuzzy graph constraint, the constrained variable is a function, f, and R is its fuzzy graph (figure 4). A fuzzy graph constraint is represented as

$$F \text{ isfg } \left(\sum_i A_i \times B_{j(i)} \right),$$

in which the fuzzy sets A_i and $B_{j(i)}$, with j dependent on i, are the granules of X and Y, respectively, and $A_i \times B_{j(i)}$ is the Cartesian product of A_i and $B_{j(i)}$. Equivalently, a fuzzy graph may be expressed as a collection of fuzzy if-then rules of the form

$$\text{if } X \text{ is } A_i \text{ then } Y \text{ is } B_{j(i)}, \quad i = 1, \ldots, m; \; j = 1, \ldots, n$$

For example:

$$F \text{ isfg } (small \times small + medium \times large + large \times small)$$

may be expressed as the rule set:

$$\text{if } X \text{ is } small \text{ then } Y \text{ is } small$$
$$\text{if } X \text{ is } medium \text{ then } Y \text{ is } large$$
$$\text{if } X \text{ is } large \text{ then } Y \text{ is } small$$

Such a rule set may be interpreted as a description of a perception of f.

A bimodal constraint involves a combination of two modalities: probabilistic and possibilistic. More specifically, in the generalized constraint

$$X \text{ isbm } R,$$

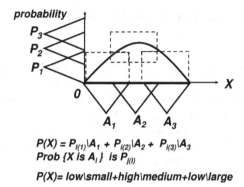

$$P(X) = P_{l(1)}\backslash A_1 + P_{l(2)}\backslash A_2 + P_{l(3)}\backslash A_3$$
$$\text{Prob}\ \{X\ is\ A_i\}\ is\ P_{j(i)}$$

$$P(X)= low\backslash small+high\backslash medium+low\backslash large$$

Fig. 5. Bimodal distribution: perception-based probability distribution.

X is a random variable, and R is what is referred to as a bimodal distribution, P, of X, with P expressed as

$$P : \sum_i P_{j(i)} \setminus A_i \ ,$$

in which the A_i are granules of X, and the $P_{j(i)}$, with j dependent on i, are the granules of probability (figure 5). For example, if X is a real-valued random variable with granules labeled *small, medium,* and *large* and probability granules labeled *low, medium,* and *high*, then

$$X\ isbm\ (low\backslash small\ +\ high\backslash medium\ +\ low\backslash large)$$

which means that

Prob(X is *small*) is *low*
Prob(X is *medium*) is *high*
Prob(X is *large*) is *low*

In effect, the bimodal distribution of X may be viewed as a description of a perception of the probability distribution of X. As a perception of likelihood, the concept of a bimodal distribution plays a key role in perception-based calculus of probabilistic reasoning [46].

The concept of a bimodal distribution is an instance of combination of different modalities. More generally, generalized constraints may be combined and propagated, generating generalized constraints that are composites of other generalized constraints. The set of all such constraints together with deduction rules – rules that are based on the rules governing generalized-constraint propagation – constitutes the generalized-constraint language (GCL). An example of a generalized constraint in GCL is

$$(X\ isp\ A)\ \text{and}\ ((X,\ Y)\ is\ B),$$

where A is the probability distribution of X and B is the possibility distribution of the binary variable (X, Y). Constraints of this form play an important role in the Dempster-Shafer theory of evidence [30].

3 The Concepts of Precisiability and Precisiation Language

Informally, a proposition, p, in a natural language, NL, is precisiable if its meaning can be represented in a form that lends itself to computation and deduction. More specifically, p is precisiable if it can be translated into what may be called a precisiation language, PL, with the understanding that the elements of PL can serve as objects of computation and deduction. In this sense, mathematical languages and the languages associated with propositional logic, first-order and higher-order predicate logics, modal logic, LISP, Prolog, SQL, and related languages may be viewed as precisiation languages. The existing PL languages are based on bivalent logic. As a direct consequence, the languages in question do not have sufficient expressive power to represent the meaning of propositions that are descriptors of perceptions. For example, the proposition "All men are mortal" can be precisiated by translation into the language associated with first-order logic, but "Most Swedes are tall" cannot.

The principal distinguishing feature of PNL is that the precisiation language with which it is associated is GCL. It is this feature of PNL that makes it possible to employ PNL as a meaning-precisiation language for perceptions. What should be understood, however, is that not all perceptions or, more precisely, propositions that describe perceptions, are precisiable through translation into GCL. Natural languages are basically systems for describing and reasoning with perceptions, and many perceptions are much too complex to lend themselves to precisiation.

The key idea in PNL is that the meaning of a precisiable proposition, p, in a natural language is a generalized constraint X isr R. In general, X, R, and r are implicit, rather than explicit, in p. Thus, translation of p into GCL may be viewed as an explicitation of X, R, and r. The expression X isr R will be referred to as the GC form of p, written as $\mathrm{GC}(p)$.

In PNL, a proposition, p, is viewed as an answer to a question, q. To illustrate, the proposition p: *Monika is young* may be viewed as the answer to the question q: *How old is Monika?* More concretely:

$$p : \text{Monika is young} \quad \rightarrow \quad p^* : \; Age(Monika) \text{ is } young$$
$$q : \text{How old is Monika?} \quad \rightarrow \quad q^* : \; Age(Monika) \text{ is } ?R$$

where p^* and q^* are abbreviations for $\mathrm{GC}(p)$ and $\mathrm{GC}(q)$, respectively.

In general, the question to which p is an answer is not unique. For example, p: *Monika is young* could be viewed as an answer to the question q: *Who is young?* In most cases, however, among the possible questions there is one that is most likely. Such a question plays the role of a default question. The GC form of q is, in effect, the translation of the question to which p is an answer. The following simple examples are intended to clarify the process of translation from NL to GCL.

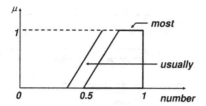

Fig. 6. Calibration of most and usually represented as trapezoidal fuzzy numbers.

$$p: \text{Tandy is much older than Dana} \rightarrow$$
$$(Age(Tandy), Age(Dana)) \text{ is } much.older$$

where *much.older* is a binary fuzzy relation that has to be calibrated as a whole rather through composition of *much* and *older*.

To deal with the example $p: Most Swedes are tall$, it is necessary to have a means of counting the number of elements in a fuzzy set. There are several ways in which this can be done, with the simplest way relating to the concept of ΣCount (sigma count). More specifically, if A and B are fuzzy sets in a space $U = \{u_1, \ldots, u_N\}$, with respective membership functions μ_A and μ_B, respectively, then

$$\Sigma\text{Count}(A) = \sum_i \mu_A(u_i) ,$$

and the relative ΣCount, that is, the relative count of elements of A that are in B, is defined as

$$\Sigma\text{Count}(A/B) = \frac{\Sigma\text{Count}(A \cap B)}{\Sigma\text{Count}(B)}$$

in which the membership function of the intersection $A \cap B$ is defined as

$$\mu_{A \cap B}(u) = \mu_A(u) \wedge \mu_B(u) ,$$

where $\wedge$ is min or, more generally, a *t*-norm [16, 28].

Using the concept of sigma count, the translation in question may be expressed as

$$p: \text{Most Swedes are tall} \rightarrow$$
$$\Sigma\text{Count}(tall.Swedes/Swedes) \text{ is } most ,$$

where *most* is a fuzzy number that defines *most* as a fuzzy quantifier [25, 41] (figure 6). A further sample translation looks as follows:

p : Usually Robert returns from work at about 6 PM.
q : When does Robert return from work?
X : Time of return of Robert from work, *Time(Return)*
R : about 6 PM (6*PM)
r : u (usuality)
p^* : Prob(*Time(Return)* is 6*PM) is *usually*.

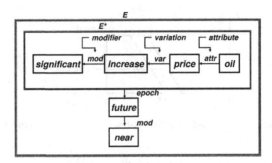

Fig. 7. Semantic network of p: It is very unlikely that there will be a significant increase in the price of oil in the near future.

A less simple example is:

> p: It is very unlikely that there will be a significant increase in the price of oil in the near future.

In this example, it is expedient to start with the semantic network representation [34] of p that is shown in figure 7. In this representation, E is the main event, and E^* is a sub-event of E:

> E : significant increase in the price of oil in the near future
> E^* significant increase in the price of oil
> Thus, *near future* is the epoch of E^*.

The GC form of p may be expressed as

$$\mathrm{Prob}(E) \text{ is } R,$$

where R is the fuzzy probability, *very unlikely*, whose membership function is related to that of *likely* by figure 8:

$$\mu_{\text{very.unlikely}}(u) = (1 - \mu_{\text{likely}}(u))^2 \,,$$

where it is assumed for simplicity that *very* acts as an intensifier that squares the membership function of its operand, and that the membership function of *unlikely* is the mirror image of that of *likely*.

Given the membership functions of *significant increase* and *near future* (figure 9), we can compute the degree to which a specified time function that represents a variation in the price of oil satisfies the conjunction of the constraints *significant increase* and *near future*. This degree may be employed to compute the truth value of p as a function of the probability distribution of the variation in the price of oil. In this instance, the use of PNL may be viewed as an extension of truth-conditional semantics [1, 5].

What should be noted is that precisiation and meaning representation are not coextensive. More specifically, precisiation of a proposition, p, assumes that the meaning of p is understood and that what is involved is a precisiation of the meaning of p.

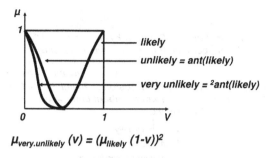

$$\mu_{very.unlikely}\ (v) = (\mu_{likely}\ (1\text{-}v))^2$$

Fig. 8. Precisiation of *very unlikely*.

4 The Concept of a Protoform and the Structure of PNL

A concept that plays a key role in PNL is that of a protoform – an abbreviation of prototypical form. Informally, a protoform is an abstracted summary of an object that may be a proposition, command, question, scenario, concept, decision problem, or, more generally, a system of such objects. The importance of the concept of a protoform derives from the fact that it places in evidence the deep semantic structure of the object to which it applies. For example, the protoform of the proposition

$$p\text{: Monika is young}$$

is

$$PF(p)\text{: } A(B) \text{ is } C\ ,$$

where A is abstraction of the attribute *Age*, B is abstraction of *Monika*, and C is abstraction of *young*. Conversely, *Age* is instantiation of A, *Monika* is instantiation of B, and *young* is instantiation of C. Abstraction may be annotated, for example, A/Attribute, B/Name, and C/Attribute.value. A few examples are shown in figure 10. Basically, abstraction is a means of generalization. Abstraction has levels, just as summarization does. For example, successive abstractions of *p: Monika is young* are $A(Monika)$ is *young*, $A(B)$ is *young*, and $A(B)$ is C, with the last abstraction resulting in the terminal protoform, or simply the protoform. With this understanding, the protoform of *p: Most Swedes are tall* is QAs are Bs, or equivalently, Count(B/A) is Q, and the protoform of *p: Usually Robert returns from work at about 6* PM, is Prob$(X$ is $A)$ is B, where X, A, and B are abstractions of "Time (Robert.returns.from work)", "About 6 PM", and "Usually." For simplicity, the protoform of p may be written as p^{**}.

Abstraction is a familiar concept in programming languages and programming systems. As will be seen in the following, the role of abstraction in PNL is significantly different and more essential because PNL abandons bivalence. The concept of a protoform has some links to other basic concepts such as ontology [4, 31, 32, 35], conceptual graph [33] and Montague grammar [27].

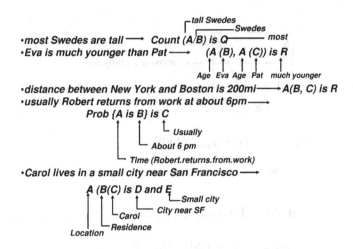

Fig. 9. Examples of translation from NL to PFL.

However, what should be stressed is that the concept of a protoform is not limited – as it is in the case of related concepts – to propositions whose meaning can be represented within the conceptual structure of bivalent logic.

As an illustration, consider a proposition, p, which was dealt with earlier:

p: It is very unlikely that there will be a significant increase in the price of oil in the near future.

With reference to the semantic network of p (figure 9), the protoform of p may be expressed as:

$$\text{Prob}(E) \text{ is } A \quad (A : \text{very unlikely})$$
$$E:\ B(E^*) \text{ is } C \quad (B : \text{epoch}; C : \text{near.future})$$
$$E^*:\ F(D) \quad\quad (F : \text{significant increase}; D : \text{price of oil})$$
$$D:\ G(H) \quad\quad (G : \text{price}; H : \text{oil})$$

Using the protoform of p and calibrations of *significant increase, near-future,* and *likely,* (figure 9), we can compute, in principle, the degree to which any given probability distribution of time functions representing the price of oil satisfies the generalized constraint, $\text{Prob}(E)$ is A. As was pointed out earlier, if the degree of compatibility is interpreted as the truth value of p, computation of the truth value of p may be viewed as a PNL-based extension of truth-conditional semantics.

By serving as a means of defining the deep semantic structure of an object, the concept of a protoform provides a platform for a fundamental mode of classification of knowledge based on protoform equivalence, or PF equivalence for short. More specifically, two objects are protoform equivalent at a specified level of summarization and abstraction if at that level they have

**E*: Epoch (Variation (Price (oil)) is significant.increase)
is near.future**

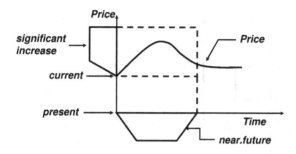

Fig. 10. Computation of degree of compatibility.

identical protoforms. For example, the propositions p: *Most Swedes are tall*, and q: *Few professors are rich*, are PF equivalent since their common protoform is *QAs are Bs* or, equivalently, Count(B/A) is Q. The same applies to propositions p: *Oakland is near San Francisco*, and q: *Rome is much older than Boston*. A simple example of PF equivalent concepts is: cluster and mountain.

A less simple example involving PF equivalence of scenarios of decision problems is the following. Consider the scenarios of two decision problems, A and B:

Scenario A:

Alan has severe back pain. He goes to see a doctor. The doctor tells him that there are two options: (1) do nothing and (2) do surgery. In the case of surgery, there are two possibilities: (a) surgery is successful, in which case Alan will be pain-free; and (b) surgery is not successful, in which case Alan will be paralyzed from the neck down. Question: Should Alan elect surgery?

Scenario B:

Alan needs to fly from San Francisco to St. Louis and has to get there as soon as possible. One option is to fly to St. Louis via Chicago, and the other is to go through Denver. The flight via Denver is scheduled to arrive in St. Louis at time a. The flight via Chicago is scheduled to arrive in St. Louis at time b, with $a < b$. However, the connection time in Denver is short. If the connection flight is missed, then the time of arrival in St. Louis will be c, with $c > b$. Question: Which option is best?

The common protoform of A and B is shown in figure 11. What this protoform means is that there are two options, one that is associated with a certain gain or loss and another that has two possible outcomes whose probabilities may not be known precisely.

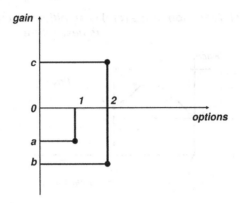

Fig. 11. Protoform equivalence of scenarios A and B.

The protoform language, PFL, is the set of protoforms of the elements of the generalized-constraint language, GCL. A consequence of the concept of PF equivalence is that cardinality of PFL is orders of magnitude lower than that of GCL or, equivalently, the set of precisiable propositions in NL. As will be seen in the sequel, the low cardinality of PFL plays an essential role in deduction.

The principal components of the structure of PNL (figure 12) are (1) a dictionary from NL to GCL; (2) a dictionary from GCL to PFL (figure 13); (3) a multiagent, modular deduction database, DDB; and (4) a world knowledge database, WKDB. The constituents of DDB are modules, with a module consisting of a group of protoformal rules of deduction, expressed in PFL (figure 14), that are drawn from a particular domain, for example, probability, possibility, usuality, fuzzy arithmetic [14], fuzzy logic, search, and so on. For example, a rule drawn from fuzzy logic is the compositional rule of inference, expressed in figure 14 where $A \circ B$ is the composition of A and B, defined in the computational part, in which μ_A, μ_B, and $\mu_{A \circ B}$ are the membership functions of A, B, and $A \circ B$, respectively. Similarly, a rule drawn from probability is shown in figure 15, where D is defined in the computational part.

The rules of deduction in DDB are, basically, the rules that govern propagation of generalized constraints. Each module is associated with an agent whose function is that of controlling execution of rules and performing embedded computations. The top-level agent controls the passing of results of computation from a module to other modules. The structure of protoformal, that is, protoform based, deduction is shown in figure 16. A simple example of protoformal deduction is shown in figure 17.

The world knowledge database, WKDB, consists of propositions that describe world knowledge, for example, Parking near the campus is hard to find on weekdays between 9 and 4; Big cars are safer than small cars; If A/person works in B/city then it is likely that A lives in or near B; If A/person is at home at time t then A has returned from work at t or earlier, on the under-

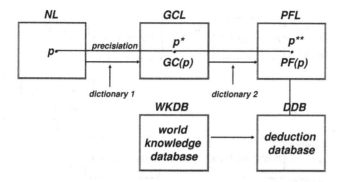

Fig. 12. Basic structure of PNL.

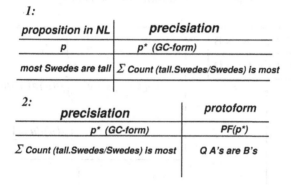

Fig. 13. Structure of PNL: dictionaries.

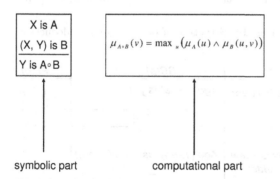

Fig. 14. Structure of protoform-based deduction.

standing that A stayed home after returning from work. Much, perhaps most, of the information in WKDB is perception based.

World knowledge – and especially world knowledge about probabilities – plays an essential role in almost all search processes, including searching the

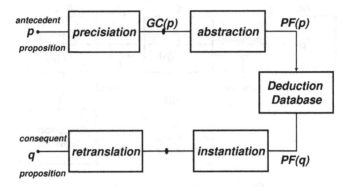

Fig. 15. Rule drawn from probability.

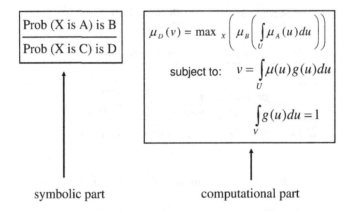

symbolic part   computational part

Fig. 16. Structure of protoform-based deduction.

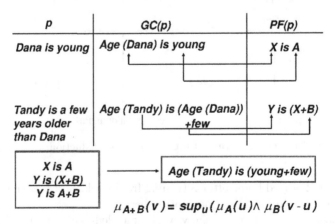

Fig. 17. Example of protoformal reasoning.

Web. Semantic Web and related approaches have contributed to a significant improvement in performance of search engines. However, for further progress it may be necessary to add to existing search engines the capability to operate on perception-based information. It will be a real challenge to employ PNL to add this capability to sophisticated knowledge-management systems such as the Web Ontology Language (OWL) [32], Cyc [19], WordNet [8], and ConceptNet [20].

An example of PFL-based deduction in which world knowledge is used is the so-called Robert example. A simplified version of the example is the following: The initial data set is the proposition (perception) p: *Usually Robert returns from work at about 6* PM. The question is q: *What is the probability that Robert is home at 6:15* PM?

The first step in the deduction process is to use the NL to GCL dictionary for deriving the generalized-constraint forms, $GC(p)$ and $GC(q)$, of p and q, respectively. The second step is to use the GCL to PFL dictionary to derive the protoforms of p and q. The forms are:

p^*: Prob($Time(Robert.returns.from.work)$ is about 6 PM) is *usually*
q^*: Prob($Time(Robert$ is home) is 6:15 PM) is $?E$

and

$$p^{**} : \text{Prob}(X \text{ is } A) \text{ is } B$$
$$q^{**} : \text{Prob}(Y \text{ is } C) \text{ is } ?D$$

The third step is to refer the problem to the top-level agent with the query: Is there a rule or a chain of rules in DDB that leads from p^{**} to q^{**}? The top-level agent reports a failure to find such a chain but success in finding a proximate rule of the form

$$\frac{\text{Prob}(X \text{ is } A) \text{ is } B}{\text{Prob}(X \text{ is } C) \text{ is } D}$$

The fourth step is to search the world knowledge database, WKDB, for a proposition or a chain of propositions that allow Y to be replaced by X. A proposition that makes this possible is (A/person is in B/location) at T/time if A arrives at B before T, with the understanding that A stays at B after arrival.

The last step involves the use of the modified form of q^{**}: Prob(X is E) is $?D$, in which E is "before 6:15 PM". The answer to the initial query is given by the solution of the variational problem associated with the rule that was described earlier (figure 15):

$$\frac{\text{Prob}(X \text{ is } A) \text{ is } B}{\text{Prob}(X \text{ is } C) \text{ is } D}$$

The value of D is the desired probability.

What is important to observe is that there is a tacit assumption that underlies the deduction process, namely, that the chains of deduction are short. This assumption is a consequence of the intrinsic imprecision of perception-based information. Its further implication is that PNL is likely to be effective, in the main, in the realm of domain-restricted systems associated with small universes of discourse.

5 PNL as a Definition Language

As we move further into the age of machine intelligence and automated reasoning, a problem that is certain to grow in visibility and importance is that of definability – that is, the problem of defining the meaning of a concept or a proposition in a way that can be understood by a machine.

It is a deeply entrenched tradition in science to define a concept in a language that is based on bivalent logic [10, 11, 13]. Thus defined, a concept, C, is bivalent in the sense that every object, X, is either an instance of C or it is not, with no degrees of truth allowed. For example, a system is either stable or unstable, a time series is either stationary or nonstationary, a sentence is either grammatical or ungrammatical, and events A and B are either independent or not independent.

The problem is that bivalence of concepts is in conflict with reality. In most settings, stability, stationarity, grammaticality, independence, relevance, causality, and most other concepts are not bivalent. When a concept that is not bivalent is defined as if it were bivalent, the ancient Greek sorites (heap) paradox comes into play. As an illustration, consider the standard bivalent definition of independence of events, say A and B. Let $P(A), P(B)$, and $P_A(B)$ be the probabilities of A, B, and B given A, respectively. Then A and B are independent if and only if $P_A(B) = P(B)$.

Now assume that the equality is not satisfied exactly, with the difference between the two sides being ϵ. As ϵ increases, at which point will A and B cease to be independent?

Clearly, independence is a matter of degree, and furthermore the degree is context dependent. For this reason, we do not have a universally accepted definition of *degree of independence* [17].

One of the important functions of PNL is that of serving as a definition language. More specifically, PNL may be employed as a definition language for two different purposes: first, to define concepts for which no general definitions exist, for example, causality, summary, relevance, and smoothness; and second, to redefine concepts for which universally accepted definitions exist, for example, linearity, stability, independence, and so on. In what follows, the concept of independence of random variables will be used as an illustration.

For simplicity, assume that X and Y are random variables that take values in the interval $[a, b]$. The interval is granulated as shown in figure 18, with S, M, and L denoting the fuzzy intervals *small*, *medium*, and *large*.

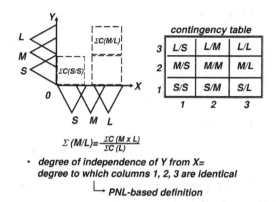

$$\Sigma(M/L) = \frac{\Sigma C\,(M \times L)}{\Sigma C\,(L)}$$

- **degree of independence of Y from X=**
  degree to which columns 1, 2, 3 are identical
  ↳ **PNL-based definition**

Fig. 18. PNL-based definition of statistical independence.

Using the definition of relative ΣCount, we construct a contingency table, C, of the form shown in figure 18, in which an entry such as ΣCount(S/L) is a granulated fuzzy number that represents the relative ΣCount of occurrences of Y, which are small, relative to occurrences of X, which are large.

Based on the contingency table, the degree of independence of Y from X may be equated to the degree to which the columns of the contingency table are identical. One way of computing this degree is, first, to compute the distance between two columns and then aggregate the distances between all pairs of columns. PNL would be used for this purpose

An important point that this example illustrates is that, typically, a PNL-based definition involves a general framework with a flexible choice of details governed by the context or a particular application. In this sense, the use of PNL implies an abandonment of the quest for universality, or, to put it more graphically, of the one-size-fits-all modes of definition that are associated with the use of bivalent logic.

Another important point is that PNL suggests an unconventional approach to the definition of complex concepts. The basic idea is to define a complex concept in a natural language and then employ PNL to precisiate the definition.

More specifically, let U be a universe of discourse and let C be a concept that I wish to define, with C relating to elements of U. For example, U is a set of buildings, and C is the concept of tall building. Let $p(C)$ and $d(C)$ be, respectively, my perception and my definition of C. Let $I(p(C))$ and $I(d(C))$ be the intensions of $p(C)$ and $d(C)$, respectively, with intension used in its logical sense [5, 10], that is, as a criterion or procedure that identifies those elements of U that fit $p(C)$ or $d(C)$. For example, in the case of tall buildings, the criterion may involve the height of a building.

Informally, a definition, $d(C)$, is a good fit or, more precisely, is cointensive, if its intension coincides with the intension of $p(C)$. A measure of goodness of fit is the degree to which the intension of $d(C)$ coincides with that of $p(C)$. In this sense, cointension is a fuzzy concept. As a high-level definition language, PNL makes it possible to formulate definitions whose degree of cointensiveness is higher than that of definitions formulated through the use of languages based on bivalent logic.

6 Concluding Remarks

Existing theories of natural languages are based, anachronistically, on Aristotelian logic – a logical system whose centerpiece is the principle of the excluded middle: Truth is bivalent, meaning that every proposition is either true or not true, with no shades of truth allowed.

The problem is that bivalence is in conflict with reality – the reality of pervasive imprecision of natural languages. The underlying facts are (a) a natural language, NL, is, in essence, a system for describing perceptions; and (b) perceptions are intrinsically imprecise, reflecting the bounded ability of sensory organs, and ultimately the brain, to resolve detail and store information.

PNL abandons bivalence. What this means is that PNL is based on fuzzy logic – a logical system in which everything is, or is allowed to be, a matter of degree.

Abandonment of bivalence opens the door to exploration of new directions in theories of natural languages. One such direction is that of precisiation. A key concept underlying precisiation is the concept of a generalized constraint. It is this concept that makes it possible to represent the meaning of a proposition drawn from a natural language as a generalized constraint. Conventional, bivalent constraints cannot be used for this purpose. The concept of a generalized constraint provides a basis for construction of GCL – a language whose elements are generalized constraints and their combinations. Within the structure of PNL, GCL serves as a precisiation language for NL. Thus, a proposition in NL is precisiated through translation into GCL. Not every proposition in NL is precisiable. In effect, the elements of PNL are precisiable propositions in NL.

What should be underscored is that in its role as a high-level definition language, PNL provides a basis for a significant enlargement of the role of natural languages in scientific theories.

References

[1] K. Allan. *Natural Language Semantics*. Blackwell Publishers, 2001.
[2] J. Barwise and R. Cooper. Generalized Quantifiers and Natural Language. *Linguistics and Philosophy*, 4(1):159–209, 1981.
[3] A. W. Biermann and B. W. Ballard. Toward Natural Language Computation. *American Journal of Computational Linguistics*, 6(2):71–86, 1980.

[4] O. Corcho, M. Fernandez-Lopez, and A. Gomez-Perez. Methodologies, Tools and Languages for Building Ontologies. Where is their Meeting Point? *Data and Knowledge Engineering*, 46(1):41–64, 2003.

[5] M. J. Cresswell. *Logic and Languages*. Methuen, London, 1973.

[6] E. Davis. *Representations of Common-sense Knowledge*. Morgan Kaufmann, San Francisco, 1990.

[7] D. Dubois and H. Prade. Approximate and Commonsense Reasoning: From Theory to Practice. In *Proceedings of the Foundations of Intelligent Systems*, pages 19–33, Berlin, 1996. Springer.

[8] C. Fellbaum, editor. *WordNet: An Electronic Lexical Database*. MIT Press, Cambridge, Mass., 1998.

[9] N. E. Fuchs and U. Schwertel. Reasoning in Attempto Controlled English. In *Lecture Notes in Computer Science*, pages 174–188, Berlin, 2003. Workshop on Principles and Practice of Semantic Web Reasoning (PP-SWR 2003), Springer.

[10] T. F. Gamat. *Language, Logic and Linguistics*. University of Chicago Press, Chicago, 1996.

[11] G. Gerla. Fuzzy Metalogic for Crisp Logics. In V. Novak and I. Perfilieva, editors, *Discovering the World with Fuzzy Logic: Studies in Fuzziness and Soft Computing*, pages 175–187. Physica-Verlag, Heidelberg, 2000.

[12] P. Hajek. *Metamathematics of Fuzzy Logic: Trends in Logic*, volume 4. Kluwer Academic Publishers, Dordrecht, The Netherlands, 1998.

[13] P. Hajek. Many. In V. Novak and I. Perfilieva, editors, *Discovering the World with Fuzzy Logic: Studies in Fuzziness and Soft Computing*, pages 302–304. Physica-Verlag, Heidelberg, 2000.

[14] A. Kaufmann and M. M. Gupta. *Introduction to Fuzzy Arithmetic: Theory and Applications*. Van Nostrand, New York, 1985.

[15] E. Klein. A Semantics for Positive and Comparative Adjectives. *Linguistics and Philosophy*, 4(1):1–45, 1980.

[16] P. Klement, R. Mesiar, and E. Pap. Triangular Norms – Basic Properties and Representation Theorems. In V. Novak and I. Perfilieva, editors, *Discovering the World with Fuzzy Logic: Studies in Fuzziness and Soft Computing*, pages 63–80. Physica-Verlag, Heidelberg, 2000.

[17] G. J. Klir. Uncertainty-Based Information: A Critical Review. In V. Novak and I. Perfilieva, editors, *Discovering the World with Fuzzy Logic: Studies in Fuzziness and Soft Computing*, pages 29–50. Physica-Verlag, Heidelberg, 2000.

[18] W. G. Lehnert. *The Process of Question Answering – A Computer Simulation of Cognition*. Lawrence Erlbaum Associates, Hillsdale, New Jersey, 1978.

[19] D. B. Lenat. CYC: A large-scale investment in knowledge infrastructure. *Communications of the ACM*, 38(11):32–38, 1995.

[20] H. Liu and P. Singh. Commonsense Reasoning in and over Natural Language. In *Proceedings of the Eighth International Conference on Knowledge-Based Intelligent Information and Engineering Systems*,

Brighton, U. K., 2004. KES Secretariat, Knowledge Transfer Partnership Centre.

[21] B. Macias and S. G. Pulman. A Method for Controlling the Production of Specifications in Natural Language. *The Computing Journal*, 38(4):310–318, 1995.

[22] I. Mani and M. T. Maybury, editors. *Advances in Automatic Text Summarization*. MIT Press, Cambridge, Mass., 1999.

[23] D. A. McAllester and R. Givan. Natural Language Syntax and First-Order Inference. *Artificial Intelligence*, 56(1):1–20, 1992.

[24] J. McCarthy. *Formalizing Common Sense*. Ablex Publishers, Norwood, New Jersey, 1990.

[25] R. Mesiar and H. Thiele. On T-Quantifiers and S-Quantifiers. In V. Novak and I. Perfilieva, editors, *Discovering the World with Fuzzy Logic: Studies in Fuzziness and Soft Computing*, pages 310–318. Physica-Verlag, Heidelberg, 2000.

[26] V. Novak. Fuzzy Logic, Fuzzy Sets, and Natural Languages. *International Journal of General Systems*, 20(1):83–97, 1991.

[27] B. Partee. *Montague Grammar*. Academic Press, New York, 1976.

[28] W. Pedrycz and F. Gomide. *Introduction to Fuzzy Sets*. MIT Press, Cambridge, Mass., 1998.

[29] P. Peterson. On the Logic of Few, Many and Most. *Journal of Formal Logic*, 20(1-2):155–179, 1979.

[30] G. Shafer. *A Mathematical Theory of Evidence*. Princeton University Press, Princeton, New Jersey, 1976.

[31] B. Smith and C. Welty. What Is Ontology? Ontology: Towards a New Synthesis. In *Proceedings of the Second International Conference on Formal Ontology in Information Systems*, New York, 2002. ACM.

[32] M. K. Smith, C. Welty, and D. McGuinness. OWL Web ontology language guide. W3C working draft 31. In M. K. Smith, C. Welty, and D. McGuinness, editors, *W3C Working Draft*. World Wide Web Consortium (W3C), Cambridge, Mass., 2003.

[33] J. F. Sowa. *Conceptual Structures: Information Processing in Mind and Machine*. Addison-Wesley, Reading, Mass., 1984.

[34] J. F. Sowa. *Principles of Semantic Networks: Explorations in the Representation of Knowledge*. Morgan Kaufmann, San Francisco, 1991.

[35] J. F. Sowa. Ontological Categories. In L. Albertazzi, editor, *Shapes of Forms: From Gestalt Psychology and Phenomenology to Ontology and Mathematics*, pages 307–340. Kluwer Academic Publishers, Dordrecht, The Netherlands, 1999.

[36] J. Sukkarieh. Mind your Language! Controlled Language for Inference Purposes. Oral presentation, May 2003. Paper presented at the Joint Conference of the Eighth International Workshop of the European Association for Machine Translation and the Fourth Controlled Language Applications Workshop, Dublin, Ireland.

[37] R. Sun. *Integrating Rules and Connectionism for Robust Commonsense Reasoning*. John Wiley, New York, 1994.

[38] R. R. Yager. Deductive Approximate Reasoning Systems. *IEEE Transactions on Knowledge and Data Engineering*, 3(4):399–414, 1991.

[39] L. A. Zadeh. Probability Measures of Fuzzy Events. *Journal of Mathematical Analysis and Applications*, 23:421–427, 1968.

[40] L. A. Zadeh. A Computational Approach to Fuzzy Quantifiers in Natural Languages. *Computers and Mathematics*, 9:149–184, 1983.

[41] L. A. Zadeh. Syllogistic Reasoning in Fuzzy Logic and its Application to Reasoning with Dispositions. In *Proceedings International Symposium on Multiple-Valued Logic*, pages 148–153, Los Alamitos, Calif., 1984. IEEE.

[42] L. A. Zadeh. Outline of a Computational Approach to Meaning and Knowledge Representation Based on the Concept of a Generalized Assignment Statement. In M. Thoma and A. Wyner, editors, *Proceedings of the International Seminar on Artificial Intelligence and Man-Machine Systems*, pages 198–211, Heidelberg, 1986. Springer.

[43] L. A. Zadeh. Toward a Theory of Fuzzy Information Granulation and its Centrality in Human Reasoning and Fuzzy Logic. *Fuzzy Sets and Systems*, 90(2):111–127, 1997.

[44] L. A. Zadeh. From Computing with Numbers to Computing with Words – From Manipulation of Measurements to Manipulation of Perceptions. *IEEE Transactions on Circuits and Systems*, 45(1):105–119, 1999.

[45] L. A. Zadeh. Toward a Logic of Perceptions Based on Fuzzy Logic. In V. Novak and I. Perfilieva, editors, *Discovering the World with Fuzzy Logic: Studies in Fuzziness and Soft Computing*, pages 4–25. Physica-Verlag, Heidelberg, 2000.

[46] L. A. Zadeh. Toward a Perception-Based Theory of Probabilistic Reasoning with Imprecise Probabilities. *Journal of Statistical Planning and Inference*, 105(1):233–26, 2002.

Appendix

The Tall Swedes Problem (Version 2)

In the following version of the tall Swedes problem, a^* denotes "approximately a.":

> Swedes more than twenty years of age range in height from 140 centimeters to 220 centimeters. Over 70^* percent are taller than 170^* centimeters; less than 10^* percent are shorter than 150^* centimeters; and less than 15 percent are taller than 200^* centimeters. What is the average height of Swedes over twenty?

Fuzzy Logic Solution

Consider a population of Swedes over twenty, i.e. $S = \{\text{Swede}_1, \text{Swede}_2, \ldots, \text{Swede}_N\}$, with h_i, $i = 1, \ldots, N$, being the height of $S_i \in S$.

The datum "Over 70^* percent of S are taller than 170^* centimeters", constrains the h_i in $h = (h_i, \ldots, h_N)$. The constraint is precisiated through translation into GCL. More specifically, let X denote a variable taking values in S, and let $X|(h(X)$ is $\geq 170^*)$ denote a fuzzy subset of S induced by the constraint $h(X)$ is $\geq 170^*$. Then

Over 70^* percent of S are taller than 170^* $\rightarrow$
(GCL): $\frac{1}{N}\Sigma\text{Count}(X|h(X)$ is $\geq 170^*)$ is $\geq 0.7^*$

where ΣCount is the sigma count of Xs that satisfy the fuzzy constraint $h(X)$ is $\geq 170^*$. Similarly,

Less than 10^* percent of S are shorter than 150^* $\rightarrow$
(GCL): $\frac{1}{N}\Sigma\text{Count}(X|h(X)$ is $\leq 150^*)$ is ≤ 0.1

and

Less than 15^* percent of S are taller than 200^* $\rightarrow$
(GCL): $\frac{1}{N}\Sigma\text{Count}(X|h(X)$ is $\geq 200^*)$ is ≤ 0.15

A general deduction rule in fuzzy logic is the following – in this rule, X is a variable that takes values in a finite set $U = \{u_1, u_2, \ldots, u_N\}$, and $a(X)$ is a real-valued attribute of X, with $a_i = a(u_i)$ and $a = (a_i, \ldots, a_N)$:

$$\frac{\frac{1}{N}\Sigma\text{Count}(X|a(X) \text{ is } C) \text{ is } B}{Av(X) \text{ is } ?D}$$

where $Av(X)$ is the average value of X over U. Thus, computation of the average value, D, reduces to the solution of the nonlinear programming problem

$$\mu_D(v) = \max_{a_i} \mu_B\left(\frac{1}{N}\sum_i \mu_i(a_i)\right)$$

subject to

$$v = \frac{1}{N}\sum_i a_i \quad \text{(average value)}$$

where μ_D, μ_B and μ_C are the membership functions of D, B, and C, respectively. To apply this rule to the constraints in question, it is necessary to form their conjunction. Then, the fuzzy logic solution of the problem may be reduced to the solution of the nonlinear programming problem

$$\mu_D(v) = \max_{h_i} \left(\mu_{\geq 0.7^*} \left(\frac{1}{N}\sum_i \mu_{\geq 170^*}(h_i) \right) \wedge \right.$$

$$\left. \mu_{\leq 0.1^*} \left(\frac{1}{N}\sum_i \mu_{\leq 150^*}(h_i) \right) \wedge \mu_{\leq 0.15^*} \left(\frac{1}{N}\sum_i \mu_{\geq 200^*}(h_i) \right) \right)$$

subject to

$$v = \frac{1}{N}\sum_i h_i$$

Note that computation of D requires calibration of the membership functions of $\leq 170^*$, $\leq 0.7^*$, $\leq 150^*$, $\leq 0.1^*$, $\geq 200^*$, and $\leq 0.15^*$.

On the Issue of Linguistic Approximation

George J. Klir[1] and Kari Sentz[2]

[1] Binghamton University (SUNY)
gklir@binghamton.edu
[2] Los Alamos National Laboratory
ksentz@lanl.gov

1 Introduction

The purpose of this paper is twofold. First, we discuss the various issues of linguistic approximation in the context of intelligent systems. Second, we address one of the approximation issues in particular, which seems to be somewhat neglected in the literature.

In general, *intelligent systems* are defined in the literature as human-made systems that are capable of achieving highly complex tasks in human-like, intelligent way. The qualifier "human-like" in this definition is crucial for distinguishing the area of intelligent systems within the broad field of artificial intelligence. In intelligent systems, the *human mind* is viewed as a *role model* and the aim is to understand and emulate its various cognitive capabilities that allow human beings to perform remarkably complex tasks.

Two of the most exemplary capabilities of the human mind are the capability of using perceptions in purposeful ways and the capability of approximating perceptions by statements in natural language. Understanding these capabilities and emulating them by machines is the crux of intelligent systems. To construct intelligent systems, we need to develop appropriate methodological tools for dealing with perceptions in machines. As has recently been argued by Zadeh [22, 23], a feasible way to deal with perceptions in machines is to approximate them by statements in natural language and then to use fuzzy logic to represent these statements and deal with them as needed. This approach to developing perception-based machines, which is currently a subject of active research, is referred to in the literature as *computing with words* [18, 24]. This evocative term was coined by Zadeh [22]; to capture its meaning, we can hardly do better than to quote from his more recent paper:

"Computing, in its usual sense, is centered on manipulation of numbers and symbols. In contrast, computing with words is a methodology in which the objects of computation are words and propositions drawn from a natural language [...]. Computing with words is inspired by the remarkable human capability to perform a wide variety of physical and mental tasks without any measurements and any computations. Familiar examples of such tasks are parking a car, driving in heavy traffic, playing golf, riding a bicycle, understanding speech, and summarizing a story. Underlying this remarkable capability is the brain's crucial ability to manipulate perceptions – perceptions of distance, size, weight, color, speed, time, direction, force, number, truth, likelihood, and other characteristics of physical and mental objects. Manipulation of perceptions plays a key role in human recognition, decision and execution processes. As a methodology, computing with words provides a foundation for a computational theory of perceptions – a theory which may have an important bearing on how human beings make – and machines might make – perception-based rational decisions in an environment of imprecision, uncertainty and partial truth.

A basic difference between the perception and measurements is that, in general, measurements are crisp whereas perceptions are fuzzy. One of the fundamental aims of science has been and continues to be that of progressing from perceptions to measurements. Pursuit of this aim has led to brilliant successes. [...]. But alongside the brilliant successes stand conspicuous underachievements and outright failures. We cannot build robots which can move with the agility of animals or humans; we cannot automate driving in heavy traffic; we cannot translate from one language to another at the level of a human interpreter, we cannot create programs which can summarize nontrivial stories; our ability to model the behavior of economic systems leaves much to be desired; and we cannot build machines that can compete with children in the performance of a wide variety of physical and cognitive tasks.

It may be argued that underlying the underachievements and failures is the unavailability of a methodology for reasoning and computing with perceptions rather than measurements." [23, p. 105].

Computing with words can thus be viewed as an underlying methodology for computing with perceptions. It utilizes the capability of natural language to approximate perceptions. However, the meaning of statements in natural language is strongly context dependent. Once we approximate them in the context of each particular application by appropriate propositions of fuzzy logic, we can utilize all available resources of fuzzy logic to formalize approximate, human-like reasoning [1, 10, 11, 19]. Sound metamathematical foundations for this approximate reasoning have already been developed [5].

The usual outcome of reasoning with fuzzy propositions is a fuzzy set [10]. For some purpose (such as control), we need to replace this fuzzy set with a single value which, in some sense, is its best representative. This replacement (or a single-value approximation) of the given fuzzy set is called a *defuzzification* [10]. For other purposes (such as communication of intelligent machines with human beings), we need to approximate the given fuzzy set by an appropriate linguistic term represented by another fuzzy set. This latter approximation is of our interest in this article and we use the term "linguistic approximation" in this sense. Our primary approach to dealing with this approximation issue is based on quantifying information closeness between the given fuzzy set and its approximation. However, we also examine several other approaches.

Although we assume that the reader is familiar with fundamentals of fuzzy set theory and fuzzy logic, we introduce relevant concepts and symbols in section 2. The problem of linguistic approximation that is of our interest in this article is discussed in section 3. Information-based approximation is introduced in section 4 and several other approaches to linguistic approximation are examined in section 5. The various approaches are illustrated and compared by examples in sections 6 and 7. Our conclusions are presented in section 8.

2 Relevant Concepts and Notation

In this article, we consider only *standard fuzzy sets*, and we denote them by capital letters. To define a standard fuzzy set A on a universal set of concern, X, a value in the unit interval $[0, 1]$ is assigned to each element x of X. This number is viewed as the degree of membership of x in A. The function defining these assignments for all elements in X is called a *membership function* of A. According to the common practice, we use the same symbol for a membership function and the associated fuzzy set. A degenerate fuzzy set for which $A(x) \in \{0, 1\}$ for all $x \in X$ is called a *crisp set*.

The largest membership degree defined by A is called a *height* of A and we denote it by h_A. When $h_A = 1$, A is called a *normal fuzzy set*. For each $\alpha \in [0, 1]$, we define a crisp set

$$^{\alpha}A = \{x \in X | A(x) \geq \alpha\}, \tag{1}$$

which is called an α-*cut* of A. It is well established [10] that the family of α-cuts $^{\alpha}A$ for all $\alpha \in [0, 1]$ is a unique representation of A. That is, there is a one-to-one correspondence between membership functions of fuzzy sets and their α-cut representations. Any property of classical set theory can be extended to fuzzy set theory by requiring that it be preserved in the classical sense in each α-cut of a given fuzzy set. Properties of fuzzy sets that satisfy this requirement are called *cutworthy properties*.

When a fuzzy set A is defined on a finite universal set X, its sigma count, $|A|$, is defined by the formula

$$|A| = \sum_{x \in X} A(x). \tag{2}$$

The most common operations of a complement, intersection, and union of fuzzy sets, which are usually referred to in the literature as the *standard operations*, are defined for all $x \in X$ by the following formulas:

$$\overline{A}(x) = 1 - A(x) \text{ (standard complement)},$$
$$(A \cap B)(x) = \min\{A(x), B(x)\} \text{ (standard intersection)},$$
$$(A \cup B)(x) = \max\{A(x), B(x)\} \text{ (standard union)}.$$

Standard intersection and union are the only cutworthy operations among all possible intersections and unions of fuzzy sets. None of the operations of complement of fuzzy sets is cutworthy.

Given two fuzzy sets, A and B, defined on the same universal set X, A is said to be a *subset* of B if and only if

$$A(x) \leq B(x) \tag{3}$$

for all $x \in X$. The usual notation $A \subseteq B$ is used to signify that A is a subset of B. It is also useful to define a more general concept, the *degree of subsethood*, $S(A \subseteq B)$, by the formula:

$$S(A \subseteq B) = \frac{|A \cap B|}{|A|} \tag{4}$$

The family of all fuzzy subsets of X is called the *fuzzy power set* of X and it is denoted by $\mathbb{F}(X)$.

Fuzzy sets that are defined on the set $\mathbb{R}$ of real numbers (i.e. $X = \mathbb{R}$) have a special significance in this article. They are interpreted as *fuzzy intervals* if they satisfy the following requirements:

1. they are normal fuzzy sets;
2. their supports are bounded;
3. their α-cuts are closed intervals of real numbers for all $\alpha \in [0, 1]$.

When $A(x) = 1$ for exactly one $x \in \mathbb{R}$, then this special fuzzy interval A is called a *fuzzy number*. A convenient way of expressing any fuzzy interval A is the canonical form

$$A(x) = \begin{cases} f_A(x) & \text{when } x \in [a, b), \\ 1 & \text{when } x \in [b, c], \\ g_A(x) & \text{when } x \in (c, d], \\ 0 & \text{otherwise,} \end{cases} \tag{5}$$

where $x \in \mathbb{R}$, f_A is a real-valued function that is nondecreasing and continuous from the right, g_A is a real-valued function that is nonincreasing and continuous from the left, and a, b, c, d are real numbers such that $a \leq b \leq c \leq d$. When $b = c$, A is a fuzzy number. Given any fuzzy interval A in this canonical form, its α-cuts are expressed for all $\alpha \in (0, 1]$ by the formula

$$^{\alpha}A = \begin{cases} [f_A^{-1}(\alpha), g_A^{-1}(\alpha)] & \text{when } \alpha \in (0, 1), \\ [b, c] & \text{when } \alpha = 1, \end{cases} \tag{6}$$

where f_A^{-1} and g_A^{-1} are inverse (or pseudoinverse) functions of f_A and g_A, respectively.

For fuzzy intervals, equation (2) is not directly applicable. However, it is meaningful to modify the definition of the sigma count by replacing the summation in (2) by integration. That is,

$$|A| = \int_X A(x)dx \tag{7}$$

when A is a fuzzy interval. This modified definition is then used in (4) when A and B are fuzzy intervals.

A *linguistic variable* [20, 21] is a variable whose states are fuzzy intervals (or fuzzy numbers) rather than real numbers. Each of these fuzzy intervals represents the meaning of a linguistic term in the context of a particular application. More specifically, a linguistic variable consists of the following components:

1. a *name*, which should adequately capture its meaning;
2. a *base variable*, which is an ordinary real-valued variable with its range of values (a closed interval of real numbers);
3. a set of *linguistic terms* that refer to values of the base variable;
4. a *semantic rule*, which assigns to each linguistic terms its meaning, in a given application context, by an appropriate fuzzy interval (or a fuzzy number) defined on the range of the base variable.

As is discussed in Section 3, linguistic variables play a crucial role in the kind of linguistic approximation that is of our concern in this article.

Each fuzzy set involves two basic types of uncertainty: *fuzziness* and *nonspecificity* [9]. While fuzziness is *linguistic uncertainty*, nonspecificity is *information-based uncertainty*. In this article, we develop an approach to linguistic approximation that is based on the information-based uncertainty: *nonspecificity*. For fuzzy sets on a finite universal set X, the measure of nonspecificity is a generalization of the classical Hartley measure for crisp sets [9]. Given a normal fuzzy set A on a finite universal set X, *the generalized Hartley measure of nonspecificity* of A, $GH(A)$, is defined by the formula

$$GH(A) = \int_0^1 \log_2 |{}^\alpha A| d\alpha, \tag{8}$$

provided that the measurement units are *bits*. One bit of uncertainty is equivalent to the uncertainty regarding the truth value of one elementary proposition. This measure is well justified and its uniqueness has been proven [9].

For normal fuzzy sets defined on an infinite universal set X (a convex subset of the n-dimensional Euclidean space $\mathbb{R}^n$ for some $n \geq 1$), a *generalized Hartley-like measure of nonspecificity*, GHL, is also well established [9, 16]. It is defined by the formula

$$GHL(A) = \int_0^1 \min_{t \in T} \log_2 [\prod_{i=1}^n (1 + \mu({}^\alpha A_{i_t})) + \mu({}^\alpha A) - \prod_{i=1}^n \mu({}^\alpha A_{i_t}))]d\alpha, \tag{9}$$

where A is a convex fuzzy subset of X, T is the set of all isometric transformations from one orthogonal system to another, A_{i_t} is the i-th projection of A in coordinate system t, and μ denotes the Lebesgue measure. When $n = 1$, which is the case of our interest in this article, equation (9) reduces to

$$GHL(A) = \int_0^1 \log_2[1 + \mu(^\alpha A)]d\alpha. \tag{10}$$

3 Linguistic Approximation of Fuzzy Sets

The problem of linguistic approximation that is of our interest in this article can be described, in general, as follows. A linguistic variable v is given whose base variable x ranges over the interval $X = [\underline{x}, \overline{x}]$. We assume that v is associated with a set of linguistic terms, $L = \{L_1, L_2, \ldots, L_n\}$, and that these linguistic terms are represented by a family of specific fuzzy intervals $F = \{F_1, F_2, \ldots, F_n\}$ defined on X. Fuzzy intervals in F are usually referred to as *granules*. We assume that $n \geq 2$ and that any linguistic term L_i is represented by a fuzzy interval (granule) F_i for each $i \in \mathbb{N}_n$, where $\mathbb{N}_n = \{1, 2, \ldots, n\}$. We also assume that F is a fuzzy partition of X so that

$$\sum_{i=1}^n F_i(x) = 1 \tag{11}$$

for all $x \in X$, and that $F_i \prec F_{i+1}$ for all $i \in \mathbb{N}_{n-1}$, where $\prec$ denotes the partial ordering of fuzzy intervals [10]. Given now a fuzzy set A expressing information about the actual value of variable x (i. e. A is a fuzzy interval on X), our aim is to determine the most descriptive linguistic approximation of A in terms of the linguistic terms available in L. To illustrate possible variations of this approximation problem, let us consider the following cases:

1. the approximation is required to have the form "x is L_i" for some particular $L_i \in L$,
2. the approximation may be in the form "x is $\mathrm{EXPR}(L_1, L_2, \ldots, L_n)$", where EXPR denotes a linguistic expression containing two or more linguistic terms in L represented by fuzzy intervals in F that are combined by appropriate logical connectives;
3. the approximation may be in the form "x is $m(L_i)$" for some particular $L_i \in L$, where m denotes an appropriate modifier describing some linguistic hedge;
4. the approximation may be in a form that is some combination of the forms described in variations 2 and 3.

Variation 1 is clearly the simplest one. At the same time, it is also the most restrictive one. Variations 2 and 3 allow us to obtain more expressive linguistic approximations, and the expressiveness of variation 4 is clearly superior to both of them. When none of the variations is sufficient for obtaining an adequate approximation of A by granules in F, the only possibility is to refine (or enlarge in some other way) the set of granules in F and the associated set of linguistic terms L.

Although we primarily address variation 1 in this article (see sections 4-6), we discuss some possible enhancements of this basic way for dealing with linguistic approximation as well (see section 7).

Throughout this article, we assume that A is a fuzzy interval. This is a reasonable assumption with one exception. The result of a fuzzy inference may be a fuzzy set on $\mathbb{R}$ that is not normal. Let us denote this fuzzy set by B and let us assume that $h_B \leq 1$. When $h_B < 1$, then we need to convert B to its normal counterpart A. As is now well established [8], there is only one way to make this conversion from B to A without changing evidence conveyed by B: for all $x \in X$,

$$A(x) = B(x) + 1 - h_B. \tag{12}$$

Clearly, when B is normal ($h_B = 1$), then $A = B$.

4 Most Informative Approximation

The simplest problem (variation 1) of the linguistic approximation of our interest may be formulated as follows: Given a fuzzy set A and a family of fuzzy sets $F = \{F_i | i \in \mathbb{N}_n\}$, all defined on the same universal set X, determine a particular set F_i from the family that *best describes* A. The term "best describes" may of course be interpreted in different ways. If we interpret it as "F_i is the *most informative* granule in F about A", then our choice should be based on measuring for each F_i the relevant amount of information about A, and choose the one with the highest amount.

As is well known [9], information is measured by a reduction of uncertainty. In our case, uncertainty is measured by the generalized Hartley-like measure defined by equation (10) or, when we deal with a discrete approximation of X, by the generalized Hartley measure defined by equation (8).

Using the generalized Hartley-like measure, we define the *degree of informativeness* of F_i about A by the formula

$$I(F_i, A) = 1 - \frac{\int_0^1 \log_2[1 + \mu({}^\alpha F_i \cup {}^\alpha A) - \mu({}^\alpha F_i \cap {}^\alpha A)]d\alpha}{\int_0^1 \log_2[1 + \mu({}^\alpha F_i \cup {}^\alpha A)]d\alpha}. \tag{13}$$

Two desirable properties of this degree of informativeness can readily be observed:

- $I(F_i, A) = 0$ if and only if $F_i \cap A = \emptyset$;
- $I(F_i, A) = 1$ if and only if $F_i = A$.

In addition to (13), we also consider an alternative definition of the degree of informativeness that is based on the *information metric distance* introduced for possibility theory by Higashi and Klir [6]. In terms of the generalized Hartley-like measure defined by (10), the metric distance between F_i and A, $D(F_i, A)$, has the form

$$D(F_i, A) = 2 \cdot GHL(F_i \cup A) - GHL(F_i) - GHL(A). \qquad (14)$$

Its minimum, $D_{\min}(F_i, A) = 0$, is obtained for $F_i = A$. Its maximum, $D_{\max}(F_i, A)$, is obtained for $F_i \cap A = \emptyset$ and has the value

$$D_{\max}(F_i, A) = 2 \cdot GHL(F_i + A) - GHL(F_i) - GHL(A), \qquad (15)$$

where

$$GHL(F_i + A) = \int_0^1 \log_2[1 + \mu(^\alpha F_i) + \mu(^\alpha A)] d\alpha. \qquad (16)$$

Information closeness is then defined by the difference $D_{\max}(F_i, A) - D(F_i, A)$, and its normalized version,

$$I'(F_i, A) = 1 - \frac{2\,GHL(F_i \cup A) - GHL(F_i) - GHL(A)}{2\,GHL(F_i + A) - GHL(F_i) - GHL(A)}, \qquad (17)$$

may be viewed as the *degree of informativeness* of F_i about A and vice versa.

5 Other Approaches to Linguistic Approximation

In this section, we explore four alternative approaches to linguistic approximation. We employ the same notation and deal with the same issues that are introduced in section 3. The difference from the information-based approach is that the degree of informativeness defined by (13) or (17) is replaced with alternative definitions of the degree of approximation.

5.1 Approximate Equality of Fuzzy Sets

In this approach, we utilize the concept of *approximate equality* of two fuzzy sets. This concept emerged in the context of fuzzy relation equations [4], where it has been employed for dealing with approximate solvability [10]. Given two fuzzy sets, A and B, their *degree of equality*, $E(A, B)$, is usually defined by the formula

$$E(A, B) = \min\{S(A \subseteq B), S(A \supseteq B)\}, \qquad (18)$$

where the function S is defined by equation (4). Three desirable properties of E for linguistic approximation can be readily recognized:

- $E(A, B) = 0$ if and only if $A \cap B = \emptyset$;
- $E(A, B) = 1$ if and only if $A = B$;
- if $A \subseteq B \subseteq C$, then $E(A, B) \geq E(A, C)$ and $E(B, C) \geq E(A, C)$.

5.2 Tolerance Relation

The problem of linguistic approximation can be formulated in terms of a *fuzzy tolerance relation*, T, defined on the fuzzy power set $F(X)$. Defining T by the formula

$$T(F_i, A) = 1 - \frac{\int_X |F_i(x) - A(x)| dx}{\int_X [F_i(x) + A(x)] dx}, \tag{19}$$

where $|F_i(x) - A(x)|$ denotes here the absolute value of the difference, we may interpret $T(F_i, A)$ as the *degree of tolerance* to approximate fuzzy set A by fuzzy set F_i.

We can see that $T(F_i, A) = 0$ iff $F_i \cap A = \emptyset$ and $T(F_i, A) = 1$ iff $F_i = A$. Moreover, the degree of approximation decreases with the difference $|F_i(x) - A(x)|$, which is a desirable property.

5.3 Max-Min Ratio

In this approach, the degree of approximation, $M(F_i, A)$, is defined by the formula

$$M(F_i, A) = \frac{\int_X \min\{F_i(x), A(x)\} dx}{\int_X \max\{F_i(x), A(x)\} dx} \tag{20}$$

The operations min and max represent here, respectively, the standard operations of intersection and union on fuzzy sets. Clearly, $M(F_i, A) = 0$ iff $F_i \cap A = \emptyset$ and $M(F_i, A) = 1$ iff $F_i = A$.

Equation (20) may be modified by replacing the min operation with another *t*-norm or with an averaging operation, *Ave*, such that

$$Ave(0, a) = Ave(a, 0) = 0 \tag{21}$$

for any $a \in [0, 1]$. Replacing, for example, min with the geometric average, we would obtain an alternative degree of approximation, $G(F_i, A)$, given by the formula

$$G(F_i, A) = \frac{\int_X \sqrt{F_i(x) \cdot A(x)} dx}{\int_X \max\{F_i(x), A(x)\} dx} \tag{22}$$

5.4 Closeness of Fuzzy Intervals

This approach to linguistic approximation is based on the concept of a normalized distance between fuzzy intervals [17]. Using fuzzy intervals of our concern, $F_i \in F$ and A (both defined on $X = [\underline{x}, \overline{x}]$), the *degree of closeness* between F_i and A, $C(F_i, A)$, is defined by the formula

$$C(F_i, A) = 1 - \frac{\int_0^1 [|\underline{f}_i(\alpha) - \underline{a}(\alpha)| + |\overline{f}_i(\alpha) - \overline{a}(\alpha)|] d\alpha}{2(\overline{x} - \underline{x})}. \tag{23}$$

As can be easily shown, the term

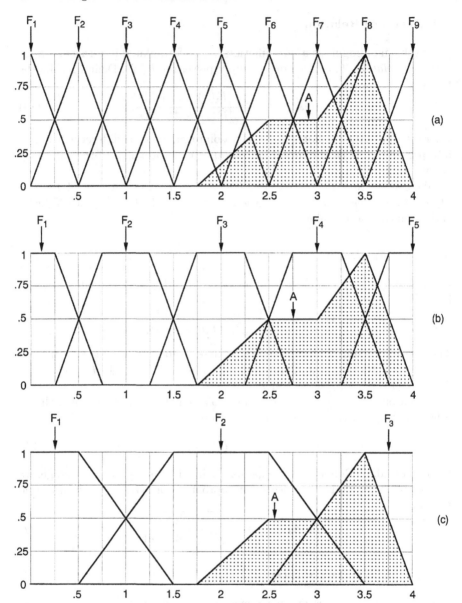

Fig. 1. Three families of fuzzy sets.

$$D(F_i, A) = \int_0^1 [|\underline{f}_i(\alpha) - \underline{a}(\alpha)| + |\overline{f}_i(\alpha) - \overline{a}(\alpha)|]d\alpha \qquad (24)$$

in this formula is a *metric distance* between F_i and A. The term $2(\overline{x} - \underline{x})$ in equation (23) is the largest value of $D(F_i, A)$, which is obtained when F_i

and A collapse to real numbers $\bar{x}$ and $\underline{x}$. The value of $C(F_i, A)$ expresses thus the opposite of the distance between F_i and A, i.e. their closeness. It is obvious that $C(F_i, A) = 1$ iff $F_i = A$. However, $F_i \cap A = \emptyset$ does not imply $C(F_i, A) = 0$, which is a deficiency of this approach.

6 Examples

The purpose of this section is to examine and compare the proposed approaches to linguistic approximation. We begin with three linguistic variables defined on the same numerical base variable whose range is $[0, 4]$. The linguistic variables have 9, 5, and 3 linguistic states, respectively, which are represented by the families of fuzzy sets identified in figure 1 as (a), (b), and (c). Linguistic states that these fuzzy sets represent are:

1. *close to 0, close to 0.5, close to 1, ..., close to 4*;
2. *very small, small, medium, large, very large*;
3. *small, medium, large.*

Also shown in figure 1 is fuzzy set A whose approximation in terms of the linguistic terms available in each case is to be determined.

First, we need formal descriptions of the fuzzy sets defined by their graphs in figure 1. From the graphs, we obtain the following formulas:

1. For all $i \in \mathbb{N}_9$,

$$F_i(x) = \begin{cases} 2x + 2 - i & \text{when } x \in [\max\{0, (i-2)/2\}, (i-1)/2] \\ i - 2x & \text{when } x \in [(i-1)/2, \min\{4, i/2\}] \\ 0 & \text{otherwise,} \end{cases}$$

$$^{\alpha}F_i = [\max\{0, -1 + (i + \alpha)/2\}, \min\{4, (i - \alpha)/2\}]$$

2. For all $i \in \mathbb{N}_5$,

$$F_i(x) = \begin{cases} \min\{1, 2x - 2i + 3.5\} & \text{when } x \in \\ & [\max\{0, i - 1.75\}, \max\{0, i - 1.25\}] \\ 1 & \text{when } x \in \\ & [\max\{0, i - 1.25\}, \min\{4, i - 0.75\}] \\ \min\{1, 2i - 2x - 0.5\} & \text{when } x \in \\ & \min\{4, i - 0.75\}, \min\{4, i - 0.25\}] \\ 0 & \text{otherwise} \end{cases}$$

$$^{\alpha}F_i = [\max\{0, i - 1.75 + 0.5\alpha\}, \min\{4, i - 0.25 - 0.5\alpha\}]$$

Table 1. Degrees of linguistic approximations for the three situations (a), (b), and (c) in figure 1.

(a)	F_4	F_5	F_6	F_7	F_8	F_9	(b)	F3	F4	F5	(c)	F2	F3
(13)	.004	.096	.186	.245	.385	.096	(10)	.121	.413	.164	(10)	.154	.645
(17)	.012	.118	.323	.392	.593	.123	(17)	.166	.682	.270	(17)	.247	.653
(18)	.015	.132	.324	.373	.471	.118	(18)	.235	.666	.262	(18)	.281	.706
(19)	.020	.180	.440	.507	.640	.190	(19)	.242	.687	.313	(19)	.367	.727
(20)	.010	.099	.282	.340	.471	.105	(20)	.138	.530	.220	(20)	.225	.571

3. For all $i \in \mathbb{N}_3$,

$$
F_i(x) = \begin{cases}
\min\{1, x - 2i + 3.5\} & \text{when } x \in \\
& [\max\{0, 2i - 3.5\}, \max\{0, 2i - 2.5\}] \\
1 & \text{when } x \in \\
& [\max\{0, 2i - 2.5\}, \min\{4, 2i - 1.5\}] \\
\min\{1, 2i - x - 0.5\} & \text{when } x \in \\
& [\min\{4, 2i - 1.5\}, \min\{4, 2i - 0.5\}] \\
0 & \text{otherwise}
\end{cases}
$$

$$
{}^{\alpha}F_i = [\max\{0, 2i + \alpha - 3.5\}, \min\{4, 2i - \alpha - 0.5\}]
$$

Moreover,

$$
A(x) = \begin{cases}
(x - 1.75)/1.5 & \text{when } x \in [1.75, 2.5) \\
0.5 & \text{when } x \in [2.5, 3) \\
x - 2.5 & \text{when } x \in [3, 3.5) \\
8 - 2x & \text{when } x \in [3.5, 4] \\
0 & \text{otherwise}
\end{cases}
$$

$$
{}^{\alpha}A = \begin{cases}
[1.5\alpha + 1.75, 4 - 0.5\alpha] & \text{when } \alpha \in (0, 0.5] \\
[\alpha + 2.5, 4 - 0.5\alpha] & \text{when } \alpha \in (0.5, 1]
\end{cases}
$$

The various degrees of linguistic approximations of the given fuzzy set A by granules shown in figure 1 are given in table 1. The three sets of granules are identified in both figure 1 and table 1 as (a), (b), and (c). Each row in the tables refers to an equation that defines the respective type of the degree of approximation. Columns in the tables correspond to granules in the three groups. Only nonzero degrees of approximation are shown in the tables. For example, the degree of approximation for F_1 is equal to zero in each of the three sets of granules and for each of the considered definitions of the approximation degree. We do not show degrees of approximation based on equation (23) since this definition is ill suited for our purpose. For F_1 in group (c), for example, we have $C(F_1, A) = 0.484$, which is not acceptable in our context.

We can see from table 1 that the same granule in each group has the highest degree of approximation in terms of all the considered definitions: F_8 in group (a); F_4 in group (b); and F_3 in group (c). These unique maxima are also in these cases compatible with our intuition. The granule with the next highest degree of approximation is also unique in each group. Moreover, the order of granules in each group by their degrees of approximation is almost the same for all the considered definition. Hence, there does not seem to be any significant differences between the five approaches that are based on equations (13) and (17)-(20).

Let $Apr(F_i, A)$ denote the degree of approximation of A by F_i based on one of the five considered definitions or, possibly, some other definition. Regardless of which of the definitions is used, the fundamental question is: Knowing values of $Apr(F_i, A)$ for a given fuzzy set A and all granules F_i in a given set F, which of the granules in F, if any, should be chosen as the best approximation of A? We believe that a particular granule $F_k \in F$ is acceptable as a sound approximation of A provided that the following requirements be satisfied:

- $Apr(F_k, A) > Apr(F_i, A)$ for all $i \neq k$;
- $Apr(F_k, A) \geq \gamma$, where γ is a specified minimum acceptable degree of approximation (a practical value of γ should be close to 1);
- $Apr(F_k, A) - \max_{i \neq k}\{Apr(F_i, A)\} \geq \delta Apr(F_k, A)$,

where δ is a specified minimum acceptable fraction of $Apr(F_k, A)$ by which the approximation degree of F_k is required to exceed the next highest approximation degree for granules in F (a practical value should be close to 0.5).

When some of these requirements are violated for chosen values γ and δ, we need to invoke some of the other variations of linguistic approximation within F or, possibly, deal with the approximation by modifying F.

7 Approximation Enhancement

It is important to recognize that we always deal with linguistic approximation in a given application context. We may thus assume that each granule $F_i \in F$ was properly constructed to represent the meaning of the associated linguistic term $L_i \in L$ in this application context. If, under this assumption, a particular granule $F_k \in F$ is found to be an acceptable approximation of a given fuzzy set A (according to the requirements stated in section 6), then the corresponding linguistic term captures well the information conveyed about the base variable x by A. If, however, none of the granules in F is an acceptable approximation of A, we need to construct additional granules by combining or modifying the given granules in F. It is crucial that each constructed granule represents properly the meaning of some linguistic term, as understood in the given application context.

In this paper, we consider only two methods of constructing meaningful new granules. The first method is based on taking appropriate unions of two or more contiguous granules in F, the second one is based on modifying individual granules in F appropriately to account for some linguistic hedges.

In order to describe the first method, we begin by considering unions $F_i \cup F_{i+1}$ of two contiguous granules for all $i \in \mathbb{N}_{n-1}$ and assigning each of them to the linguistic expression "L_i or L_{i+1}". To represent adequately the usual meaning of these linguistic expressions in natural language, the chosen operation of union (t-conorm) cannot be arbitrary. It must be such that $F_i \cup F_{i+1}$ is again a fuzzy interval. When F is a fuzzy partition, as is assumed here, this requirement is satisfied by using the bounded sum defined for each $x \in X$ by the formula

$$(F_i \cup F_{i+1})(x) = \min[1, F_i(x) + F_{i+1}(x)]. \tag{25}$$

In a more general case, when F is not a fuzzy partition, the requirement is satisfied by using the drastic union [10]. If desirable, we can take unions of more than two contiguous granules. In the extreme case, when we take the union of all of them, we obtain of course the whole universal set X.

As an example, consider the granules in figure 1(b). Using the information closeness defined by equation (13), A is best approximated by granule F_4 with the degree of approximation 0.413 (see table 1). We can obtain a higher degree of approximation 0.618, when we take $F_4 \cup F_5$. The linguistic expression "x is large or very large" is thus more accurate then the expression "x is large".

As another example, consider the granules in figure 1(a) and let A be defined for all $\alpha \in (0, 1]$ by its α-cut representation

$$^\alpha A = [1.25 - \sqrt{1 - \alpha}, 1.25 + \sqrt{1 - \alpha}]. \tag{26}$$

In this case, the same maximum degree of informativeness is obtained for two granules: $I(F_3, A) = I(F_4, A) = 0.545$. This suggests to consider the union of these granules, for which we obtain $I(F_3 \cup F_4) = 0.864$. The linguistic expression "x is close to 1 or 1.5" is thus by far more informative about A than either "x is close to 1" or "x is close to 1.5." The situation is illustrated in figure 2(a). We obtain similar results when A has the same location but a different shape, as shown in figure 2(b). In this case, A is defined by

$$^\alpha A = [\sqrt{\alpha} + 0.25, 2.25 - \sqrt{\alpha}]. \tag{27}$$

The relevant degrees of informativeness are $I(F_3, A) = I(F_4, A) = 0.199$ and $I(F_3 \cup F_4) = 0.565$. Again, the union of the two granules is considerably more informative about A than either of them alone.

The issue of modifying fuzzy sets to account for linguistic hedges is a complex one [7, 12]. We do not intend to investigate this issue, only to illustrate

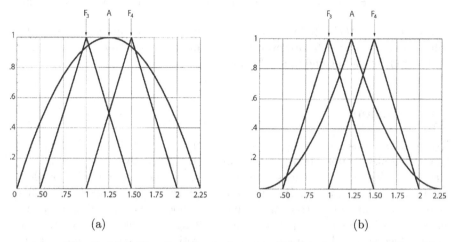

Fig. 2. Modelling the informativeness of linguistic expressions.

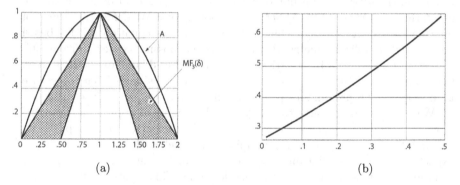

Fig. 3. A range of modified granules $MF_3(\delta)$.

that linguistic hedges play a useful role in linguistic approximation. As an example, consider again the set of granules in figure 1(a) and let

$$^{\alpha}A = [1 - \sqrt{1 - \alpha}, 1 + \sqrt{1 - \alpha}]. \tag{28}$$

We can readily determine that F_3 is the most informative granule. However, its degree of informativeness about A is only 0.272, which is rather small. This degree can be increased by using the linguistic hedge *somewhat*, which would allow us to describe the information conveyed by A as "x is somewhat close to 1." A simple way of representing this linguistic hedge is to modify F_3 to MF_3 by extending its support from $(0.5, 1.5)$ to $(0.5 - \delta, 1.5 + \delta)$, where δ is a small positive number. Considering, for example, that $\delta \in [0, 0.5]$, we obtain a range of modified granules $MF_3(\delta)$ illustrated in figure 3(a) by the shaded area. Their α-cuts are:

$$^{\alpha}MF_3(\delta) = [0.5 - \delta + (0.5 + \delta)\alpha, 1.5 + \delta - (0.5 + \delta)\alpha]. \tag{29}$$

The values of $I(MF_3(\delta), A)$ for all $\delta \in [0, 0.5]$ are shown in figure 3(b). The value is 0.659 for $\delta = 0.5$. However, we can still obtain a better approximation of A, 0.844, by taking the union $F_2 \cup F_3 \cup F_4$.

8 Conclusion

Except for a few scattered discussions in the literature [2, 3, 15, 17], the problem of linguistic approximation has not been sufficiently investigated. It is fair to say that no comprehensive methodology for dealing with the problem has been developed as yet. Our motivation for writing this paper was rather modest. We wanted to attract the attention of both linguists and mathematicians to this interesting, important, and challenging problem, and we wanted to introduce and explore an information-theoretic approach to deal with it. In this approach, which is based on some fairly recent results in generalized information theory [9], we formulate the degree of approximation as the degree of informativeness of one fuzzy set about another one. We define the degree of informativeness in two ways. One of them (defined by equation (17)) is based on the well-established metric distance between possibility distributions [6]. The other one (defined by equation (13)) is introduced here for the first time. Our conjecture is that it is also based on a metric distance. This conjecture is supported by a preliminary proof, which needs to be properly scrutinized.

We also suggested four other types of the degree of approximation. One of them (defined by equation (23)) was found deficient for our purpose and was not further pursued. The remaining types behaved in our examples as expected. It seems significant that the degree of informativeness defined by equation (13) was consistently the smallest one. This seems to suggest that the information-theoretic approach is the most guarded one. We also believe that defining the degree of approximation in information-theoretic terms is epistemologically sound.

To formulate linguistic approximation in terms of a finite family F of fuzzy intervals that form a fuzzy partition of X is somewhat restrictive. We can make it less restrictive by allowing meaningful combinations of the fuzzy intervals in F and some relevant modifiers. Assuming, for example, that $|F| = n$, that we use all possible unions of contiguous fuzzy intervals in F, and that we recognize m modifiers, the number of linguistic expressions we are capable of representing is extended from n to $mn(n+1)/2$. This is certainly a substantial extension, but it may still be restrictive for some purposes.

We can imagine an alternative approach to linguistic approximation that is not based on a finite family of predefined fuzzy intervals. Instead, the given fuzzy interval A is converted to an approximating fuzzy interval F of a certain type that has a meaningful linguistic interpretation (in context of a given application) and is the most informative one about A. The idea of a *verbal quantity* introduced by Mareš and Mesiar [13, 14] seems quite relevant for this purpose.

We envision the following steps as a possible way of developing the alternative approach in information-theoretic terms:

1. Determine the value of α for which $GHL({}^{\alpha}\!A) = GHL(A)$; ${}^{\alpha}\!A$ may be viewed as the most informative crisp set representing A.
2. Take the midpoint, $a(\alpha)$, of ${}^{\alpha}\!A$ determined in step 1 as the most informative real number representing A.
3. Use $a(\alpha)$ or some crisp interval around it (to make the linguistic interpretation more meaningful) as the core, ${}^{1}\!F$, of the constructed fuzzy interval F that is supposed to approximate A.
4. Determine the fuzzy interval F with the core, ${}^{1}\!F$ chosen in step 3 whose membership function conforms to a desirable shape and for which $GHL(F) = GHL(A)$.

We intend to investigate this fairly specific approach to linguistic approximation in the future. In the meantime, we hope that this paper will stimulate the interest of other researchers to address this extremely challenging problem.

References

[1] J. C. Bezdek, D. Dubois, and H. Prade, editors. *Fuzzy Sets in Approximate Reasoning and Information Systems*. Kluwer, Boston, 1999.

[2] A. Dvořák. On Linguistic Approximation in the Frame of Fuzzy Logic Deduction. *Soft Computing*, 3(2):111–115, 1999.

[3] F. Eshragh and E. H. Mamdani. A General Approach to Linguistic Approximation. *International Journal of Man-Machine Studies*, 11:501–519, 1979.

[4] S. Gottwald. Generalized Solvability Criteria for Fuzzy Equations. *Fuzzy Sets and Systems*, 17(3):285–296, 1985.

[5] P. Hájek. *Metamathematics of Fuzzy Logic*. Kluwer, Boston, 1999.

[6] M. Higashi and G. J. Klir. On the Notion of Distance Representing Information Closeness: Possibility and Probability Distribution. *International Journal of General Systems*, 19(2):103–115, 1983.

[7] E. E. Kerre and M. de Cock. Linguistic Modifiers: An Overview. In G. Chen et al., editor, *Fuzzy Logic and Soft Computing*. Kluwer, Boston, 1999.

[8] G. J. Klir. On Fuzzy-set Interpretation of Possibility Theory. *Fuzzy Sets and Systems*, 108(3):263–273, 1999.

[9] G. J. Klir and M. J. Wierman. *Uncertainty-Based Information: Elements of Generalized Information Theory*. Physica/Springer, Heidelberg and New York, 1998.

[10] G. J. Klir and B. Yuan. *Fuzzy Sets and Fuzzy Logic: Theory and Applications*. Prentice Hall, PTR, Upper Saddle River, NJ, 1995.

[11] G. J. Klir and B. Yuan, editors. *Fuzzy Sets, Fuzzy Logic, and Fuzzy Systems: Selected Papers by Lotfi A. Zadeh.* World Scientific, 1996.

[12] G. Lakoff. Hedges: A Study in Meaning Criteria and the Logic of Fuzzy Concepts. *Journal of Philosophical Logic,* 2:458–508, 1973.

[13] M. Mareš and R. Mesiar. Calculation over Verbal Quantities. In L. A. Zadeh and J. Kacprzyk, editors, *Computing with Words in Information/Intelligent Systems,* volume 1, pages 409–427. Physica, 1999.

[14] M. Mareš and R. Mesiar. Dual Meaning of Verbal Quantities. *Kybernetika,* 38(6):709–716, 2002.

[15] V. Novák. *Fuzzy Sets and Their Applications.* Adam Hilger, Bristol and Philadelphia, 1989.

[16] A. Ramer and C. Padet. Nonspecificity in $\mathbb{R}^n$. *International Journal of General Systems,* 30(6):661–680, 2001.

[17] J. Talašová. *Fuzzy Methods of Multicriteria Valuation and Decision Making (in Czech).* Palacký University Press, 2003.

[18] P. P. Wang, editor. *Computing with Words.* John Wiley, 2001.

[19] R. R. Yager, S. Ovchinnikov, R. M. Tong, and H. T. Nguyen, editors. *Fuzzy Sets and Applications – Selected Papers by L. A. Zadeh.* John Wiley, New York, 1987.

[20] L. A. Zadeh. The Concept of a Linguistic Variable and its Application to Approximate Reasoning. Parts 1 and 2. *Information Sciences,* 8:199–249,301–357, 1975.

[21] L. A. Zadeh. The Concept of a Linguistic Variable and its Application to Approximate Reasoning. Part 3. *Information Sciences,* 9:43–80, 1975.

[22] L. A. Zadeh. Fuzzy Logic = Computing with Words. *IEEE Transactions on Fuzzy Systems,* 4(2):103–111, 1996.

[23] L. A. Zadeh. From Computing with Numbers to Computing with Words – From Manipulation of Measurements to Manipulation of Perceptions. *IEEE Transactions on Circuits and Systems – I. Fundamental Theory and Applications,* 45(1):105–119, 1999.

[24] L. A. Zadeh and J. Kacprzyk, editors. *Computing with Words in Information/Intelligent Systems.* Physica/Springer, Heidelberg and New York, 1999.

A Semiotic Approach to Complex Systems

Harald Atmanspacher

Institut für Grenzgebiete der Psychologie und Psychohygiene
haa@igpp.de

1 Introduction

A key topic in the work of Burghard Rieger is the notion of meaning. To explore this notion, he and his collaborators developed a most sophisticated approach combining theoretical ideas and concepts of semiotics with empirical and numerical tools of computational linguistics (see [29] for a most recent comprehensive account). In the present contribution, relations of Rieger's achievements to some issues of interest in the physics and philosophy of complex systems will be addressed.

The notion of meaning will first be introduced in the framework of the Cartesian distinction between mind and matter or, more precisely, between states and properties of mental and material systems as well as their dynamics. In this dualistic framework, meaning can be formulated in terms of a reference relation between a descriptive term (in a model or a theory) and an object (or a set of objects) in the material world.

Such a reference relation can be considered from a semiotic point of view first proposed by Peirce. Within the semiotic triad of syntactics, semantics, and pragmatics, the notion of meaning forces us to leave a purely syntactic level of discussion and include questions of semantics and pragmatics.

The significance of the semiotic triad in a relatively new field of modern physics, the study of complex systems, will then be outlined. The crucial point is that the concept of complexity can be defined in a way permitting a straightforward semiotic interpretation. This interpretation may be conversely used to discuss basic methodological cornerstones of traditional physics in view of some of its novel developments.

Some speculative perspectives toward an implicit, non-dualistic kind of meaning without explicit reference relations conclude this contribution. This reflects Peirce's conviction that semiotics is ultimately holistic. In terms of Rieger's approach, it reflects the idea that mental processes "not only cut across the distinction of mind and matter but can even be considered to underlie and allow for this distinction" ([29], p. 352).

H. Atmanspacher: *A Semiotic Approach to Complex Systems*, StudFuzz **209**, 79–91 (2007)
www.springerlink.com   © Springer-Verlag Berlin Heidelberg 2007

2 The Cartesian Distinction

In the night from November 10 to 11, 1619, close to the German city of Ulm, the French soldier René Descartes had a series of dreams.[1] According to virtually all biographers of Descartes, these dreams were of key influence for his future philosophical insights and achievements. Waking up the morning of November 11, he was left with the persistent question of how he could ever know for sure whether he was indeed awake or still dreaming. Major parts of his philosophy were motivated by the crucial question of how to reliably distinguish between dream and reality, including the most fundamental issue of the reality of one's own existence. Descartes' solution of this problem is reflected in his famous "cogito ergo sum", one of the most-cited quotations of European philosophy.

Roughly speaking, this solution rests on Descartes' proposal to distinguish material and mental domains of reality, leading to Cartesian dualism as an essential part of his philosophy. Of course, this dualistic stance does not characterize his thinking exhaustively. Descartes' philosophical writings are very rich, they are partially incoherent, and they cover much more than the split of matter and mind. Nevertheless, it is correct to speak of a "Cartesian" distinction at least insofar as some of Descartes' successors, notably those who contributed to the development of the natural sciences, compactified and simplified Descartes' thoughts considerably. For this reason the notion of a Cartesian distinction, or Cartesian cut [27], should be considered as a central term of "scientific Cartesianism".

Since Descartes' time, the Cartesian distinction turned out to be an extremely powerful tool for reducing the arbitrariness inherent in the allegorical and speculative schemes of late scholasticism and Renaissance neoplatonism; it provided the possibility of a rational, consistent description of reality. In Descartes' terminology, the Cartesian distinction splits the entirety of reality into a material component (*res extensa*) and a non-material component (*res cogitans*) [14]. These labels, literally translated, characterize the realms of "extended substance" and "thinking substance". The notion of extension in *res extensa* refers to the fact that material reality is extended in its spatial location and in its temporal duration (although Descartes himself did not put much emphasis on the latter). The notion of cognition in *res cogitans* is probably best characterized as referring to conscious activity in general rather than "thinking" in the narrow sense of cognitive capabilities.

---

[1]Baillet [6] reported this series of three dreams in his biography of Descartes. Especially the second and third dream contain elements clearly referring to the issue of distinguishing between realistic chains of events and unrealistic chains of events as they are typical for dreams. Of course, much more material is contained in Baillet's report which has been analyzed in a number of accounts. For a detailed interpretation of Descartes' dreams, together with an overview concerning other accounts, see [35].

The Cartesian distinction can therefore be regarded as a conceptual border between a material and a mental domain. Without any reasonable doubt this is of central significance for the world view Western science and philosophy have developed. On the other hand it is obvious that this cut is nothing more than a conceptual tool – it is itself not an object in the material world, but belongs to the non-material world of *res cogitans*. Although Descartes himself thought he had *discovered* the cut as an ontological fact superior to the realms it separates, it is today much more appropriate to say that he *invented* it. So the question arises as to whether there might be modes of mental activity (i.e., operations within *res cogitans*) reducing the relevance of the Cartesian distinction, or even avoiding it completely, at least in its rigorous interpretation of a prescribed and impenetrable border.

This question receives additional motivation by the fact that this border is simply not recognized during many kinds of mental activity as it operates in practice. Who does explicitly and consciously distinguish between what he sees (as a fact) and what he thinks he sees (as his model of this fact)? Is it possible to make this distinction at all, and, if yes, how can it be cognitively realized? Moreover – a bit apart from everyday experience, but still close to the subject of this article – each abstract scientific model contains terms which refer to objects in the material world of concrete, empirical, material facts. The corresponding relation of reference is crucial for the possibility to check the validity of a model [4]. Reference relations of this kind express the interpretation or, more colloquially, the meaning of conceptual terms with respect to objects in the material world. In this sense they are relations across the Cartesian distinction.

In other words, issues of meaning are primary candidates for the connection of material and mental domains of reality, and they are of fundamentally relational character. In the following section, another, more recent approach to address meaning will be outlined.

3 The Semiotic Approach

The father of present-day semiotics is Charles Sanders Peirce. He developed semiotics as a theory of signs which is always embedded in a framework of relations. In [26], he says:

"A *Sign*, or *Representamen*, is a First which stands in such a genuine triadic relation to a Second, called its *Object*, as to be capable of determining a Third, called its *Interpretant*, to assume the same triadic relation to its Object in which it stands itself to the same Object. The triadic relation is *genuine*, that is its three members are bound together by it in a way that does not consist in any complexus of dyadic relations." [26, vol. 2, §274].

This quotation expresses the basically holistic significance that Peirce ascribed to the semiotic triad. Nevertheless, contemporary semiotic approaches often, maybe even typically, distinguish the semiotic areas of syntactics, semantics and pragmatics. Syntactics deals with (grammatical or stylistic) interrelations among signs, e.g. in a code. Semantics deals with interrelations between signs and what they designate, i.e. their meaning. And pragmatics addresses relations between signs and their users. Correspondingly, one can conceptually distinguish syntactic, semantic and pragmatic information [25]. In contrast to Peirce's quotation above, their demarcation may be justified from an abstract, analytical viewpoint where signs can be considered without their relational context. From a phenomenological ("lifeworld") point of view, concrete signs are never context-free: concreteness requires context. In this sense, the notion of concreteness entails some type of holism, lifting the conceptual separation inherent in the distinction between syntax, meaning, and usage.

With respect to models or theories, the syntactic component can be considered to refer to their formal codification, the semantic component addresses their interpretation, and the pragmatic component comprises their range of applications. From this example, it is evident that a formal codification without interpretation and application is possible only in an abstract sense. The concrete development of a theoretical concept is never isolated from its meaning and, ultimately, usage. Every element of syntax is inseparably linked to semantic and pragmatic aspects.

Nevertheless, the history of semiotic aspects of information has shown that it can be methodologically helpful to distinguish them analytically. For particular technical problems of communication through noisy channels, aspects of syntactic information were extensively studied in the influential publication by Shannon [32], which explicitly omits any reference whatsoever to meaning-related or pragmatic issues. Shannon-type information is purely syntactic insofar as it quantifies by which amount a message carrying information reduces the uncertainty of a receiver as compared to his/her state before receiving that message. Weaver's contribution in [32] already pointed out that this syntactical component of information requires extension to semantic and pragmatic aspects (for more details see [3]).

Shortly after Shannon's work, Bar Hillel and Carnap [8] proposed to quantify semantic information based on a receiver's ability to draw logical consequences from a message. If a message contains a huge amount of syntactic information, which is not or cannot be understood by its receiver, then he/she cannot draw conclusions from it. Yet the problem remains how to evaluate, or operationalize, an understanding of information. Clearly, self-reports may be insufficient for this purpose, not only since they may be mistaken, but more importantly for the reason that they can hardly be normalized and, thus, are incomparable.

At this point, the significance of pragmatic information becomes clear. If semantic information, i.e. meaning, is understood, then it triggers action, e.g. changes efficiency, or leaves some other imprint on the behavior of its receiver. (In this sense, focusing on pragmatic information resembles a particular kind of behaviorism.) A corresponding concept has been proposed by von Weizsäcker [36] and further developed in [16, 23].

It relies on the two notions of *primordiality* ("Erstmaligkeit") and *confirmation* ("Bestätigung"). Weizsäcker argued that a (redundant) message that does nothing but confirm the prior knowledge of a receiver will not change its structure or behavior. On the other hand, a message providing only material completely unrelated (primordial) to any prior knowledge of the receiver will also not change its structure or behavior, simply because it will not be understood. In both cases, the pragmatic information of the message vanishes. A maximum of pragmatic information is assigned to a message that transfers an optimum mixture of primordiality and confirmation to its receiver. For the limiting case of complete confirmation, purely syntactic Shannon information and pragmatic information vanish coincidentally. If primordiality is added, Shannon information increases monotonically.

4 Concepts of Complexity

In recent decades, complexity has become an extremely popular notion covering a huge variety of different kinds of behavior. From a scientific point of view, such a colorful concept is useful only in combination with a clear-cut definition. However, there is a plenitude of different concepts of complexity. A systematic orientation among them requires a reasonable classification. There are several approaches that can be found in the literature: two of them are (i) the distinction of structural and dynamical measures [37] and (ii) the distinction of deterministic and statistical measures [13]. Another, epistemologically inspired scheme (iii) assigns ontic and epistemic levels of description to deterministic and statistical measures, respectively [2, 31].

In addition to these approaches, a purely phenomenological criterion for classification can be given by the functional behavior in which a complexity measure is related to measures of randomness.[2] Within such an approach (for an early reference see [38]), there are two classes of complexity measures: (iv) those for which complexity increases monotonically with randomness and those with a globally convex behavior as a function of randomness (cf. Fig. 1). It turns out that classifications according to (ii) and (iii) distinguish measures of complexity in precisely the same manner as (iv) does: deterministic or ontic measures behave monotonically, and statistical or epistemic measures are convex. In other words: deterministic (ontic) measures are essentially measures of randomness, whereas statistical (epistemic) measures are not.

The class of monotonic measures of complexity contains, e.g., algorithmic complexity [22], various kinds of Rényi information [7] (among them Shan-

---

[2]It is worth mentioning that randomness itself is a concept that is anything but finally clarified. In the framework of the present paper we use the notion of randomness in the broad sense of an entropy.

non's information [32]), multifractal scaling indices [20], or dynamical entropies [21]. The class of convex measures of complexity contains, e.g., effective measure complexity [17], ϵ-machine complexity [13], fluctuation complexity [10], and variance complexity [5]. See also [15, 24] for further discussion.

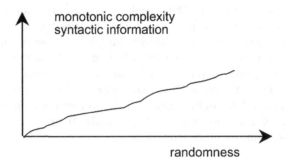

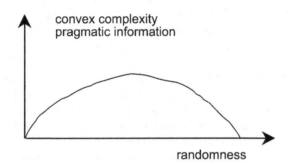

Fig. 1. Schematic illustration of two different classes of complexity measures, corresponding to different information measures and distinguished by their functional dependence on randomness.

A most intriguing additional difference (v) between both classes can be recognized if one focuses on the way statistics is implemented in each of these measures. The crucial point is that convex measures, in contrast to monotonic measures, are *meta*-statistically formalized, i.e. effectively represent (in one or another way) second-order statistics in the sense of "statistics of statistics". Fluctuation complexity is the standard deviation (second-order) of a net mean information flow (first-order); effective measure complexity is the convergence rate (second-order) of a difference of entropies (first-order); ϵ-machine complexity is the Shannon information with respect to machine states (second-order) that are constructed as a compressed description of a data stream (first-order); and variance complexity is based on the variance (second-order) of the mean of many individual variances (first-order) of a distribution of data. To our knowledge, there is no monotonic complexity measure providing such a two-level statistical structure. Although it would be desirable to have a theorem for the corresponding relationship between convex complexity measures and their two-level statistical structure, such a theorem is not yet available.

5 Complexity and Information

Since so many complexity measures bear an intimate relation to information theoretical concepts, it is interesting to see whether first-order and second-order complexity measures can be related to corresponding information measures. For this purpose, let us now consider some examples.

Applying a proper algorithm in order to generate a regularly alternating, periodic pattern, the corresponding generation process is obviously recurrent after the first steps, i.e., after the generation of the first elements of the pattern. Considering the entire generation process as a process of information transmission, it presents only confirmation of its first time steps once they have passed by. In this sense, a regular pattern, exhibiting no complexity, corresponds to a process of information transmission that has vanishing pragmatic information (or "meaning") after an initial transient phase (the first time steps). This applies to both notions of complexity, the deterministic as well as the statistical one.

For a completely random pattern the situation is more involved, since deterministic complexity and statistical complexity lead to different viewpoints. Deterministically, a random pattern is generated by an incompressible algorithm which contains as many steps as the pattern contains elements. The process of generating the pattern is not recurrent within the length of the algorithm. This means that it never ceases to produce elements that are unpredictable, except under the assumption that the entire algorithm was known *a priori*. Such knowledge, however, would imply that the pattern itself were known, since the algorithm is nothing but an incompressible description of it. Hence, the process generating a random pattern can be interpreted as a transmission of information completely lacking confirmation, and consequently with vanishing pragmatic information.

As a consequence, there is indeed a strong conceptual similarity between complexity measures and information measures. Pragmatic information is as convex as second-order complexity, and syntactic information is as monotonic as first-order complexity (compare again Fig. 1). In this context, it is worthwhile to mention that quite a number of authors have emphasized that the concept of meaning, reference, or intentionality is essential to a definition of complexity [1, 2, 11, 12, 17, 19]. For instance, Grassberger wrote:

"Complexity in a very broad sense is a *difficulty* of a *meaningful task*. More precisely, the complexity of a pattern, a machine, an algorithm, etc. is the difficulty of the most important task related to it. [...] As a consequence of our insistence on *meaningful* tasks, the concept of complexity becomes *subjective*. We really cannot speak of the complexity of a pattern without reference to the observer. [...] A unique definition with a universal range of applications does not exist. Indeed one of the most obvious properties of a complex object is that there is no *unique* most important task related to it." [18].

This quotation can be assessed in more detail if the two classes of complexity measures and associated information measures as discussed above are taken into account. Since monotonic, first-order measures of complexity are related to purely syntactic information, they can only be used to characterize systems in a way disregarding meaning. If meaning is to be considered explicitly, one has to proceed to semantic or pragmatic information and associated convex, second-order measures of complexity. Corresponding definitions of complexity provide the validity domain to which Grassberger's quotation applies.

In this respect, a conceptual framework relating second-order complexity measures to the notion of second-order models of complex systems has been outlined recently [4]. This approach is motivated by the idea that any reference relation between models and data, e.g. meaning, can only be explicitly addressed from the perspective of a meta-model, or second-order model. This move implies interesting consequences, some of which are explored in [4].

Two points should be stressed at the end of this section. First, the fact that monotonic complexity is not related to meaning does not imply that corresponding measures are useless or ill-defined. It is obvious that there are many interesting applications of first-order complexity measures, and their benefit is that they do not lead to the complications which second-order complexity entails. Second, it should be kept in mind that, in contrast to syntactic information, semantic and/or pragmatic information are not defined as precisely as desirable. Hence their relation to second-order complexity cannot be demonstrated as clearly as the relation between monotonic complexity and syntactic information. Nevertheless, their common feature of convexity is prominent enough to conjecture an intimate connection between convex complexity and semantic/pragmatic information.

6 Implicit Meaning without Explicit Reference

Complexity is a concept that has its origin in the study of physical properties of material systems. Meaning, on the other hand, originates in human concerns. It has become a topic of philosophy and, more recently, cognitive science, and is discussed as pertaining to a non-material domain. From this viewpoint, the concept of meaning is prior to the complexity of the brain as the material carrier of mental states.

From a material perspective, however, the complexity of a system is prior to its capability to constitute and understand meaning. In fact, it seems plausible to expect that a certain degree and kind of complexity is a precondition for the capability of a system to constitute and understand meaning. Although it is still unclear what the exact criteria are in this respect, it would certainly be far too anthropocentric to fix them such as to exclude non-human beings.[3] It is even an open question to what extent meaning might be a reasonable concept for non-living systems. Atlan [1] has proposed distinguishing different types of complexity and to assign the notion of meaning only to a specific one among them. Approaches like those of von Weizsäcker [36] or Crutchfield and Young [13] do not restrict the notion of meaning in this manner.

Focusing back on the convexity of both second-order complexity and pragmatic information, it is remarkable how the perspectives of physics (complexity) and of cognitive science (meaning) show an explicit *complementarity* [2, 4]. As Casti states [11], "the impression of complexity often appears as something like the expression of an experience of meaning". And Sheriff, interpreting Peirce, writes similarly:

"We might say that the unlimited complexity of the object that the representamen [sign] denotes is the "external", and the indefinite continuity of consciousness that the interpretant of the sign signifies is the "inward" view of a sign." [33].

A complementarity relation between two (or more) concepts typically indicates that the respective concepts share important features at a level of description underlying that at which the complementarity relation applies. With respect to the notions of complexity and meaning this can be taken as a hint to look for a common ground at which they are *implicitly* embedded and from which they emerge as *explicitly* different concepts under particular conditions. A top candidate for such conditions is the need for distinctions in order to gain epistemic access. In this sense, the Cartesian distinction can be regarded as a tool that generates the complementary concepts of complexity and meaning, which are unseparated without that distinction.

Another way to look at scenarios like this, motivated by physical examples, uses the terminology of symmetry breakings and contextual representations [28]. Insofar as symmetries (also called invariances) express indistinguishability, breaking symmetries means nothing else than introducing distinctions.[4] As

---

[3]Nevertheless, notions of meaning intended to apply beyond human beings (e.g., animals or AI systems) are often configured by analogies or similarities with our everyday notion of meaning.

[4]For instance, consider the homogeneity and isotropy of a property of a system in space. These two terms express translational and rotational symmetry in such a way that the considered property is indistinguishable with respect to translations and rotations in space. Breaking the translational symmetry generates distinguishable (local) positions, breaking the rotational symmetry generates distinguishable (local) directions.

Primas [28] has discussed in detail, symmetry breakings, leading to emergent properties, always require contextual conditions to be fixed. Such contexts are often introduced by the environment of a system, allowing a contextual representation of its states and properties that is different from a less specific (more general) representation of the system without the chosen context.

Introducing new contexts and breaking symmetries is, therefore, a viable approach to understand the emergence of complexity in physical systems. Stepping back from the material domain of physics, it is tempting to use the same idea to describe the emergence of this domain as such versus its non-material counterpart. This means to apply the notion of a distinction as a, or even as *the*, basic tool to achieve epistemic access, i.e. gather knowledge.[5] In this way, the Cartesian distinction of material and mental domains of reality plays a significant role for the distinction of complexity and meaning.

To the same extent as the distinction between complexity and meaning is blurred, we have to face a reality in which mind and matter are not as unrelated as they appear from the viewpoint of traditional science and its established methodology. Possible modifications with respect to cornerstones of scientific methodology have been proposed and discussed elsewhere [4]. Particularly interesting in this context is the notion of reproducibility, a basic requirement for using empirical facts and data to reject or confirm models and theories. This presupposes a well-defined reference relation between theoretical terms and empirical data, which can be addressed in a second-order modeling framework as it is necessary for complex mind-matter systems. For more details, see [4].

Questioning the unrestricted assumption of a perfect Cartesian distinction, it becomes problematic to develop or maintain a clear-cut understanding of complexity and meaning in terms of reference relations between separate mental and material domains. This leads to the question of how meaning could be conceived without an explicit decomposition of the semiotic triad, i.e. within the holistic framework of Peirce's original ideas.

This has been and is a central issue in Rieger's work. His starting point within the field of linguistics, as far as I can reconstruct it (see, e.g., [30]), received major input from the development of "situation semantics" (see, e.g., [9]). This approach emphasizes the difference between abstract and concrete reference relations as discussed above. It focuses on the concrete aspects in terms of the embodiment and situatedness (and related concepts) of cognitive systems. Their environmental constraints serve as contexts of different degrees of generality, thus leading to nested systems of corresponding contextual knowledge representations.

Explicitly emphasizing the concrete side of semiosis, the traditional paradigm of cognitive information processing becomes *semiotic* in the sense of

---

[5]In a pronounced way, Spencer-Brown proposed such a procedure as the basis of all cognitive activity in [34]: "We take as given the idea of distinction and the idea of indication, and that we cannot make an indication without drawing a distinction."

Peirce's original intention. Using fairly sophisticated formal instruments and concepts, Rieger developed a wonderfully refined framework allowing him to define and model the constitution and understanding of meaning on the empirical basis of natural language structures. Readers interested in details will especially enjoy sec. 5 in [29], which is compactly summarized in Fig. 6 of [29].

It is evident that modeling concrete aspects of semiotics requires a second-order approach insofar as any model of those concrete aspects is inevitably abstract. Therefore, a natural extension of semiotic cognitive information processing models is the concrete implementation of these models in terms of "agents" capable of constituting and understanding meaning. Indeed, successful first steps into this direction have been reported in sec. 7 of [29]. Although these first steps are still fairly simple, they allow us to hope for further insight into the fascinating problem of how meaning can emerge from an implicit, holistic domain to an explicit reference relation.

I do not know a comparably viable and promising approach to address the problem of meaning, as related to mind-matter relations in general, with such detailed knowledge and broad relevance. Rieger's work is so attractive because it combines the merits of being philosophically informed, conceptually convincing, formally elaborated and empirically grounded. He has initiated and achieved continuing progress concerning our understanding of the notion of meaning – one of the most difficult and most interdisciplinary topics of consciousness research. Independent of the hustle and bustle of contemporary "scientific business" and all its ramifications, this work will endure.

References

[1] H. Atlan. Intentional Self-organization in Nature and the Origin of Meaning. In C. Rossi and E. Tiezzi, editors, *Ecological Physical Chemistry*, pages 311–331. Elsevier, Amsterdam, 1991.

[2] H. Atmanspacher. Complexity and Meaning as a Bridge Across the Cartesian Cut. *Journal of Consciousness Studies*, 1:168–181, 1994.

[3] H. Atmanspacher. Cartesian cut, Heisenberg cut, and the concept of complexity. *World Futures*, 49:333–355, 1997.

[4] H. Atmanspacher and R. G. Jahn. Problems of Reproducibility in Complex Mind-matter Systems. *Journal of Scientific Exploration*, 17:243–270, 2003.

[5] H. Atmanspacher, C. Räth, and G. Wiedenmann. Statistics and Metastatistics in the Concept of Complexity. *Physica A*, 234:819–829, 1997.

[6] A. Baillet. *La Vie de M. Descartes*. Daniel Horthemels, Paris, 1691.

[7] J. Balatoni and A. Rényi. Remarks on Entropy. *Publ. Math. Inst. Hung. Acad. Sci.*, 9:9–40, 1956.

[8] Y. Bar Hillel and R. Carnap. Semantic Information. *British Journal of the Philosophy of Science*, 4:147–157, 1953.

[9] J. Barwise and J. Perry. *Situations and Attitudes*. MIT Press, Cambridge, 1983.

[10] J. E. Bates and H. Shepard. Measuring Complexity Using Information Fluctuations. *Physics Letters A*, 172:416–425, 1993.

[11] J. Casti. The Simply Complex: Trendy Buzzword or Emerging New Science? *Bulletin of the Santa Fe Institute*, 7:10–13, 1992.

[12] J. P. Crutchfield. Knowledge and Meaning ... Chaos and Complexity. In L. Lam and V. Naroditsky, editors, *Modeling Complex Phenomena*, pages 66–101. Springer, Berlin, 1992.

[13] J. P. Crutchfield and K. Young. Inferring Statistical Complexity. *Physical Review Letters*, 63:105–108, 1989.

[14] R. Descartes. Meditationes De Prima Philosophia. In C. Adam and P. Tannery, editors, *Œuvres de Descartes*. Cerf, Paris, 1897/1913.

[15] D. P. Feldman and J. P. Crutchfield. Measures of Statistical Complexity: Why? *Physics Letters A*, 238:244–252, 1998.

[16] D. Gernert. Measurement of Pragmatic Information. *Cognitive Systems*, 1:169–176, 1985.

[17] P. Grassberger. Toward a Quantitative Theory of Self-generated Complexity. *International Journal of Theoretical Physics*, 25:907–938, 1986.

[18] P. Grassberger. Problems in Quantifying Self-generated Complexity. *Helv. Phys. Acta*, 62:489–511, 1989.

[19] H. Haken. *Information and Self-Organization*, chapter Sect. 1.6. Springer, Berlin, second, enlarged edition, 2000.

[20] T. C. Halsey, M. H. Jensen, L. P. Kadanoff, I. Procaccia, and B. I. Shraiman. Fractal Measures and their Singularities: The Characterization of Strange Sets. *Physical Review A*, 33:1141–1151, 1986.

[21] A. N. Kolmogorov. A New Metric Invariant of Transitive Dynamical Systems and Automorphisms in Lebesgue Spaces. *Doklady Akademii Nauk SSSR*, 119:861–864, 1958. See also Ya. G. Sinai. On the notion of entropy of a dynamical system *Doklady Akademii Nauk SSSR*, 124:768, 1959.

[22] A. N. Kolmogorov. Three Approaches to the Quantitative Definition of Complexity. *Problems in Information Transmission*, 1:3–11, 1965.

[23] K. Kornwachs and W. von Lucadou. Pragmatic Information as a Non-classical Concept to Describe Cognitive Processes. *Cognitive Systems*, 1:79–94, 1985.

[24] P. T. Landsberg and J. S. Shiner. Disorder and Complexity in an Ideal Non-equilibrium Fermi Gas. *Physics Letters A*, 245:228–232, 1998.

[25] C. W. Morris. Foundations of the Theory of Signs. In O. Neurath, R. Carnap, and C. W. Morris, editors, *International Encyclopedia of Unified Science*, volume I/2, pages 77–137. University of Chicago Press, Chicago, 1955.

[26] C. S. Peirce. *The Collected Papers (1931-1935)*, volume 1-6 edited by C. Hartshorne and P. Weiss; volume 7-8 edited by A. W. Burks. Harvard University Press, Cambridge, 1958.

[27] H. Primas. The Cartesian Cut, the Heisenberg Cut, and Disentangled Observers. In K. V. Laurikainen and C. Montonen, editors, *Symposia on the Foundations of Modern Physics. Wolfgang Pauli as a Philosopher*, pages 245–269. World Scientific, Singapore, 1993.

[28] H. Primas. Emergence in Exact Natural Sciences. *Acta Polytechnica Scandinavica*, Ma 91:83–98, 1998.

[29] B. B. Rieger. Semiotic Cognitive Information Processing: Learning to Understand Discourse. A Systemic Model of Meaning Constitution. In R. Kühn, R. Menzel, W. Menzel, U. Ratsch, M. M. Richter, and I. O. Stamatescu, editors, *Perspectives on Adaptivity and Learning*, pages 347–403. Springer, Berlin, 2002.

[30] B. B. Rieger and C. Thiopoulos. Situations, Topoi and Dispositions. In J. Retti and K. Leidlmair, editors, *KI-Informatik-Fachberichte*, volume 208, pages 365–375. Springer, Berlin, 1989.

[31] E. Scheibe. *The Logical Analysis of Quantum Mechanics*. Pergamon, Oxford, 1973.

[32] C. E. Shannon and W. Weaver. *The Mathematical Theory of Communication*. University of Illinois Press, Urbana, 1949.

[33] J. K. Sheriff. *Charles Peirce's Guess at the Riddle*. Indiana University Press, Bloomington, 1994.

[34] G. Spencer Brown. *Laws of Form*, chapter 1. George Allen and Unwin, London, 1969.

[35] M. L. von Franz. Der Traum des Descartes. In M. L. von Franz, editor, *Träume*, pages 137–224. Daimon, Zürich, 1985.

[36] E. von Weizsäcker. Erstmaligkeit und Bestätigung als Komponenten der pragmatischen Information. In E. von Weizsäcker, editor, *Offene Systeme*, volume I, pages 83–113. Klett-Cotta, Stuttgart, 1974.

[37] R. Wackerbauer, A. Witt, H. Atmanspacher, J. Kurths, and H. Scheingraber. A Comparative Classification of Complexity Measures. *Chaos, Solitons, & Fractals*, 4:133–173, 1994.

[38] W. Weaver. Science and Complexity. *American Scientist*, 36:536–544, 1968.

Part II

Models of Semantic Spaces

On the Mathematics of Semantic Spaces

Peter Gritzmann

Technische Universität München
gritzman@ma.tum.de

Summary. We study a generalization of the models of semantic spaces introduced by Rieger. The focus will be on the following aspects. We show to what extent different choices of conceptual freedom leads to dramatically different behaviour. For instance, the linguistic differentiation process introduced by Rieger is highly dependent on the underlying metric space. Also, we introduce certain invariants that may be seen as leading to new approaches for identifying meaning and relevance. In particular, we study a normalized limiting process in Rieger's original model that may help to identify certain key elements of corpora. Also, we show how sensitivities in defining associated measurements like dependency trees might be used to identify linguistic relevance.

1 Introduction

Computer oriented text analysis systems are based on formal descriptions of the relevant objects and underlying relations. Rieger [23] introduces a formal model of some high dimensional *semantic space* whose elements reflect the words and their usage correlation in sample texts; see also [26]. This space can then be used to analyse dependencies with the goal to understand the constitution of meaning.

The present paper begins with a short concise introduction of Rieger's basic model of semantic spaces. Then we give a generalized model that enables us to identify the sensitivity of Rieger's original model with respect to some of its assumptions. In particular we will show how already the use of different norms leads to a dramatically different behaviour of Rieger's linguistic differentiation. Also, we study a limit process in Euclidean space that may be used for identifying semantic strength and relevance of word tokens in text corpora.

Since semantic spaces are generally of very high dimension we briefly investigate the possibility of 'grasping structure' via visualization by means of appropriate mappings of low-dimensional range. The underlying mathematical task is that of near isometric embeddings into lower dimensional spaces.

P. Gritzmann: *On the Mathematics of Semantic Spaces*, StudFuzz **209**, 95–115 (2007)
www.springerlink.com

Since due to information theoretical bounds this approach is somewhat limited we then study certain associated measurements like radii, largest simplices and various kinds of associated trees with the goal to identify certain key features of the underlying spaces. In particular, the identification of the degree of dependence on the intrinsic context leads to a new concept of corpus immanent contextuality and intrinsic relevance.

The present paper should be seen as a mathematical reflection of certain computer linguistic concepts. Its main goal is to raise questions that might stimulate the linguistic discourse. In order to make the paper more easily accessible some of the basic mathematical background is therefore included.

2 Semantic Spaces

2.1 Rieger's Model

We will now briefly describe the model introduced by Rieger that is based on a 2-fold linguistic differentiation; for more details, examples and applications we refer to the original texts by Rieger [21, 22, 23, 24, 25, 26].

Let $n \in \mathbb{N}$, and let

$$W = \{w_1, \ldots, w_n\}$$

denote a set of semiotic entities called *words*. W will be referred to as the underlying *vocabulary*. Further, let $m \in \mathbb{N}$ and let

$$T = \{T_1, \ldots, T_m\}$$

be distinct texts over the vocabulary W. T is called *corpus*. By n_{ij} we denote the number of occurrences of w_i in T_j, and we set

$$N_i = \sum_{j=1}^{m} n_{ij},$$

counting the total number of occurrences of w_i in T. The *size* t_j of T_j is the number of words in T_j counted with multiplicity, hence the *size* of the corpus T is given by

$$t = \sum_{j=1}^{m} t_j.$$

Then $n_{ij}^* = N_i t_j / t$ gives the expected number of occurrences of w_i in a text T_j. For $i, k = 1, \ldots, n$ the correlation coefficient

$$\kappa(w_i, w_k) = \frac{\sum_{j=1}^{m}(n_{ij} - n_{ij}^*)(n_{kj} - n_{kj}^*)}{\left(\sum_{j=1}^{m}(n_{ij} - n_{ij}^*)^2\right)^{\frac{1}{2}} \left(\sum_{j=1}^{m}(n_{kj} - n_{kj}^*)^2\right)^{\frac{1}{2}}},$$

is a measure of occurrence regularities of the pair (w_i, w_k). Now let for $i = 1, \ldots, n$

$$x_i = \big(\kappa(w_i, w_1), \ldots, \kappa(w_i, w_n)\big)^T.$$

Of course, $x_1, \ldots, x_n \in \mathbb{R}^n$. Then Rieger applies a *linguistic differentiation* twice to obtain first the vectors $y_1, \ldots, y_n \in \mathbb{R}^n$ by setting

$$y_i = \big(\|x_i - x_1\|_{(2)}, \ldots, \|x_i - x_n\|_{(2)}\big)^T$$

for $i = 1, \ldots, n$, where $\| \ \|_{(2)}$ denotes the Euclidean norm, subsequently arriving at the vectors $z_1, \ldots, z_n \in \mathbb{R}^n$ defined by

$$z_i = \big(\|y_i - y_1\|_{(2)}, \ldots, \|y_i - y_n\|_{(2)}\big)^T$$

for $i = 1, \ldots, n$, The vectors $x_1, \ldots, x_n$ are called *corpus points* and can be thought of as characterizing the *use* of the words $w_1, \ldots, w_n$ in $\mathcal{T}$. The vectors $y_1, \ldots, y_n$ measure the differences in the occurrence regularities and express the *intrinsic context* in which the words are used in $\mathcal{T}$; they are called *meaning points*, [23]. Clearly, two meaning points y_i and y_j will be the nearer, the more similar the occurrences of the corresponding corpus points x_i and x_j are, i.e., the more closely semantically related the corresponding words w_i and w_j are, and will therefore cluster according to the main subjects of the texts in $\mathcal{T}$. Hence the vectors $z_1, \ldots, z_n$ can be used to define a corpus inherent notion of synonymity. In fact, this representation is obtained without any external knowledge, thus reflects the 'relative meaning' of semiotic entities in the given natural language texts.

2.2 A Generalized Metric Model

In the following two subsections we generalize Rieger's model in order to show how its specific choices for the underlying degrees of freedom affects its analytic and computational properties. Of course, the decision which assumptions are regarded most suitable has to be based on the specific application one has in mind.

Since in the above linguistic context metric properties of finite point sets carry information about occurrence regularities of the words in texts, our main subject of study are the finite metric spaces introduced below. We will, however, show how this model can be further generalized so as to accommodate for instance asymmetric linguistic similarity relations.

Recall that a *metric* in $\mathbb{R}^n$ is a functional

$$d(\ ,\ ) : \mathbb{R}^n \times \mathbb{R}^n \longrightarrow \mathbb{R}$$

with the following three properties for each $x, y, z \in \mathbb{R}^n$:

$$d(x, y) \geq 0 \qquad \text{and } d(x, y) = 0 \Leftrightarrow x = y,$$
$$d(x, y) = d(y, x)$$
$$d(x, y) \leq d(x, z) + d(z, y).$$

The last inequality is called *triangle inequality*. With any such metric d, $(\mathbb{R}^n, d)$ becomes a metric space, and it is a priorily not evident which metric is suited best for linguistic applications.

Metrics of specific importance are those derived from a norm $\| \ \|$, a functional

$$\| \ \| : \mathbb{R}^n \longrightarrow \mathbb{R}$$

with the following properties for each $x, y \in \mathbb{R}^n$ and $\lambda \in \mathbb{R}$:

$$\begin{aligned}
\|x\| &\geq 0 \qquad \text{and } \|x\| = 0 \Leftrightarrow x = 0, \\
\|\lambda x\| &= |\lambda| \|x\| \\
\|x + y\| &\leq \|x\| + \|y\|.
\end{aligned}$$

In fact, every norm $\| \ \|$ induces a metric d by

$$d(x, y) = \|x - y\|.$$

In the original model of Rieger, the Euclidean norm $\| \ \|_{(2)}$ was used. However, as we will see below other metrics might be reasonable as well, particularly ℓ_p-metrics $d_{(p)}$ that are induced by the ℓ_p-norms $\| \ \|_{(p)}$ defined by

$$\|x\|_{(p)} = \left(\sum_{l=1}^{n} |\xi_l|^p \right)^{\frac{1}{p}} \qquad \text{for } p \in [1, \infty[$$

$$\|x\|_{(\infty)} = \max_{l=1,\ldots,n} |\xi_l|,$$

for all $x = (\xi_1, \ldots, \xi_n)^T \in \mathbb{R}^n$.

Now, let $V^{(0)}$ be a family[1] of n points in $\mathbb{R}^n$,

$$V^{(0)} = \{v_1^{(0)}, v_2^{(0)}, \ldots, v_n^{(0)}\} \subset \mathbb{R}^n,$$

and let $d(,)$ be a metric in $\mathbb{R}^n$. Then

$$S^{(0)} = \left(V^{(0)}, d \right)$$

is a finite metric space, our ground space. In Rieger's model, $S^{(0)}$ is the space $\left(\{x_1, \ldots, x_n\}, d_{(2)} \right)$. In order to define the iteration process it is important to note that (in order not to have to deal with equivalence classes of different embeddings of the finite metric spaces into $(\mathbb{R}^n, d)$) we fix the order in which the points are considered, i.e., we actually deal with the n-tuple

$$\bar{V}^{(0)} = \left(v_1^{(0)}, v_2^{(0)}, \ldots, v_n^{(0)} \right) \in \mathbb{R}^n \times \cdots \times \mathbb{R}^n.$$

Now, for $k \in \mathbb{N}$ and $i = 1, \ldots, n$ let

---

[1]Note that the difference between a family and a set is that in the former repetition of elements is allowed.

$$v_i^{(k)} = \alpha^{(k)}\big(d(v_i^{(k-1)}, v_1^{(k-1)}), \ldots, d(v_i^{(k-1)}, v_n^{(k-1)})\big),$$

where $\alpha^{(k)}$ is a suitable positive real used for appropriate normalization. Of course, we then set

$$V^{(k)} = \{v_1^{(k)}, v_2^{(k)}, \ldots, v_n^{(k)}\} \subset \mathbb{R}^n$$

(or

$$\bar{V}^{(k)} = \big(v_1^{(k)}, v_2^{(k)}, \ldots, v_n^{(k)}\big) \in \mathbb{R}^n \times \cdots \times \mathbb{R}^n,$$

to emphasise the particular order of points[2]) and

$$S^{(k)} = \big(V^{(k)}, d\big).$$

$V^{(k)}$ is called the kth *metric derivative* and $S^{(k)}$ the kth *derived space*.

In most cases we will keep the vocabulary W fixed. In Section 5 it will however be necessary to derive the corresponding sets with respect to different (sub-)vocabularies. If needed we will then indicate that vocabulary W is referred to by the suffix W, i.e., we write $V_W^{(k)}$ and $S_W^{(k)}$.

In Rieger's model the iteration is performed up to $k = 2$ (and linguistically interpreted) with respect to the Euclidian metric with normalization factor 1. In fact, $S^{(1)}$ and $S^{(2)}$ are the spaces $\big(\{y_1, \ldots, y_n\}, d_{(2)}\big)$ and $\big(\{z_1, \ldots, z_n\}, d_{(2)}\big)$, respectively.

2.3 Abstract Operators and Gauge Functionals

In Subsection 2.2 we gave an extension of Rieger's model based on more general metric spaces, and the main emphasis of the present paper will actually be on metric models with metrics induced by norms. In the present subsection we will, however, show how semantic spaces can be further generalized. In particular, we will show how Minkowski geometry can be used to accommodate asymmetric distances; see [3, 27] for comprehensive studies of the theory of convex bodies.

Let us begin with a rather abstract (but still $\mathbb{R}^n$-based) generalization. Let $\mathcal{P}_n$ denote the n-fold Cartesian product of $\mathbb{R}^n$, i.e., the set of n-tuples of vectors from $\mathbb{R}^n$; hence

$$\mathcal{P}_n = \mathbb{R}^n \times \cdots \times \mathbb{R}^n.$$

Further, let $\mathcal{C}_n$ denote the set of all corpora composed of n words. In order to define a more general linguistic differentiation procedure, we need first a function

$$f : \mathcal{C} \to \mathcal{P}_n$$

which actually produces for a given corpus on n words n (generalized) corpus points. Subsequently we apply operators

---

[2]In the following we will not further stress the difference between $V^{(k)}$ and $\bar{V}^{(k)}$.

$$g_k : \mathcal{P}_n \to \mathcal{P}_n$$

for $k \in \mathbb{N}$ to define the linguistic differentiation. In the metric model, $f(\mathcal{T}) = \bar{V}^{(0)}$, while $g_k(\bar{V}^{(k-1)}) = \bar{V}^{(k)}$.

Of course, in order to make any linguistic sense, one needs to impose appropriate assumptions on f and the operators g_k. We will, however, not try to produce an axiomatic approach here.[3] Rather, we will give an example close to the metric model of Subsection 2.2 that allows to incorporate intrinsic asymmetry without loosing other basic features of distances.

So, let again

$$V^{(0)} = \{v_1^{(0)}, v_2^{(0)}, \ldots, v_n^{(0)}\} \subset \mathbb{R}^n.$$

Now, let B be a compact convex subset of $\mathbb{R}^n$ containing the origin in its interior. Then we define the *gauge functional*

$$\gamma_B : \mathbb{R}^n \to [0, \infty[$$

for each $x \in \mathbb{R}^n$ by

$$\gamma_B(x) = \min\{\lambda : \lambda \in [0, \infty[\wedge x \in \lambda B\}.$$

Note that the minimum is actually attained since 0 is an interior point of B and B is compact. We can then replace the metric used in Subsection 2.2 by the functional

$$\varphi_B : \mathbb{R}^n \times \mathbb{R}^n \to \mathbb{R}$$

defined by

$$\varphi_B(x, y) = \gamma_B(x - y)$$

to define a linguistic differentiation operator. In fact, we set for $k \in \mathbb{N}$ and $i = 1, \ldots, n$

$$v_i^{(k)} = \alpha^{(k)}\left(\varphi_B(v_i^{(k-1)}, v_1^{(k-1)}), \ldots, \varphi_B(v_i^{(k-1)}, v_n^{(k-1)})\right),$$

where $\alpha^{(k)}$ is again a suitable positive real used for appropriate normalization. Note that φ_B is in general not a metric, since the functional is not symmetric. However, it is still *positive definite*, i.e., for $x, y \in \mathbb{R}^n$

$$\varphi_B(x, y) \geq 0 \quad \text{and} \quad \varphi_B(x, y) = 0 \Leftrightarrow x = y,$$

and due to the convexity of B obeys the *triangle inequality*

$$\varphi_B(x, y) \leq \varphi_B(x, z) + \varphi_B(z, y)$$

for $x, y, z \in \mathbb{R}^n$. However, φ_B is *symmetric* i.e., $\varphi_B(x, y) = \varphi_B(y, x)$ for $x, y \in \mathbb{R}^n$ if and only if $B = -B$.

---

[3] In fact, the reason for including this paragraph is just to indicate that the framework is sufficiently general to allow for all variations that have been identified as being linguistically relevant so far.

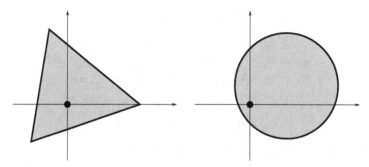

Fig. 1. Two compact convex set B that contain the origin in their interior such that φ_B is not symmetric.

Again, this is just one example deliberately constructed 'close to the metric case' showing how easily certain properties (here asymmetric distance measures) can be accommodated that may seem linguistically attractive. While, in the following, we will concentrate on the metric case since all features we want to emphasise in this paper occur already there, it should be clear that questions analogous to those studied subsequently can be investigated in much greater generality.

3 Metric Dependence and Asymptotic Behaviour

It seems natural, to study the above iteration scheme in greater generality. In particular, one can ask for the behaviour of $S^{(k)}$ for $k \to \infty$. If the starting points are obtained from texts as in Section 2.1 (and the normalization is performed correctly) then limits might encode some kind of asymptotic relevance of the semiotic entities.

This aspect will be elaborated on in Section 3.2. In Section 3.1 we first show, how dramatically the scheme changes if the maximum norm is used instead of the Euclidean norm in the metric model of Subsection 2.2.

3.1 Using the Maximum Norm

In the following we show how the iteration scheme behaves for the maximum norm

$$\|x\|_{(\infty)} = \max_{i=1,\ldots,n} |\xi_i|,$$

a norm that is quite common and often used whenever a leveling of the effects of the single coordinates is less desirable than a more radical worst case behaviour.

Theorem 3.1 *Let* $x_1, \ldots, x_n \in \mathbb{R}^n$, *and set*

$$\begin{aligned}
\eta_{ij} &= \|x_i - x_j\|_{(\infty)} &\text{for } i,j \in \{1, \ldots, n\}, \\
y_i &= (\eta_{i1}, \ldots, \eta_{in})^T &\text{for } i \in \{1, \ldots, n\}, \\
\zeta_{ij} &= \|y_i - y_j\|_{(\infty)} &\text{for } i,j \in \{1, \ldots, n\}.
\end{aligned}$$

Then

$$\zeta_{ij} = \eta_{ij} \quad \text{for } i,j \in \{1, \ldots, n\}.$$

Proof. By the triangle inequality we have for $i,j \in \{1, \ldots, n\}$

$$\begin{aligned}
\zeta_{ij} = \|y_i - y_j\|_{(\infty)} &= \max_{l=1,\ldots,n} |\eta_{il} - \eta_{jl}| \\
&= \max_{l=1,\ldots,n} \left| \|x_i - x_l\|_{(\infty)} - \|x_j - x_l\|_{(\infty)} \right| \\
&\leq \|x_i - x_j\|_{(\infty)} = \eta_{ij}.
\end{aligned}$$

On the other hand, specifying l to i (or to j), we see that

$$\max_{l=1,\ldots,n} \left| \|x_i - x_l\|_{(\infty)} - \|x_j - x_l\|_{(\infty)} \right| \geq \|x_i - x_j\|_{(\infty)} = \eta_{ij},$$

which proves the assertion. □

Theorem 3.1 means that in the case of the maximum norm, the differentiation process generating $S^{(1)}, S^{(2)}, S^{(3)}, \ldots$ becomes static after the first step.

Corollary 3.2 *Let for* $i \in \{1, \ldots, n\}$ *and* $k \in \mathbb{N}_0$, $v_i^{(k)}$ *be the ith vector of the kth metric derivative generated by using the metric* $d_{(\infty)}$. *Then*

$$v_i^{(1)} = v_i^{(2)} = v_i^{(3)} = \ldots$$

for $i \in \{1, \ldots, n\}$.

3.2 Asymptotic Behaviour of Euclidean Derivatives

In Rieger's model linguistic abstraction is employed only twice. Since in general, Euclidean iterates $v_i^{(k)}$ vary with every k, it seems reasonable to study the corresponding discrete dynamical system, and particularly to understand its asymptotic for $k \to \infty$. As an example, we investigate here explicitly what exactly happens in the first nontrivial case, that of $n = 3$. Suppose we have executed the kth step of the linguistic differentiation. Of course, if

$$v_1^{(k)} = v_2^{(k)} = v_3^{(k)}$$

everything is trivial and for $l \in \mathbb{N}$ with $l \geq k + 1$ the families $V^{(l)}$ all contain just the zero vector three times. So we assume that not all three vectors coincide. We normalize the vectors in such a way that the largest coefficient is 1, i.e., in the notation of Subsection 2.2

$$\alpha^{(k)} = \left(\max\{ \|v_i^{(k-1)} - v_j^{(k-1)}\|_{(2)} : i, j = 1, \ldots, n \} \right)^{-1}.$$

Regarding $v_1^{(k)}, v_2^{(k)}, v_3^{(k)}$ as the column vectors of a 3×3-matrix $M^{(k)}$, this matrix is of the form

$$M^{(k)} = \begin{pmatrix} 0 & \mu & \nu \\ \mu & 0 & \tau \\ \nu & \tau & 0 \end{pmatrix},$$

with suitable real entries μ, ν and τ. Suppose that $1 = \mu \geq \nu \geq \tau \geq 0$. Then,

$$M^{(k+1)} = \alpha^{(k+1)} \begin{pmatrix} 0 & \sqrt{2\mu^2 + (\nu - \tau)^2} & \sqrt{2\nu^2 + (\mu - \tau)^2} \\ \sqrt{2\mu^2 + (\nu - \tau)^2} & 0 & \sqrt{2\tau^2 + (\mu - \nu)^2} \\ \sqrt{2\nu^2 + (\mu - \tau)^2} & \sqrt{2\tau^2 + (\mu - \nu)^2} & 0 \end{pmatrix}.$$

Now it follows readily that

$$\sqrt{2\mu^2 + (\nu - \tau)^2} \geq \sqrt{2\nu^2 + (\mu - \tau)^2} \geq \sqrt{2\tau^2 + (\mu - \nu)^2},$$

thus

$$\alpha^{(k+1)} = \frac{1}{\sqrt{2\mu^2 + (\nu - \tau)^2}}.$$

Hence, if we number the starting vectors $v_1^{(0)}$, $v_2^{(0)}$ and $v_3^{(0)}$ in such a way that

$$\|v_1^{(0)} - v_2^{(0)}\|_{(2)} \geq \|v_1^{(0)} - v_3^{(0)}\|_{(2)} \geq \|v_2^{(0)} - v_3^{(0)}\|_{(2)}$$

we have

$$\|v_1^{(l)} - v_2^{(l)}\|_{(2)} \geq \|v_1^{(l)} - v_3^{(l)}\|_{(2)} \geq \|v_2^{(l)} - v_3^{(l)}\|_{(2)}$$

for every $l \in \mathbb{N}$. Of course, in view of the normalization,

$$\|v_1^{(l)} - v_2^{(l)}\|_{(2)} = 1$$

for every $l \in \mathbb{N}$. Next one checks that

$$\frac{\sqrt{2\nu^2 + (\mu - \tau)^2}}{\sqrt{2\mu^2 + (\nu - \tau)^2}} \geq \nu.$$

Therefore the sequence

$$\left(\|v_1^{(l)} - v_3^{(l)}\|_{(2)} \right)_{l \in \mathbb{N}}$$

is increasing (and bounded by 1), hence convergent. It is now easy to see that

$$\|v_1^{(l)} - v_3^{(l)}\|_{(2)} \to 1 \quad \text{as } l \to \infty.$$

Unfortunately, the sequence

$$\|v_2^{(l)} - v_3^{(l)}\|_{(2)}$$

is in general not monotonous. However, a careful analysis shows that exactly two different limits are possible, 0 or 1. Hence we have the following theorem.

Theorem 3.3 *For $n = 3$, the linguistic differentiation sequence (normalized to maximum coefficient 1, unless all entries are 0) converges. In terms of the corresponding matrices the limits are*[4]

$$\begin{pmatrix} 0\,0\,0 \\ 0\,0\,0 \\ 0\,0\,0 \end{pmatrix}, \quad \begin{pmatrix} 0\,1\,1 \\ 1\,0\,0 \\ 1\,0\,0 \end{pmatrix}, \quad \begin{pmatrix} 0\,1\,0 \\ 1\,0\,1 \\ 0\,1\,0 \end{pmatrix}, \quad \begin{pmatrix} 0\,0\,1 \\ 0\,0\,1 \\ 1\,1\,0 \end{pmatrix}, \quad \begin{pmatrix} 0\,1\,1 \\ 1\,0\,1 \\ 1\,1\,0 \end{pmatrix}.$$

For general n it is not clear whether the limit of $V^{(k)}$ always exists and what the possible limits are. In fact, computer experiments conducted by A. Dattasharma support the conjecture that the limits exist, yet show a somewhat chaotic behaviour of the interation points at the beginning of the process. This indicates that the particular values of $V^{(2)}$ may not really exhibit invariant structural information but the limits might. Hence a more detailed study is needed to understand what exactly the linguistic significance of $V^{(k)}$ and the limits of the corresponding sequences is. For instance, a distance tending to 1 might signify a pair of semantically independent keywords of a text corpus.

4 Embeddings and Measurements

Since even relatively short texts can have a large vocabulary, semantic spaces typically have a very large dimension. Hence, direct inspection may be difficult. In the following we pursue two approaches to deal with this difficulty. First we study the possibility of near isometric embeddings into spaces of much lower dimension. Then we investigate various associated measurements that can be seen as providing access to certain linguistic features.

4.1 Embeddings and Projections

For a direct study of semantic spaces via visualization it is desirable to find low-dimensional subspaces into which $S^{(k)}$ can be embedded without changing its metric properties too much. If fact, since the information is encoded in the distances of the meaning points such an embedding must be 'nearly isometric'. Theorem 3.1 shows that with respect to semiotic differentiation ℓ_∞ behaves differently than ℓ_2. In fact this can also been interpreted in terms of embeddings.

---

[4]Note, however, that the three matrices in the middle are equivalent in the sense that they differ only in the order of the underlying vectors.

Lemma 4.1 *Based on $V^{(0)}$ let $V^{(1)}$ be produced with respect to the Euclidean norm (and normalization factor 1). Then $V^{(1)}$ is the isometric embedding of $(V^{(0)}, d_{(2)})$ into ℓ_∞.*

Proof. On the one hand

$$d_{(2)}(x_i, x_j) \leq \max_{l=1,\ldots,n} \left| d_{(2)}(x_i, x_l) - d_{(2)}(x_j, x_l) \right|,$$

while on the other hand by the triangle inequality

$$d_{(2)}(x_i, x_j) \geq \left| d_{(2)}(x_i, x_l) - d_{(2)}(x_j, x_l) \right|$$

for every $l = 1, \ldots, n$. Hence for $i, j = 1, \ldots, n$

$$d_{(2)}(v_i^{(0)}, v_j^{(0)}) = d_{(\infty)}(v_i^{(1)}, v_j^{(1)}),$$

proving the assertion.   □

The following fundamental *Johnson-Lindenstrauss Lemma* shows that it is in principle possible to embed finite subsets A of Euclidean space into subspaces of rather small, in fact logarithmic dimension with arbitrarily small distortion; actually *orthogonal projections* $\Pi(A; U)$ on a suitable linear subspace U of $\mathbb{R}^n$ suffice, see [18], [8, 9], and [19].

Theorem 4.2 *Let $k, m, n \in \mathbb{N}$, $v_1, \ldots, v_m \in \mathbb{R}^n$, and $\varepsilon \in]0, \frac{1}{2}[$. If $m \leq n^2 \sqrt{\ln n}/2$ and*

$$k \geq 4(\varepsilon^2/2 - \varepsilon^3/3)^{-1} \ln n,$$

then there exists a k-dimensional subspace U such that for the linear map $f = \sqrt{\frac{n+k}{k}} \Pi(; U)$,

$$(1 - \varepsilon)\|v_i\|_2^2 \leq \|f(v_i)\|_{(2)}^2 \leq (1 + \varepsilon)\|v_i\|_{(2)}^2$$

for all $i = 1, \ldots, m$.

When applied to Euclidean semantic spaces on n words the Johnson-Lindenstrauss Lemma shows that one could in principle study all metric properties already after suitably projecting $V^{(0)}$ into some essentially $\ln(n)$-dimensional space. Of course, in order to avoid a blow up of the embedding error, for questions concerning limits one would need to do the asymptotics in the original space and project at the end rather than trying to understand the limit process in the projected space. For various theoretical and algorithmic results on embeddings and projections see [5] and [6].

4.2 Radii

In comparing semantic spaces that originate in different text corpora, the dimension and other geometric features of the *polytope*[5]

$$P = \text{conv}\left(V^{(k)}\right)$$

may be of some interest for certain k. Figure 2 shows P (in three of the five cases) according to Theorem 3.3.

 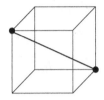 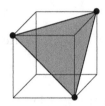

Fig. 2. Limit spaces and corresponding polytopes in Euclidean 3-space.

The main purpose of this section is to give some examples of geometric measurements that may lend themselves to a linguist interpretation. The question of algorithmic accessibility of these measurements is of course important in view of the fact that in the underlying high dimensions the semantic spaces cannot be fully analysed in practice. For more information about the following functionals and some of their relatives see [13], [14], and [15].

Let $\mathbb{M} = (\mathbb{R}^n, \|\ \|)$ denote a normed space. A j-dimensional affine subspace of $\mathbb{M}$ is called j-flat, and a j-ball of radius ρ in $\mathbb{M}$ is a set of the form

$$(q + \rho\mathbb{B}) \cap F = \{x \in F : \|x - q\| \leq \rho\}$$

for some j-flat F in $\mathbb{M}$ and point $q \in F$. A j-simplex S in $\mathbb{R}^n$ is a particular polytope in $\mathbb{R}^n$ that is the convex hull of $j + 1$ affinely independent points of $\mathbb{R}^n$; see Subsection 4.3. As an example, consider the three simplices depicted in Figure 2, (from left to right) a 0-simplex, a 1-simplex and a 2-simplex.

The following notions will be used only for polytopes of the form $P = \text{conv}\left(V^{(k)}\right)$ since those are relevant in the context of semantic spaces. For general results see [12].

For $1 \leq j \leq n$, the *inner j-radius* $r_j(P)$ of a polytope $P \subset \mathbb{M}$ is the maximum of the radii of the j-balls contained in P.

The *outer j-radius* $R_j(P)$ of P measures how well P can be approximated, in a minimax sense, by an $(n - j)$-flat. Specifically, $R_j(P)$ is the minimum

---

[5]Note that each point of P corresponds to a 'potential meaning point' that is constructed by 'averaging' some actual meaning points v. In particular, the edges of P can be interpreted as the locus of all 'potential meaning points' that are obtained by weighing the corresponding vertices with a total weight of 1.

of the positive numbers ρ such that $\mathbb{M}$ contains an $(n-j)$-flat F for which $P \subset F + \rho\mathbb{B}$.

The numbers $r_n(P)$ and $R_n(P)$ are respectively the radius of a largest n-ball contained in P and of a smallest n-ball containing P. They are called, respectively, the *inradius* and the *circumradius* of P. The number $2r_1(P)$ is the *diameter* of P, the maximum distance that is realized between two points of P. The number $2R_1(P)$ is the *width* of P, the smallest of the distances between pairs of parallel supporting hyperplanes of P.

It is quite conceivable that when applied to the polytopes $P = \text{conv}\left(V^{(k)}\right)$ these radii and the corresponding optimal flats may lead to some linguistic interpretation. For instance, a small circumball may indicate that the underlying text corpus is thematically quite focussed. Also, a large diameter might be seen as an indication for the existence of more than one key theme. A large inradius might be characteristic for a greater semantic diversity of the underlying text corpus while a small width of P indicates some specific linguistic dependencies. In particular, the orthogonal projection of $V^{(k)}$ on a hyperplane that (in the sense of width computation) best approximates P may then carry essentially the same linguistic information as the whole space.

It turns out that the algorithmic difficulties in computing radii vary dramatically with the norm that is used; see [13]. Even in the innocently looking case of n-simplices, particularly relevant in the context of semantic spaces, computations may be easy when performed with respect to one norm yet difficult with respect to some other.[6] It is for instance true that computing the width is hard in Euclidean spaces (and also in ℓ_∞-spaces) but easy in ℓ_1-spaces, [13].

Hence whenever linguistic studies involve the actual computation of parameters it is important to identify which assumptions are mandatory and which are up to discussion. If, for instance, the use of the Euclidean norm is linguistically binding then of course one has to do the computations in Euclidean spaces and might have to cope with computational intractabilities (as it is the case for the width). If, however, from a linguistical point of view various different norms may be acceptable then, of course, one should do the computations in such spaces that render them most easy.

4.3 Largest j-Simplices and Heterogeneity

In order to detect larger substructures of maximal heterogeneity that are based on the metric properties of $V^{(k)}$ one might want to identify, say, $j+1$ words,

---

[6]The notion of algorithmic difficulty can be made precise within the realm of computational complexity theory; see [10], [17] for precise definitions and various results on computational complexity. A problem is regarded as algorithmically *easy* if there exists an algorithm solving it on every instance whose running time is bounded by a polynomial in the binary size of the input. Problems belonging to the large class of so called NP-hard problems are regarded as algorithmically *hard*.

that are 'as independent as possible'.[7] A possible measure for the common heterogeneity may be the (j-dimensional) volume of the convex hull of the corresponding vectors. Hence one is longing to find a j-dimensional simplex S in the polytope $P = \text{conv}\left(V^{(k)}\right)$ of maximum j-volume. The algorithmic properties of this problem have been studied in [16]; see also [15]. Essentially it turns out that this problem can be solved efficiently if j (or $n-j$) is constant but is NP-hard if j is 'in between', e.g. if $j = \gamma(n) = \max\{1, \lfloor \mu n^{1/k} \rfloor\}$ for some $k \in \mathbb{N}$ and fixed rational μ with $0 < \mu \leq 1$, and also if $j = \gamma(n) = n - \mu n^{1/k}$ for some $k \in \mathbb{N}$ and fixed positive rational μ.

4.4 Trees

The semantic spaces $S^{(k)}$ can naturally be regarded as weighted complete graphs $G = (N, E; \omega)$. The points of $V^{(k)}$ constitute the nodes of G, the edge set E consists of each pair of different nodes, and the weight $\omega(e)$ of an edge e is some function of the distance $d(v_i, v_j)$ of the points corresponding to its nodes.[8] For background material on graphs see [7].

Rieger [23, Chapter 9] and Mehler [20, Chapter 10] study the use of minimum spanning trees and dependency trees of G to derive linguistic information. A *tree* is a connected acyclic graph, a *spanning tree* of G is a tree that contains all nodes of G. The *weight* of a graph is the sum of the weights of its edges. Hence a *minimum spanning tree* in G is a connected acyclic subgraph of G that contains all vertices of G and is of minimal weight among all such spanning trees.[9] In particular, subtrees rooted at some specific node (corresponding to some specific word) and containing some other specified node as a leaf are used to derive so called *dispositional dependency structures*. In fact, the combinatorial distance of two nodes within a minimal spanning tree is regarded as a measure for the semantic relevance of a relation of the corresponding words. The notion of *criteriality* then assigns an information theoretical numerical value that is based on the metric distances.

Since the set of acyclic subgraphs of a given graph carries the structure of a *matroid* simple greedy algorithms are available for computing minimum spanning trees. In particular, the classical algorithm of Prim starts with an arbitrary node and successively adds a minimum weight edge (and its new node) that connects a node that has already been reached with one that had not been reached previously; see e.g. [4] or [1]. (For a popular introduction to combinatorial optimization see [11].)

A *dependency tree* of G rooted at some node v is built by ordering all nodes according to increasing distance from v and then successively adding

---

[7]Another approach for measuring heterogeneity based on so called inference trees is suggested in [20, Chap. 10.3].

[8]Of course, if needed one could study more general weights that need not come from a metric. In particular, asymmetry can be modelled with the aid of weighted *directed graphs*.

[9]In the case of weighted directed graphs one is led to the notion of *arborescence*.

an edge of minimal weight connecting the node under consideration with some node previously considered. Clearly, dependency trees are spanning trees but in general not minimum spanning trees.

In the following examples and subsequent analysis we will use as the weight $\omega(e)$ of an edge e simply the Euclidean distance $d_{(2)}(v_i, v_j)$ of the points corresponding to its nodes. It should be clear, however, that similar properties hold for a great variety of different weight functions.[10]

Figure 3 depicts a set of six points in the Euclidean plane which has the property that no matter where the root is located, no dependency tree is a minimal spanning tree. Specifically, no dependency tree contains all four minimum length edges that are present in any minimum spanning tree since either the right most or the left most vertex is the unique point furthest from the root.

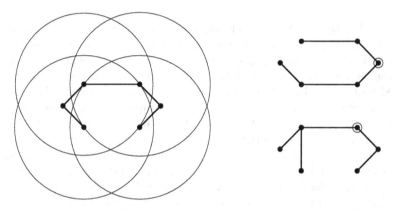

Fig. 3. No dependency tree (right: two such trees with marked roots) is a minimum spanning tree (left).

Of course, dependency trees depend on the root, which, in the linguistic interpretation is seen as an interpretation perspective. The special order in which new nodes and edges are added reflects the specific perspective seen from the corresponding word.

In [23] and [20] it is suggested to derive linguistic dependency information by using the distances within spanning trees and dependency trees, more precisely the length of the (uniquely determined) shortest path connecting two nodes v_1 and v_2 within the produced tree. Of course, this distance does depend on the specific tree that is used and does in general not coincide with the shortest path in G between v_1 and v_2.[11] Naturally, such a minimum

---

[10]As a matter of fact in [20, Chapter 10] the weights are defined differently, reversing the order of lengths.

[11]In fact, due to the metric nature of the weight function this shortest path in G is always the edge connecting v_1 and v_2 itself.

spanning tree or dependency tree based analysis of linguistic relations has to cope with the fact that these trees need not be unique. Figure 4 depicts three topologically quite different minimum spanning trees (which are also dependency trees for the central root) connecting seven points in the Euclidean plane.

Fig. 4. Minimum spanning trees or dependency trees need not be unique.

Note that the regular $(n-1)$-simplex can be embedded in a hyperplane of Euclidean n-space, in fact as the convex hull of the standard coordinate vectors. Hence the complete graph on n vertices and unit length edges shows that the number of different minimum spanning trees or dependency trees (for a fixed root) of an Euclidean semantic space on n points might be as large as n^{n-2}; see e.g. [2]. For texts with only 10 words we might therefore have already up to 100 million different such trees, and it does not seem obvious how to prefer one over the other.

One might argue that a semantic space with all distances being the same is on the one hand unrealistic and on the other hand does not exhibit any kind of metric discrimination anyway, this lack being just reflected by the 'arbitrariness' of the associated trees. However, the problem lies deeper. For, if one 'wiggles' the points of the uniform distance semantic space slightly (and generically) the minimum spanning tree as well as the dependency trees (one for each root) become unique with lengths still arbitrarily close to $n-1$. Hence, the associated tree approaches bear some disturbing inherent instability. Changing only one word in a large corpus might lead to a dramatically different minimum spanning tree or dependency tree hence seems to support a quite different linguistic relation.

5 Intrinsic Relevance

While human information processing is highly context sensitive, a very important fact of Rieger's model of automatic meaning detection is that of context independence in the sense that only intrinsic features of the underlying texts are considered. On the other hand, the derived intrinsic notion of meaning is highly sensitive to the corpus immanent textuality. In the next paragraph we will show this in more detail while Subsection 5.2 will lead to a measure of relevance that is specifically based on this sensitivity.

5.1 Corpus Immanent Contextuality

We are now interested in the corpus immanent contextuality in the following sense. Suppose we are still given the same corpus $\mathcal{T}$ but we are only interested in the linguistic relations within a (nontrivial) subset B of the vocabulary W. This may be the case if we have already identified the relevant key words of the texts and want to study their linguistic relation or if we want to restrict the analysis to a clearly defined semantic theme.[12] It is then natural to delete all words of $W \setminus B$ from all texts and compute the restricted semantic space $S_B^{(k)}$. The obvious question is of course whether the main linguistic features of $S_B^{(k)}$ are the same as those of the space obtained from $S_W^{(k)}$ after ignoring the irrelevant vectors and coordinates, i.e., after deleting the vectors corresponding to $W \setminus B$ and projecting the remaining vectors orthogonally on the linear space U of all vectors whose coordinates correspond to the words of B. We denote this space by $S_{(W|B)}^{(k)}$.

Let us perform a simple computation for one step of the differentiation procedure. We assume (without loss of generality) that the words of W are ordered so that the first $|B|$ belong to B. Hence with $r = |B|$, the vectors $v_1^{(0)}, \ldots, v_r^{(0)}$ correspond to the words of B while the others are related to $W \setminus B$. Further, let for $i = 1, \ldots, n$

$$b_i^{(0)} \in \mathbb{R}^r \times \{0\}^{n-r}, \quad c_i^{(0)} \in \{0\}^r \times \mathbb{R}^{n-r} \quad \text{such that} \quad v_i^{(0)} = b_i^{(0)} + c_i^{(0)}.$$

Clearly, performing the differentiation within $S_B^{(0)}$ we obtain $S_B^{(1)}$ consisting of r vectors with r coordinates of the form

$$\|b_i^{(0)} - b_j^{(0)}\|_{(2)}.$$

The resulting vectors of $S_{(W|B)}^{(1)}$, on the other hand, have coordinates

$$\left(\|b_i^{(0)} - b_j^{(0)}\|_{(2)}^2 + \|c_i^{(0)} - c_j^{(0)}\|_{(2)}^2 \right)^{\frac{1}{2}}.$$

To give an example how dramatically different the metric spaces $S_B^{(1)}$ and $S_{(W|B)}^{(1)}$ can be even in the case $r = n - 1$, let $\epsilon \in\,]0, 1/2[$, $\eta \in\,]0, \infty[$ and set[13]

$$v_1^{(0)} = \sqrt{1 - \epsilon}\, e_1 + \eta e_n,$$

and for $i = 2, \ldots, n - 1$

---

[12] Also such a restriction may be desirable in order to perform computations that due to the underlying algorithmic intractability cannot be executed for the whole set W.

[13] Note that for the purpose of comparing $S_B^{(1)}$ and $S_{(W|B)}^{(1)}$ the nth vector is irrelevant.

$$v_i^{(0)} = \sqrt{1 - \epsilon^i}\, e_i,$$

where $e_1, \ldots, e_n$ denote the standard unit vectors of $\mathbb{R}^n$. The different outcome of the computation of the jth coordinate of the first vector is

$$\left(2 - \epsilon - \epsilon^j\right)^{\frac{1}{2}} \quad \text{vs.} \quad \left(2 - \epsilon - \epsilon^j + \eta^2\right)^{\frac{1}{2}},$$

while all other vectors are independent of η, having coordinates

$$\left(2 - \epsilon^i - \epsilon^j\right)^{\frac{1}{2}}.$$

Since $0 < 2\epsilon < 1$ we have for $t, i, j \geq 2$

$$\epsilon - \epsilon^t > \epsilon^i - \epsilon^j,$$

hence the minimum spanning tree in $S_B^{(1)}$ is composed of all $r - 1$ edges containing $v_1^{(1)}$. However, different choices of η lead to topologically different minimum spanning trees in $S_{(W|B)}^{(1)}$. Clearly, a similar 'contextuality' holds for dependency trees.

This means that in order to semantically relate certain basic words B within the given corpus one cannot simply eliminate all words from $W \setminus B$ that seem irrelevant for this task and produce and analyse the derived spaces. In fact, performing the linguistic differentiation on the whole space $S_W^{(k)}$ and then ignoring the irrelevant meaning points can give fundamentally different results.

5.2 Relevance by Differentiation

On the one hand, the analysis of Subsection 5.1 depicts a potentially undesired dependency on the seemingly irrelevant. On the other hand, this dependency allows us to study relevance from the point of view of the induced changes. In fact, a single word w of W may be regarded the more relevant the greater the difference between $S_{W \setminus \{w\}}^{(k)}$ and $S_{(W|W \setminus \{w\})}^{(k)}$ is. Of course, this difference can be measured in various ways. A mathematically intriguing approach could be based again on a metric, this time on $\mathbb{R}^{n-1} \times \cdots \times \mathbb{R}^{n-1}$, where each of the $n - 1$ copies of $\mathbb{R}^{n-1}$ contains the difference of the corresponding two vectors of $S_{W \setminus \{w\}}^{(k)}$ and $S_{(W|W \setminus \{w\})}^{(k)}$, respectively. But, of course, all measurements discussed before can also be used.

For instance, the relevance of a word w for some other word v could be identified by the fact that the (closest) dependency trees rooted at v are different in $S_{W \setminus \{w\}}^{(k)}$ and $S_{(W|W \setminus \{w\})}^{(k)}$. One might then continue and regard a word w most relevant for a corpus if the number of different words v for which it is relevant in the above sense is highest. Obviously, a similar approach might be based on partial trees with different tree depths and any other notion of dependency.

So, there is a lot of freedom in devising notions of intrinsic relevance along the lines outlined above, and it might be worthwhile to conduct some detailed studies based on real world corpora to prove (or disprove) the linguistic significance of this concept and its relatives.

6 Further Extensions and Concluding Remarks

Rieger's model is based on pair correlations of words. Suppose we wanted to extend the generalized model by encoding occurrence regularities of subsets of W of higher cardinality. Say, we wanted to consider for each $w \in W$ all subsets of cardinality s with $1 \le s \le n - 2$. In full analogy to the case of pair correlations this would mean to associate with every element $w \in W$ a vector with

$$q = \binom{n}{s}$$

components, one for each subset of W of cardinality s. One step of Rieger's original differentiation would then immediately lead to n vectors now again in $\mathbb{R}^n$. One could, however, vary the linguistic differentiation so as to stay with n vectors in $\mathbb{R}^q$ in various ways. Since the Euclidean distance of two points is the 1-dimensional volume of their convex hull, a direct generalization of Rieger's differentiation would be obtained by taking as the updated component the s-volume of the convex hull of the union of $\{w\}$ and the s vectors whose indices correspond to the elements of the s-element subset associated with this component.

It would be interesting to see what kind of additional linguistic information could be derived from such a generalized derivation.

Clearly this and other possible extensions as well as some of the new concepts introduced before are at present based more on their mathematical appeal than on their proven linguistic significance. A natural next step would therefore be to conduct practical studies to see which of the suggestions are linguistically most promising. It should be clear that in the case of linguistic relevance further mathematical studies may lead to extended structural insight into the relevant semantic spaces that might yield new algorithms for automatic semantic text processing.

References

[1] R. K. Ahuja, T. L. Magnanti, and J. B. Orlin. *Network Flows: Theory, Algorithms and Applications.* Prentice Hall, Englewood Cliffs, N. J., 1993.

[2] M. Aigner. *Diskrete Mathematik.* Vieweg Studium: Aufbaukurs Mathematik. Vieweg, 5 edition, 2004.

[3] T. Bonnesen and W. Fenchel. *Theorie der konvexen Körper*. Springer, Berlin, 1974. Translation: *Theory of convex bodies*, BCS Associates, Moscow, Idaho (USA), 1987.

[4] K. H. Borgwardt. *Optimierung, Operations Research, Spieltheorie: Mathematische Grundlagen*. Birkhäuser Verlag, Basel, 2001.

[5] T. Burger. *Optimal Orthogonal Projections*. Dissertation, Universität Trier, Trier (Germany), 1997.

[6] T. Burger and P. Gritzmann. Finding Optimal Shadows of Polytopes. *Discrete Comput. Geom.*, 24:219–240, 2000.

[7] R. Diestel. *Graphentheorie*. Springer, Berlin, 1997.

[8] P. Frankl and H. Maehara. The Johnson-Lindenstrauss Lemma and the Sphericity of Some Graphs. *J. Comb. Theory, Ser. B*, 44(3):355–362, 1988.

[9] P. Frankl and H. Maehara. Some Geometric Applications of the Beta Distribution. *Ann. Inst. Stat. Math.*, 42(3):463–474, 1990.

[10] M. R. Garey and D. S. Johnson. *Computers and Intractability: A Guide to the Theory of NP-Completeness*. Freeman, San Francisco, 1979.

[11] P. Gritzmann and R. Brandenberg. *Das Geheimnis des kürzesten Weges*. Springer, Heidelberg, 2 edition, 2003.

[12] P. Gritzmann and V. Klee. Inner and Outer j-Radii of Convex Bodies in Finite-Dimensional Normed Spaces. *Discrete Comput. Geom.*, 7(3):255–280, 1992.

[13] P. Gritzmann and V. Klee. Computational Complexity of Inner and Outer j-radii of Polytopes in Finite-dimensional Normed Spaces. *Math. Program.*, 59A(2):163–213, 1993.

[14] P. Gritzmann and V. Klee. Mathematical Programming and Convex Geometry. In P. M. Gruber and J. M. Wills, editors, *Handbook of Convex Geometry*, volume A, pages 627–674. Elsevier, North-Holland, Amsterdam, 1993.

[15] P. Gritzmann and V. Klee. Computational Convexity. In J. E. Goodman and J. O'Rourke, editors, *Handbook of Discrete and Computational Geometry*, CRC Press Series on Discrete Mathematics and Its Application, pages 693–718. Chapman & Hall/CRC, Boca Raton, Florida, 2 edition, 2004.

[16] P. Gritzmann, V. Klee, and D. Larman. Largest j-simplices in n-polytopes. *Discrete Comput. Geom.*, 13(3-4):477–515, 1995.

[17] D. S. Johnson. A Catalog of Complexity Classes. In J. van Leeuwen, editor, *Handbook of Theoretical Computer Science*, volume A, pages 67–161. Elsevier and MIT Press, Cambridge, Mass., Amsterdam, 1990.

[18] W. B. Johnson and J. Lindenstrauss. Extensions of Lipschitz Mappings into a Hilbert Space. *Contemp. Math.*, 26:189–206, 1984.

[19] J. Matoušek. Bi-Lipschitz Embeddings into Low-dimensional Euclidean spaces. *Commentat. Math. Univ. Carol.*, 31(3):589–600, 1990.

[20] A. Mehler. *Textbedeutung. Zur prozeduralen Analyse und Repräsentation struktureller Ähnlichkeiten von Texten*. Peter Lang, Frankfurt M., 2000.

[21] B. B. Rieger. Fuzzy Structural Semantics. On a Generative Model of Vague Natural Language Meaning. In R. Trappl, P. Hanika, and F. R. Pichler, editors, *Progress in Cybernetics and Systems Research*, vol. V, pages 495–503. Wiley & Sons, Washington, New York, London, 1979.

[22] B. B. Rieger. Unscharfe Semantik natürlicher Sprache. Zum Problem der Repräsentation und Analyse vager Wortbedeutungen. In J. H. Scharff, editor, *Naturwissenschaftliche Linguistik. Leopoldina Symposion 1976*, volume 54(245) of *Nova Acta Leopoldina. Abhandlungen der Deutschen Akademie der Naturforscher Leopoldina*, pages 251–276. J. Ambrosius Barth, Halle/Saale, 1981.

[23] B. B. Rieger. *Unscharfe Semantik: Die empirische Analyse, quantitative Beschreibung, formale Repräsentation und prozedurale Modellierung vager Wortbedeutungen in Texten*. Peter Lang, Frankfurt a. M., 1989.

[24] B. B. Rieger. Distributed Semantic Representation of Word Meanings. In J. D. Becker, I. Eisele, and F. W. Mündemann, editors, *Parallelism, Learning, Evolution. Evolutionary Models and Strategies*, volume 565 of *Lecture Notes in Artificial Intelligence*, pages 243–273. Springer, Berlin, Heidelberg, New York, 1991.

[25] B. B. Rieger. Situation Semantics and Computational Linguistics: Towards Informational Ecology. A Semiotic Perspective for Cognitive Information Processing Systems. In K. Kornwachs and K. Jacoby, editors, *Information. New Questions to a Multidisciplinary Concept*, pages 285–315. Akademie-Verlag, Berlin, 1996.

[26] B. B. Rieger. Semiotic Cognitive Information Processing: Learning to Understand Discourse. A systematic Model of Meaning Constitution. In R. Kühn, R. Menzel, W. Menzel, U. Ratsch, M. M. Richter, and I.-O. Stamatescu, editors, *Adaptivity and Learning: An Interdisciplinary Debate*, pages 347–403. Springer, Berlin, Heidelberg, New York, 2003.

[27] R. Schneider. *Convex Bodies: The Brunn-Minkowski Theory*. Cambridge University Press, Cambridge, 1993.

Models of Semantic Spaces

Edda Leopold

Hochschule für Angewandte Wissenschaften Hamburg
leopold@mt.haw-hamburg.de

1 Introduction

This contribution gives an overview about different approaches to semantic spaces. It is not a exhaustive survey, but rather a personal view on different approaches which use metric spaces for the representation of meanings of linguistic units. The aim is to demonstrate the similarities of apparently different approaches and to inspire the generalisation of semantic spaces tailored to the representation of texts to arbitrary semiotic artefacts.

I assume that the primary purpose of a semiotic system is communication. A semiotic system $\tilde{S}$ consists of signs s. Signs fulfil a communicative function $f(s)$ within the semiotic system in order to meet the communicative requirements of system users. There are different similarity relations between functions of signs. In its most general form a semantic space can be defined as follows:

Definition 1 Let $\tilde{S}$ be a semiotic system, (S, d) a metric space and $r : \tilde{S} \to S$ a mapping from $\tilde{S}$ to S. A semantic space (S, d) is a metric space whose elements are representations of signs of a semiotic system, i.e. for each $x \in S$ there is a $s \in \tilde{S}$ such that $r(s) = x$. The inverse metric $(d(x, y))^{-1}$ quantifies some functional similarity of the signs $r^{-1}(x)$ and $r^{-1}(y)$ in $\tilde{S}$.

Semantic spaces can quantify functional similarities in different respects. If the semiotic system is a natural language, the represented units are usually words or texts – but semantic spaces can also be constructed from other linguistic units like syllables or sentences. The constructions of semantic spaces leads to a notion of semantic distance which often cannot easily be made explicit. Some constructions (like the one described in section 6) yield semantically transparent dimensions.

The definition of a semantic space is not confined to linguistic units. Anything that fulfils a function in a semiotic system can be represented in a semantic space. The calculation of a semantic space often involves a reduction

E. Leopold: *Models of Semantic Spaces*, StudFuzz **209**, 117–137 (2007)
www.springerlink.com   © Springer-Verlag Berlin Heidelberg 2007

of dimensionality and the spaces described in this paper will be ordered with decreasing dimensionality and increasing semantic transparency. In the following section, the basic notations, which are used in the subsequent sections, will be introduced. Section 3 roughly outlines the fuzzy linguistic paradigm. Sections 4 and 5 briefly describe the methods of latent semantic indexing and probabilistic latent semantic indexing. In section 6 I show how previously trained classifiers can be used in order to construct semantic spaces.

2 Notations

In order to harmonise the presentation of the different approaches I will use the following notations: A text corpus $\mathcal{C}$ consists of a number of D different textual units referred to as *documents* $d_j, j = 1, \ldots, D$. Documents can be complete texts, such as articles in a newspaper, short news e.g. the Reuters newswire corpus, or even short text fragments like paragraphs or text blocks of a constant length.

Each document consists of a (possibly huge) number of *terms*. The entire number of different term-types in $\mathcal{C}$ (i.e. the size of the vocabulary of $\mathcal{C}$) is denoted by W and the number of occurrences of a given term w_i in a given document d_j is denoted by $f(w_i, d_j)$. The definition of what is considered as a term may vary, terms can be lemmas, words as they occur in the running text (i.e. strings separated by blanks), tagged words such as in [18], strings of syllables as in [24], or even a mixture of lemmas and phrases as in [23]. The methods described below are independent from what is considered as a term in a particular application. It is merely assumed that a corpus consists of a set of documents and each of these documents consist of a set of terms[1]. The *term-document matrix* A of $\mathcal{C}$ is a $W \times D$ matrix with W rows and D columns, which is defined as

$$A = (f(w_i, d_j))_{i=1,\ldots,W, j=1,\ldots,D} \tag{1}$$

or more explicitly

$$A = \begin{pmatrix} a_{11} & a_{12} & \cdots & a_{1D} \\ a_{21} & a_{22} & \cdots & a_{2D} \\ \vdots & & \ddots & \vdots \\ a_{W1} & a_{W2} & \cdots & a_{WD} \end{pmatrix}, \qquad \text{where } a_{ij} := f(w_i, d_j) \tag{2}$$

The entry in the ith row and the jth column of the term-document matrix indicates how often term w_i appears in document[2] d_j. The rows of A represent

---

[1] Actually the assumption is even weaker: the methods simply focus on the co-occurrences of documents and terms, no matter if one is contained in the other.

[2] It should be noticed here that in many cases the term-document matrix does not contain the term-frequencies $f(w, d)$ themselves but a transformation of them like e.g. $\log f(w, d)$ or tfidf.

terms and its columns represent documents. In the so-called "bag-of-words" representation, document d_j is represented by the jth column of A, which is also called the word-frequency vector of document d_j and denoted by $\mathbf{x}_j$. The sum of the frequencies in the j-th row of A is denoted by $f(d_j)$, which is also called the *length* of document d_j. The length of corpus $\mathcal{C}$ is denoted by L. Clearly

$$f(d_j) = \sum_{i=1}^{W} f(w_i, d_j) \quad \text{and} \quad L = \sum_{j=1}^{D} f(d_j) \tag{3}$$

The ith row of A indicates how the term w_i is spread over the documents in the corpus. The rows of A are linked to the notion of *polytexty*, which was defined by Köhler [14] as the number of contexts in which a given term w_i occurs. Köhler noted that polytexty can be operationalised by the number of *texts* the term occurs in i.e. the number of non-zero entries of the i-th row. The ith column of A is therefore called *vector of polytexty* of term w_i and the vector of the respective relative frequencies is named *distribution of polytexty*. The sum over the frequencies in the ith column, i.e. the total number of occurrences of term w_i in the corpus $\mathcal{C}$, is denoted by

$$f(w_i) = \sum_{j=1}^{D} f(w_i, d_j) \ . \tag{4}$$

The polytexty measured in terms of non-zero entries in a row of the term-document matrix is also called *document-frequency* denoted as df. The so-called inverse document frequency, which was defined by [30] as $idf = (\log df)^{-1}$, is widely used in the literature on automatic text processing in order to tune term-frequencies according to the thematic relevance of a term. Other term weighting schemes, e.g. the redundancy used by [18] consider the entire vector of polytexty rather than solely the number of non-zero elements. A summary of different weighting schemes is given in [20].

Matrix transposition, subsequently indicated by a superscript $\cdot^T$, exchanges columns and rows of a matrix. So the transposed term-document matrix is defined as

$$A^T = (f(w_j, d_i))_{i=1,\dots,D, j=1,\dots,W} = \begin{pmatrix} a_{11}^t & a_{12}^t & \cdots & a_{1W}^t \\ a_{21}^t & a_{22}^t & \cdots & a_{2W}^t \\ \vdots & & \ddots & \vdots \\ a_{D1}^t & a_{W2}^t & \cdots & a_{DW}^t \end{pmatrix}, \tag{5}$$

where $a_{ij}^t := f(w_j, d_i)$

It is easy to see that the matrix transposition is inverse to itself, i.e. $(A^T)^T = A$. All algorithms presented below are symmetrical in documents and terms, i.e. they can be used to estimate semantic similarity of terms as well as of documents depending on whether A or A^T is considered.

There are various measures for judging the similarity of documents. Some measures – the so-called association measures – disregard the term frequencies and just perform set-theoretical operations on the document's term sets. An example of an association measure is the *matching coefficient*, which simply counts the number of terms that two documents have in common [31].

Other measures take advantage of the vector space model and consider the entire term-frequency vectors of the respective documents. One of the most often used similarity measure, which is also mathematically convenient, is the cosine measure [20, 30] defined as

$$\cos(\mathbf{x}_i, \mathbf{x}_j) \quad = \quad \frac{\sum_k^W f(w_k, d_i) f(w_k, d_j)}{\sqrt{\sum_k^W f(w_k, d_i)^2 \sum_k^W f(w_k, d_j)^2}} \quad = \quad \frac{\mathbf{x}_i \cdot \mathbf{x}_j}{\|\mathbf{x}_i\| \|\mathbf{x}_j\|}, \quad (6)$$

which can also be interpreted as the angle between the vectors $\mathbf{x}_i$ and $\mathbf{x}_j$ or, up to centering, as the correlation between the respective discrete probability distributions.

3 Fuzzy Linguistics

"[. . .] the investigation of linguistic problems in general, and that of word-semantics in particular, should start with more or less pre-theoretical working hypotheses, for-mulated and re-formulated for continuous estimation and/or testing against observ-able data, then proceed to incorporate its findings tentatively in some preliminary theoretical set up which may finally perhaps get formalised to become part of an encompassing abstract theory. With our objective being natural language meaning, this operational approach would have to be what I would like to call *semiotic*." [25].

Fuzzy Linguistics (FL) [25, 27, 29] aims at a spatial representation of word meanings. I.e. the units represented in the semantic space are *words* as op-posed to, in the other approaches, documents. However from a mathematical point of view there is no formal difference between semantic spaces that are constructed to represent documents and those which are intended to represent terms. One can transform one problem into the other by simply transposing the term-document matrix i.e. by considering A^T instead of A.

Rieger has calculated a semantic space of word meanings in two steps of abstraction, which are also implicitly incorporated in the other construc-tions of semantic spaces described in the sections (4) to (6). The first step of abstraction is the α-*abstraction* or more explicitly *syntagmatic abstraction* which reflects a term's usage regularities in terms of its vector of polytexty. The second abstraction step is the δ-abstraction or *paradigmatic abstraction*, which represents a word's relation to all other words in the corpus.

3.1 The Syntagmatic Abstraction

For each term w_i a vector of length W is calculated, which contains the correlations of a term's vector of polytexty with all other terms in the corpus.

$$\alpha_{i,j} = \frac{\sum_{k=1}^{D}(f(w_i, d_k) - E(f(w_i) \mid d_k))(f(w_j, d_k) - E(f(w_j) \mid d_k))}{\sqrt{\sum_{k=1}^{D}(f(w_i, d_k) - E(f(w_i) \mid d_k))^2 \sum_{k=1}^{D}(f(w_j, d_k) - E(f(w_j) \mid d_k))^2}} \quad (7)$$

where $E(f(w_i) \mid d_k) = f(w_i)\frac{f(d_k)}{L}$ is an estimator of the conditioned expectation of the frequency of term w_i in document d_j, based on all documents in the corpus. The coefficient $\alpha_{i,j}$ measures the mutual affinity ($\alpha_{i,j} > 0$) or repugnancy ($\alpha_{i,j} < 0$) of pairs of terms in the corpus [29].

Substituting $y_{i,j} = f(w_i, d_k) - E(f(w_i) \mid d_k)$, the centralised vector of polytexty of term w_i is defined as $\mathbf{y}_i = (y_{i,1}, \ldots, y_{i,D})^T$. Using this definition equation (7) can be rewritten as

$$\alpha_{i,j} = \frac{\sum_{k}^{D} y_{i,k} y_{j,k}}{\sqrt{\sum_{k}^{D} y_{i,k}^2 \sum_{k}^{D} y_{j,k}^2}} = \frac{\mathbf{y}_i \cdot \mathbf{y}_j}{\|\mathbf{y}_i\|\|\mathbf{y}_j\|}, \quad (8)$$

which is the definition of the cosine distance as defined in equation (6). The difference between the α-abstraction and the cosine distance is merely that in equation (7) the centralised vector of polytexty is considered instead of the word-frequency vector in (6). Using the notion of polytexty, one might say more abstractly that $\alpha_{i,j}$ is the correlation coefficient of the polytexty distributions of the types w_i and w_j on the texts in the corpus.

Syntagmatic abstraction realised by equation (7) refers to usage regularities in terms of co-occurrences in the same document. Documents in Rieger's works were, in general, short texts, like e.g. newspaper texts [25, 29] or small textual fragments [28]. This means that the syntagmatic abstraction solely relies on the distribution of polytexty of the respective terms.

In principle however, the approach can be generalised regarding various types of generalised syntagmatic relations. Note that documents were defined as arbitrary, disjoint subsets of a corpus. The underlying formal assumption was simply that there is a co-occurrence structure of documents and terms, which is represented in the term-document matrix. Consider for instance a syntactically tagged corpus. In such a corpus documents might be defined for example as a set of terms that all carry the same tag. The corresponding "distributions of polytexty" would describe how a term is used in different parts-of-speech and the syntagmatic abstraction $\alpha_{i,j}$ would measure the similarity of w_i and w_j in terms of part-of-speech membership.

3.2 The Paradigmatic Abstraction

The α-abstraction measures the similarities of the distribution of polytexty over all terms in the corpus. The absolute value of the similarities, however, is not solely a property of the terms themselves, but also of the corpus as a whole. That is if the corpus is confined to a small thematic domain, the documents will be more similar than in the case of a corpus that covers a wide range of themes. In order to attain a paradigmatic abstraction, which abstracts away from the thematic coverage of the corpus, the Euclidean distances to all words in the corpus are summed. This is the δ-abstraction [25, 29] given by:

$$\delta(\mathbf{y}_i, \mathbf{y}_j) = \sqrt{\sum_{n=1}^{W}(\alpha_{i,n} - \alpha_{j,n})^2}; \qquad \delta \in [0; 2\sqrt{W}] \qquad (9)$$

The δ-abstraction compensates the effect of the corpus' coverage on α. The similarity vector of each term is related to the similarity vectors of all other terms in the corpus. In this way the paradigmatic structure in the corpus is evaluated in the sense that every term is paradigmatically related to each other since every term can equally be engaged in a *occurs-in-document* relation.

So the vector $\mathbf{y}_i$, is mapped to a vector $(\delta(i, 1) \ldots \delta(i, W))$, which contains the Euclidean distance of x_i's α to all other αs generated by the corpus and is interpreted as meaning point in a semantic space [26]. Rieger concludes that in this way a semantic representation is attained that represents the numerically specified generalised paradigmatic structure that has been derived for each abstract syntagmatic usage regularity against all other in the corpus [27].

Goebl [8] uses another measurement to anchor similarity measurements of linguistic units (in his case dialectometric data sets) for the completely different purpose of estimating the centrality of dialects in a dialectal network. Let $\alpha_{i,j}$ denote the similarity of dialect x_i and x_j, and let W denote the number of dialects in the network. The centrality of x_i is given by:

$$\gamma(x_i) = \sum_{n=1}^{W}\left(\alpha_{i,n} - \frac{1}{W}\sum_{k=1}^{W}\alpha_{i,k}\right)^3 \qquad (10)$$

He argues

"The skewness of a similarity distribution has a particular *linguistic* meaning. The more symmetric a similarity distribution is, the greater the centrality of the particular local dialect in the whole network." [8].

Goebl uses (10) in order to calculate the centrality of a local dialect from the matrix $(\alpha_{i,j})_{i,j}$ of similarity measures between pairs of dialects in the network. These centrality measures are employed to draw a choropleth map of the dialectal network. Substituting the delta abstraction in (9) by the skewness in (10) would result in a measure for the centrality of a term in a term-document network: the more typical a term's usage in the corpus the larger the value of γ. Such a measure could be used as a term-weighting scheme.

Rieger's construction of a semantic space does *not* lead to a reduction of dimensionality. This was not his aim. The meaning of a term is represented by a high-dimensional vector and thus demonstrates the complexity of meaning structures in natural language. Rieger's idea to compute semantic relations from a term-document matrix and represent semantic similarities as distances in a metric space has aspects in common with pragmatically oriented approaches e.g. latent semantic analysis. The measures of the $\alpha_{i,j}$ can be written in a more condensed way as

$$B^* = A^*(A^*)^T = (\alpha_{i,j})_{i,j=1,\ldots W} \qquad (11)$$

B^* is a $W \times W$-matrix which represents the similarity of the words w_i and w_j in terms of their distribution of polytexty. The semantic similarity between words is calculated here in a way similar to the semantic similarity between words in latent semantic indexing which is described in the next section. The *similarity matrix* $B^* = A^*(A^*)^T$ however is calculated in a slightly different way. The entries of A^* are $y_{i,j} = f(w_i, d_k) - E(f(w_i) \mid d_k)$ rather than the term frequencies $f(w_i, d_j)$ themselves, as can be seen from equation (7).

More advanced techniques within the fuzzy linguistic paradigm [21] extend the concept of the semantic space to the representation of texts. The respective computations, however, are complicated and exceed the scope of this paper.

Fuzzy linguistics aims at a numerical representation of the meaning of terms. Thus the paradigmatic abstraction in equation (9) does not involve a reduction of dimensionality. This is in contrast to the principal component analysis that is performed in the paradigmatic abstraction step in latent semantic analysis. There is however a close formal relationship.

4 Latent Semantic Analysis

"In essence, and in detail, it [latent semantic analysis] assumes that the psychological similarity between any two words is reflected in the way they co-occur in small subsamples of language" [17]. (Words in square brackets added by the author.)

In contrast to fuzzy linguistics *Latent Semantic Analysis* (LSA) deals with the semantic nearness of *documents* rather than of words. The method however is symmetric and can be applied to the similarity of words as well.

LSA projects document frequency vectors into a low dimensional space calculated using the frequencies of word occurrence in each document. The relative distances between these points are interpreted as distances between the topics of the documents and can be used to find related documents, or documents matching some specified query [2]. The underlying technique of LSA was chosen to fulfil the following criteria:

1. To represent the underlying semantic structure, a model with sufficient power is needed. Since the right kind of alternative is unknown, the power of the model should be variable.
2. Both terms and documents should be explicitly represented in the model.
3. The method should be computationally tractable for large data sets. Deerwester et al. concluded that the only model which satisfied all these three criteria was the singular value decomposition (SVD), which is a well known technique in linear algebra [4].

4.1 Singular Value Decomposition

Let A be a term-document matrix as defined in section (2) with rank[3] r. The singular value decomposition of A is given by

$$A = U \Sigma V, \tag{12}$$

where $\Sigma = diag(\sigma_1, \ldots, \sigma_r)$ is a diagonal matrix with ordered diagonal elements $\sigma_1 > \cdots > \sigma_r$,

$$U = \begin{pmatrix} u_{11} & u_{12} & \cdots & u_{1r} \\ u_{21} & u_{22} & \cdots & u_{2r} \\ \vdots & & \ddots & \vdots \\ u_{W1} & u_{W2} & \cdots & u_{Wr} \end{pmatrix} \tag{13}$$

is a $W \times r$-matrix with orthonormal columns and

$$V = \begin{pmatrix} v_{11} & v_{12} & \cdots & v_{1r} \\ v_{21} & v_{22} & \cdots & v_{2r} \\ \vdots & & \ddots & \vdots \\ v_{r1} & v_{r2} & \cdots & v_{rr} \end{pmatrix} \tag{14}$$

is a $r \times r$-matrix with orthonormal rows. The diagonal elements $\sigma_1, \ldots, \sigma_r$ of the matrix Σ are singular values of A. The singular value decomposition can equivalently be written as an eigen-value decomposition of the similarity matrix

$$B = AA^T \tag{15}$$

Note that U and V are orthonormal matrices. Therefore $UU^T = I$ and $VV^T = I$, where I is the neutral element of matrix-multiplication. According to (12), the singular value decomposition of the transposed term-document matrix A^T is obtained as $A^T = V^T \Sigma U^T$. Hence $AA^T = U\Sigma VV^T \Sigma U^T = U\Sigma^2 U^T$ which is the eigen value decomposition of AA^T with eigen-values $\sigma_1^2, \ldots, \sigma_r^2$. Term frequency vectors are mapped to the latent space of artificial concepts by multiplication with $U\Sigma$, i.e. $\mathbf{x} \rightarrow \mathbf{x}^T U\Sigma$. Each of the r dimensions of the latent space may be thought of as an artificial concept, which represents common meaning components of different words and documents.

4.2 Deleting the Smallest Singular Values

A reduction of dimensionality is achieved by deleting the smallest singular values corresponding to the less important concepts in the corpus. In so doing, latent semantic analysis reduces the matrix A to a smaller K-dimensional $(K < r)$ matrix:

---

[3]In practice one can assume $r = D$, since it is very unlikely that there are two documents in the corpus with linear dependent term-frequency vectors

$$A_K = U_K \Sigma_K V_K, \tag{16}$$

Here, U_K and V_K are obtained from U and V in equation (12) by deleting respectively columns and/or rows $K + 1$ to r and the diagonal matrix is reduced to $\Sigma_K = diag(\sigma_1, \ldots, \sigma_K)$. The mapping of a term-frequency vector to the reduced latent space is now performed by $\mathbf{x} \to \mathbf{x}^T U_K \Sigma_K$. It has been found that $K \approx 100$ is a good value to chose for K [17].

LSA leads to vectors with few zero entries and to a reduction of dimensionality (K instead of W) which results in a better geometric interpretability. This implies that it is possible to compute meaningful association values between pairs of documents, even if the documents do not have any terms in common.

4.3 SVD Minimises Euclidean Distance

Truncating the singular value decomposition as described in equation (16) projects the data onto the best-fitting affine subspace of a specified dimension K. It is a well-known theoretical result in linear algebra, that there is no matrix X with $rank(X) < K$ that has a smaller Frobenius distance to the original matrix A i.e. A_K minimises

$$\|A - A_K\|_F = \sum_{i,j}^{K} (a_{i,j} - a_{i,j}^K)^2. \tag{17}$$

Interestingly, Rieger's δ-abstraction in equation (9) yields a nice interpretation of this optimality statement. The reduction of dimensionality performed by latent semantic analysis is achieved in such a way that it optimally preserves the inherent meaning (i.e. the sum of the $\delta(x_i, x_j)$). That is, the meaning points in Rieger's δ-space are changed to a minimal possible extent. Another parallel between fuzzy linguistics and LSA is that equation (7) and the corresponding matrix notation of $\alpha_{i,j}$ in equation (11) coincide with the similarity matrix in equation (15). The only difference is that the entries of A and A^* are defined in a different way. Using Rieger's terminology, one may call equation (15) a syntagmatic abstraction, because it reflects the usage regularities in the corpus. The singular value decomposition is then the paradigmatic abstraction, since it abstracts away from the paradigmatic structure of the language's vocabulary which consists of synonymy and polysemy relationships.

One objection to latent semantic indexing is that along with all other least-square methods, the property of minimising the Frobenius distance makes it suited for normally distributed data. The normal distribution however is unsuitable to model term frequency counts. Other distributions like Poisson or negative binomial are more appropriate for this purpose [20].

Alternative methods have therefore been developed [9], which assume that the term frequency vectors are multinomially distributed and therefore agree with well corroborated models on word frequency distribution developed by Chitashvili and Baayen [3]. Probabilistic Latent Semantic Analysis has advanced further in this direction.

5 Probabilistic Latent Semantic Analysis

Whereas latent semantic analysis is based on counts of co-occurrences and uses the singular value decomposition to calculate the mapping of term-frequency vectors to a low-dimensional space, *Probabilistic Latent Semantic Analysis* (PLSA; see [10, 11]) is based on a probabilistic framework and uses the maximum likelihood principle. This results in a better linguistic interpretability and makes PLSA compatible with the well-corroborated multinomial model of word frequency distributions.

5.1 The Multinomial Model

The assumption that the occurrences of different terms in the corpus are stochastically independent allows to calculate the probability of a given term frequency vector $\mathbf{x}_j = (f(w_1, d_j), \ldots, f(w_W, d_j))$. This is according to the multinomial distribution (cf. Baayen [1] and Chitashvili & Baayen [3]):

$$p(\mathbf{x}_j) = \frac{f(d_j)}{\prod_{i=1}^{W} f(w_i, d_j)!} \prod_{i=1}^{W} p(w_i, d_j)^{f(w_i, d_j)} \tag{18}$$

where $p(w_i, d_j)$ is the probability of occurrence of term w_i in document d_j which can be estimated by $p(w_i, d_j) = f(w_i, d_j)/f(d_j)$. If it is further assumed that the term-frequency vectors of the documents in the corpus are stochastically independent, the probability to observe a given term-document matrix is

$$p(A) = \prod_{j=1}^{D} \frac{f(d_j)}{\prod_{i=1}^{W} f(w_i, d_j)!} \prod_{i=1}^{W} p(w_i, d_j)^{f(w_i, d_j)} \tag{19}$$

5.2 The Aspect Model

In order to map high-dimensional term-frequency vectors to a limited number of dimensions, PLSA uses a probabilistic framework called aspect model. The aspect model is a latent variable model which associates an unobserved class variable $z_k, k = 1, \ldots, K$, with each observation, an observation being the occurrence of a word in a particular document. The latent variables z_k can be thought of as artificial concepts like the latent dimensions in LSA. Like in LSA, the number of artificial concepts K has to be chosen by the experimenter. The following probabilities are introduced: $p(d_j)$ denotes the probability that a word occurrence will be observed in a particular document, d_i, $p(w_i \mid z_k)$ denotes the conditional probability of a specific term conditioned on the latent variable z_k (i.e. the probability of term w_i given the thematic domain z_k), and finally $p(z_k \mid d_j)$ denotes a document-specific distribution over the latent variable space i.e. the distribution of artificial concepts in document d_j.

A generative model for word/document co-occurrences is defined as follows:

(1) select a document d_j with probability $p(d_j)$,
(2) pick a latent class z_k with probability $p(z_k|d_j)$, and
(3) generate word w_j with probability $p(w_i|z_k)$ [10].

Since the aspects are latent variables which cannot be observed directly, the conditioned probability $p(w_i \mid d_j)$ has to be calculated as the sum of the possible aspects:

$$p(w_i|d_j) \quad = \quad \sum_{k=1}^{K} p(w_i|z_k)p(z_k|d_j) \tag{20}$$

This implies the assumption that the conditioned probability of occurrence of aspect z_k in document d_j is independent from the conditioned probability that term w_i is used, given that aspect z_k is present [10].

In order to find the optimal probabilities $p(w_i|z_k)$ and $p(z_k|d_j)$, maximizing the probability of observing a given term-document matrix, the maximum likelihood principle is applied. The multinomial coefficient in equation (19) remains constant when the probabilities $p(w_i, d_j)$ are varied. It can therefore be omitted for the calculation of the likelihood function, which is then given as

$$\mathcal{L} = \sum_{j=1}^{D} \sum_{i=1}^{W} f(w_i, d_j) \log p(w_i, d_j) \tag{21}$$

Using the definition of the conditioned probabilities $p(w_i, d_j) = p(d_j)p(w_i \mid d_j)$ and inserting equation (20) yields

$$\mathcal{L} = \sum_{j=1}^{D} \sum_{i=1}^{W} \left(f(w_i, d_j) \log \left(p(d_j) \cdot \sum_{k=1}^{K} p(w_i \mid z_k)p(z_k \mid d_j) \right) \right) \tag{22}$$

Using the additivity of the logarithm and factoring in $f(w_i, d_j)$ gives

$$\mathcal{L} = \sum_{j=1}^{D} \left(\sum_{i=1}^{W} f(w_i, d_j) \log p(d_j) + \sum_{i=1}^{W} f(w_i, d_j) \log \sum_{k=1}^{K} p(w_i \mid z_k)p(z_k \mid d_j) \right) \tag{23}$$

Since $\sum_i f(w_i, d_j) = f(d_j)$ factoring out $f(d_j)$ finally leads to the likelihood function

$$\mathcal{L} = \sum_{j=1}^{D} f(d_j) \left(\log p(d_j) + \sum_{i=1}^{W} \frac{f(w_i, d_j)}{f(d_j)} \log \sum_{k=1}^{K} p(w_i \mid z_k)p(z_k \mid d_j) \right) \tag{24}$$

which has to be maximised with respect to the conditional probabilities involving the latent aspects z_k. Maximisation of (24) can be achieved using the EM-algorithm, which is a standard procedure for maximum likelihood estimation in latent variable models [5]. The EM-algorithm works in two steps that are iteratively repeated (see e.g. [22] for details).

Step 1 In the first step (the expectation step), the expected value $E(z_k)$ of the latent variables is calculated, assuming that the current hypothesis h_1 holds.

Step 2 In a second step (the maximisation step), a new maximum likelihood hypothesis h_2 is calculated. This assumes that the latent variables z_k equal their expected values $E(z_k)$ that have been calculated in the expectation step. Then h_1 is substituted by h_2 and the algorithm is iterated.

In the case of PLSA, the EM-algorithm is employed as follows (see [10] for details): To initialise the algorithm, generate $W \cdot K$ random values for the probabilities $p(w_i \mid z_k)$ and $D \cdot K$ random values for the probabilities $p(z_k \mid d_j)$ such that all probabilities are larger than zero and fulfil the conditions $\sum_{i,k} p(w_i \mid z_k) = 1$ and $\sum_{j,k} p(z_k \mid d_j) = 1$ respectively. The expectation step can be obtained from equation (24) by applying Bayes' formula:

$$p(z_k \mid w_i, d_j) = \frac{p(w_i \mid z_k)p(z_k \mid d_j)}{\sum_{k=1}^{K} p(w_i \mid z_k)p(z_k \mid d_j)} \qquad (25)$$

In the maximization step, the probability $p(z_k \mid w_i, d_j)$ is used to calculate the new conditioned probabilities

$$p(w_i \mid z_k) = \frac{\sum_{j=1}^{N} f(w_i, d_j)p(z_k \mid w_i, d_j)}{\sum_{k=1}^{K} \sum_{j=1}^{D} f(w_i, d_j)(z_k \mid w_i, d_j)} \qquad (26)$$

and

$$p(z_k \mid d_j) = \frac{\sum_{i=1}^{W} f(w_i, d_j)p(z_k \mid w_i, d_j)}{f(d_j)}, \qquad (27)$$

Now the conditioned probabilities $p(z_k|d_j)$ and $p(w_i|z_k)$ calculated from equation (26) and (27) are inserted into equation (25) to perform the next iteration. The iteration is stopped when a stationary point of the likelihood function is achieved. The probabilities $p(z_k \mid d_j), k = 1, \ldots, K$, uniquely define for each document a $K - 1$-dimensional point in continuous latent space.

It is reported that PLSA outperforms LSA in terms of perplexity reduction. Notably PLSA allows to train latent spaces with a continuous increase in performance. This is in contrast to LSA where the model perplexity increases when a certain number of latent dimensions is exceeded. In PLSA the number of latent dimensions may even exceed the rank of the term-document matrix [10].

The main difference between LSA and PLSA is the optimisation criterion for the mapping to the latent space which is defined by $U\Sigma$ and $p(z_k \mid d_j)$

respectively. LSA minimises the least square criterion in equation (17) and thus, implicitly assumes an additive Gaussian noise on the term-frequency data. PLSA in contrast, assumes multinomially distributed term-frequency vectors and maximises the likelihood of the aspect model. It is therefore in accordance with linguistic word frequency models. One disadvantage of PLSA is that the EM-algorithm, like most iterative algorithms, converges only locally. Therefore the solution need not be a global optimum. This is in contrast to LSA which uses an algebraic solution and ensures global optimality.

6 Classifier Induced Semantic Spaces

"[...] problems, in which the task is to classify examples into one of a discrete set of possible categories, are often referred to as *classification problems.*" [22].

The main problem in PLSA approach was to find the latent aspect variables z_k and calculate the corresponding conditioned probabilities $p(w_i|z_k)$ and $p(z_k|d_j)$. It was assumed that the latent variables correspond to some artificial concepts. It was impossible however to specify these concepts explicitly. In the approach described below, the aspect variables can be interpreted semantically. Prerequisite for such a construction of a semantic space is a semantically annotated *training corpus*. Such annotations are usually done manually according to explicitly defined annotation rules. An example of such a corpus is e.g. the news data of the German Press Agency (dpa) which is annotated according to the categories of the International Press Telecommunications Council (IPTC). These annotations inductively define the concepts z_k, or the dimensions, of the semantic space. A *classifier induced semantic space* (CISS) is generated in two steps: In the *training step* classification rules $\mathbf{x_j} \rightarrow z_k$ are inferred from the training data. In the *classification step* these decision rules are applied to possibly unannotated documents.

This construction of a semantic space is especially useful for practical applications because (1) the space is low-dimensional (up to dozens of dimensions) and thus can easily be visualised, (2) the space's dimension possesses a well defined semantic interpretation, and (3) the space can be tailored to the special requirements of a specific application. The disadvantage of classifier induced semantic spaces (CISS) is that they rely on *supervised* classifiers. Therefore manually annotated training data is required.

Classification algorithms often use an internal representation of the degree of membership. They internally calculate how much a given input vector $\mathbf{x}$, belongs to a given class z_k. This internal representation of the degree of membership can be exploited to generate a semantic space.

A Support Vector Machine (SVM) is a supervised classification algorithm that recently has been applied successfully to text classification tasks. SVMs have proven to be an efficient and accurate text classification technique [6, 7,

130 Edda Leopold

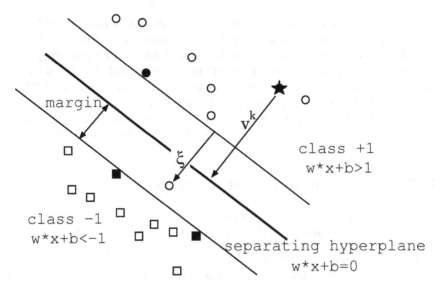

Fig. 1. Generating a CISS with a support vector machine. The SVM algorithm seeks to maximise the margin around a hyperplane that separate a positive class (marked by circles) from a negative class (marked by squares). Once an SVM is trained, $v^k = \mathbf{w}^k \mathbf{x} + b$ is calculated in the classification step. The quantity v^k measures the rectangular distance between the point marked by a star and the hyperplane. It can be used to generate a CISS.

12, 18]. Therefore Support Vector Machines appears to be the best choice for the construction of a semantic space for textual documents.

6.1 Using an SVM to Quantify the Degree of Membership

Like other supervised machine learning algorithms, an SVM works in two steps. In the first step – the *training* step – it learns a decision boundary in input space from preclassified training data. In the second step – the *classification* step – it classifies input vectors according to the previously learned decision boundary. A *single* support vector machine can only separate *two* classes – a positive class ($y = +1$) and a negative class ($y = -1$). This means that for each of the K classes z_k a new SVM has to be trained, separating z_k from all other classes.

   In the training step the following problem is solved: Given is a set of training examples $S_\ell = \{(\mathbf{x}_1, y_1), (\mathbf{x}_2, y_2), \ldots, (\mathbf{x}_\ell, y_\ell)\}$ of size $\ell \leq W$ from a fixed but unknown distribution $p(\mathbf{x}, y)$ describing the learning task. The term-frequency vectors $\mathbf{x}_i$ represent documents and $y_i \in \{-1, +1\}$ indicates whether a document has been annotated as belonging to the positive class or not. The SVM aims to find a decision rule $h_{\mathcal{L}} : \mathbf{x} \rightarrow \{-1, +1\}$ based on S_ℓ that classifies documents as accurately as possible.

The hypothesis space is given by the functions $f(\mathbf{x}) = \text{sgn}(\mathbf{wx} + b)$, where $\mathbf{w}$ and b are parameters that are learned in the training step and which determine the class separating hyperplane. Computing this hyperplane is equivalent to solving the following optimisation problem [13, 32]:

$$\text{minimise:} \quad V(\mathbf{w}, b, \boldsymbol{\xi}) = \frac{1}{2}\mathbf{ww} + C\sum_{i=1}^{\ell}\xi_i$$

$$\text{subject to: } \forall_{i=1}^{\ell} : y_i(\mathbf{wx} + b) \geq 1 - \xi_i$$
$$\forall_{i=1}^{\ell} : \xi_i \geq 0 \tag{28}$$

The constraints require that all training examples are classified correctly allowing for some outliers, symbolised by the slack variables ξ_i. If a training example lies on the wrong side of the hyperplane, the corresponding ξ_i is greater or equal to 0. The factor C is a parameter that allows one to trade off training error against model complexity. Instead of solving the above optimization problem directly, it is easier to solve the following dual optimisation problem [13, 32].

$$\text{minimise:} \quad W(\boldsymbol{\alpha}) = -\sum_{i=1}^{\ell}\alpha_i + \frac{1}{2}\sum_{i=1}^{\ell}\sum_{j=1}^{\ell}y_iy_j\alpha_i\alpha_j\mathbf{x}_i\mathbf{x}_j$$

$$\text{subject to:} \quad \sum_{\substack{i=1 \\ 0 \leq \alpha_i \leq C}}^{\ell} y_i\alpha_i = 0 \tag{29}$$

All training examples with $\alpha_i > 0$ at the solution are called support vectors. The support vectors are situated right at the margin (see the solid squares and the circle in figure (1)) and define the hyperplane. The definition of a hyperplane by the support vectors is especially advantageous in high dimensional feature spaces because a comparatively small number of parameters – the αs in the sum of equation (29) – is required.

In the classification step an unlabeled term-frequency vector is estimated to belong to the class

$$\hat{y} = \text{sgn}(\mathbf{wx} + b) \tag{30}$$

Heuristically, the estimated class membership $\hat{y}$ corresponds to whether or not $\mathbf{x}$ belongs on the lower or upper side of the decision hyperplane. Thus estimating the class membership by equation (30) consists of a loss of information since only the algebraic sign of the right-hand term is evaluated. However the value of $v = \mathbf{wx} + b$ is a real number and can be used in order to create a real valued semantic space, rather than just to estimate if $\mathbf{x}$ belongs to a given class or not.

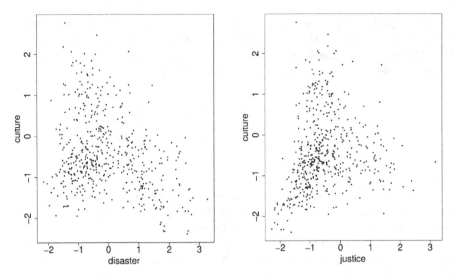

Fig. 2. A classifier induced semantic space. 17 classifiers have been trained according to the highest level of the IPTC classification scheme. The projection to two dimensions "culture" and "disaster" is displayed on the right, and the projection to "culture" and "justice" on the left. The calculation is based on 68778 documents from the "Basisdienst" of the German Press Agency (dpa) July-October 2000.

6.2 Using Several Classes to Construct a Semantic Space

Suppose there are several, e.g K, classes of documents. Each document is represented by an input vector $\mathbf{x}_j$. For each document the variable $y_j^k \in \{-1, +1\}$ indicates whether $\mathbf{x}_j$ belongs to the k-th class ($k = 1, \ldots, K$) or not. For each class $k = 1, \ldots, K$ an SVM can be learned which yields the parameters $\mathbf{w}^k$ and b^k. After the SVMs have been learned, the classification step (equation (30)) can be applied to a (possibly unlabeled) document represented by $\mathbf{x}$ resulting in a K-dimensional vector $\mathbf{v}$, whose kth component is given by $v^k = \mathbf{w}^k \cdot \mathbf{x} + b^k$. The component v^k quantifies how much a document belongs to class k. Thus the document represented by the term frequency vector $\mathbf{x_j}$ is mapped to the K-dimensional vector in the classifier induced semantic space. Each dimension in this space can be interpreted as the membership degree of the document to each of the K classes.

The relation between PLSA and CISS is given by the latent variable z_k. In the context of CISS, the latent variable z_k is interpreted as the thematic domain. This in accordance with semantic annotations in the corpus. Statistical learning theory assumes that each class k is learnable because there is an underlying conditional distribution $p(\mathbf{x}_j \mid z_k)$, which reflects the special characteristics of the class z_k. The classification rules that are learned from the training data minimise the expected error. In PLSA the aspect variables are not previously defined. The conditioned probabilities $p(w_i \mid z_k)$ and $p(z_k \mid \mathbf{x}_j)$

are chosen in such a way that they maximise the likelihood of the multinomial model.

6.3 Graphical Representation of a CISS

Self-Organising Maps (SOM) were invented in the early 80s [15]. They use a specific neural network architecture to perform a recursive regression leading to a reduction of the dimension of the data. For practical applications SOMs can be considered as a distance preserving mapping from a more than three-dimensional space to two-dimensions. A description of the SOM algorithm and a thorough discussion of the topic is given by [16].

Figure 3 shows an example of a SOM visualising the semantic relations of news messages. SVMs for the four classes 'culture', 'economy', 'politics', and 'sports' were trained by news messages from the 'Basisdienst' of the German Press Agency (dpa) April 2000. Classification and generation of the SOM was performed for the news messages of the first 10 days of April. 50 messages were selected at random and displayed as white crosses. The categories are indicated by different grey tone. Then the SOM algorithm is applied (with 100×100 nodes using Euclidean metric) in order to map the four-dimensional document representations to two dimensions admitting a minimum distortion of the distances. The grey tone indicates the topic category. Shadings within the categories indicate the confidence of the estimated class membership (dark = low confidence, bright = high confidence).

It can be seen that the change from sports (15) to economy (04) is filled by documents which cannot be assigned confidently to either classes. The area between politics (11) and economy (04), however, contains documents, which definitely belong to both classes. Note that classifier induced semantic spaces go beyond a mere extrapolation of the annotations found in the training corpus. It gives an insight into how typical a certain document is for each of the classes. Furthermore classifier induced semantic spaces allow one to reveal previously unseen relationships between classes. The bright islands in area 11 on Figure 3 show, for example, that there are messages classified as economy which surely belong to politics.

7 Conclusion

Fuzzy Linguistics, LSA, PLSA, and CISS map documents to the semantic space in a different manner. Fuzzy Linguistics computes a vector for each word which consists of the cosine distances to every other word in the corpus. Then it calculates the Euclidean distances between the vectors which gives the meaning point. Documents are represented by summing up the meaning points of the document's words.

In the case of LSA the representation of the document in the semantic space is achieved by matrix multiplication: $d_j \rightarrow \mathbf{x}_j^T U_K \Sigma_K$. The dimensions

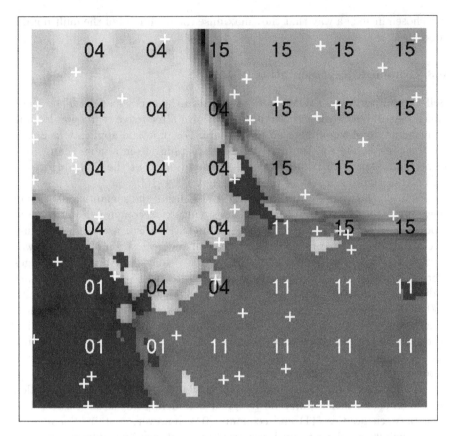

Fig. 3. A self-organising map of a classifier induced semantic space. 4 classifiers have been trained according to the highest level of the IPTC classification scheme. The shadings and numbers indicate the "true" topic annotations of the news messages. 01: culture, 04: economy, 11: politics, 15: sports. (The figure was taken from [19]).

of the semantic space correspond to the K largest eigen-values of the similarity matrix AA^T. The projection employed by LSA always leads to a global optimum in terms of the Euclidean distance between A and A_k.

PLSA maps a document to the vector of the conditional probabilities, which indicate how probable aspect z_k is, when document d_j is selected: $d_j \rightarrow (p(z_1 \mid d_j), \ldots, p(z_K \mid d_j))$. The probabilities are derived from the aspect model using the maximum likelihood principle and the assumption of multinomially distributed word frequency distributions. The the likelihood function is maximised using the EM-algorithm, which is an iterative algorithm that leads only to a local optimum.

CISS requires a training corpus of documents annotated according to their membership of classes z_k. The classes have to be explicitly defined by the human annotation rules. For each class z_k a classifier is trained, i.e. parameters $\mathbf{w}^k$ and b^k are calculated from the training data. For each document d_j the quantities $v^k = \mathbf{w}^k \cdot \mathbf{x} + b^k$ are calculated, which indicate how much d_j belongs the previously learned classes z_k. The mapping of document d_j to the semantic space is defined as $d_j \rightarrow (v_1, \ldots v_K)$. The dimensions can be interpreted according to the annotation rules.

Acknowledgements

This study is part of the project InDiGo which is funded by the German ministry for research and technology (BMFT) grant number 01 AK 915 A.

References

[1] R. H. Baayen. *Word Frequency Distributions*. Kluwer, Dordrecht, 2001.

[2] M. W. Berry, S. T. Dumais, and G. W. O'Brien. Using Linear Algebra for Intelligent Information Retrieval. *SIAM Review*, 37(4):573–595, 1995.

[3] R. J. Chitashvili and H. R. Baayen. Word Frequency Distributions. In G. Altmann and L. Hřebíček, editors, *Quantitative Text Analysis*, pages 54–135. wvt, Trier, 1993.

[4] S. Deerwester, S. T. Dumais, G. W. Furnas, T. K. Landauer, and R. Harshmann. Indexing by Latent Semantic Analysis. *Journal of the American Society for Information Science*, 41(6):391–407, 1990.

[5] A. P. Dempster, N. M. Laird, and D. B. Rubin. Maximum Likelihood from Incomplete Data Via the EM Algorithm. *Journal of the Royal Statistical Society, B*, 39:1–38, 1977.

[6] H. Drucker, D. Wu, and V. Vapnik. Support Vector Machines for Spam Categorization. *IEEE Transactions on Neural Networks*, 10:1048–1054, 1999.

[7] S. Dumais, J. Platt, D. Heckerman, and M. Sahami. Inductive Learning Algorithms and Representations for Text Categorization. In *Proceedings of the ACM-CIKM*, pages 148–155, 1998.

[8] H. Goebl. Dialectometry: A Short Overview of the Principles and Practice of Quantitative Classification of Linguistic Atlas Data. In R. Köhler and B. B. Rieger, editors, *Contributions to Quantitative Linguistics, Proceedings of the First International Conference on Quantitative Linguistics*, pages 277–315, Dordrecht, 1991. Kluwer.

[9] A. Gous. *Exponential and Spherical Subfamily Models*. PhD thesis, Stanford University, 1998.

[10] T. Hofmann. Unsupervised Learning by Probabilistic Latent Semantic Analysis. *Machine Learning*, 42:177–196, 2001.

[11] T. Hofmann and J. Puzicha. Statistical Models for Co-occurrence Data. A. I. Memo No. 1625, Massachusetts Institute of Technology, 1998.

[12] T. Joachims. Text Categorization with Support Vector Machines: Learning with Many Relevant Features. In *Proceedings of the Tenth European Conference on Machine Learning (ECML 1998)*, volume 1398 of *Lecture Notes in Computer Science*, pages 137–142, Berlin, 1998. Springer.

[13] T. Joachims. *Learning to Classify Text Using Support Vector Machines*. Kluwer, Boston, Dordrecht, London, 2002.

[14] R. Köhler. *Zur linguistischen Synergetik: Struktur und Dynamik der Lexik*. Brockmeyer, Bochum, 1986.

[15] T. Kohonen. *Content-addressable Memories*. Springer, Berlin, 1980.

[16] T. Kohonen. *Self-Organizing Maps*. Springer, Berlin, 1995.

[17] T. K. Landauer and S. T. Dumais. A Solution to Plato's Problem: The Latent Semantic Analysis Theory of Acquisition, Induction, and Representation of Knowledge. *Psychological Review*, 104(2):211–240, 1997.

[18] E. Leopold and J. Kindermann. Text Categorization with Support Vector Machines. How to Represent Texts in Input Space? *Machine Learning*, 46:423–444, 2002.

[19] E. Leopold, M. May, and G. Paaß. Data Mining and Text Mining for Science and Technology Research. In H. F. Moed, W. Glänzel, and U. Schmoch, editors, *Handbook of Quantitative Science and Technology Research*, pages 187–214. Kluwer, Dordrecht, 2004.

[20] C. D. Manning and H. Schütze. *Foundations of Statistical Natural Language Processing*. MIT Press, Cambridge, Massachusetts, 1999.

[21] A. Mehler. Hierarchical Orderings of Textual Units. In *Proceedings of the 19th International Conference on Computational Linguistics, COLING'02, Taipei*, pages 646–652, San Francisco, 2002. Morgan Kaufmann.

[22] T. M. Mitchell. *Machine Learning*. McGraw-Hill, New York, 1997.

[23] G. Neumann and S. Schmeier. Shallow Natural Language Technology and Text Mining. *Künstliche Intelligenz*, 2(2):23–26, 2002.

[24] G. Paaß, E. Leopold, M. Larson, J. Kindermann, and S. Eickeler. SVM classification using sequences of phonemes and syllables. In *Proceedings of the 6th European Conference on Principles of Data Mining and Knowledge Discovery (PKDD), Helsinki*, pages 373–384, Berlin, 2002. Springer.

[25] B. B. Rieger. Feasible Fuzzy Semantics. On Some Problems of How to Handle Word Meaning Empirically. In H.-J. Eikmeyer and H. Rieser, editors, *Words, Worlds, and Contexts. New Approaches in Word Semantics (Research in Text Theory 6)*, pages 193–209. de Gruyter, Berlin, 1981.

[26] B. B. Rieger. Definition of Terms, Word Meaning, and Knowledge Structure. On Some Problems of Semantics from a Computational View of Linguistics. In H. Czap and C. Galinski, editors, *Terminology and Knowledge Engineering. Proceedings International Congress on Terminology and Knowledge Engineering (Volume 2)*, pages 25–41, Frankfurt a. M., 1988. Indeks.

[27] B. B. Rieger. Computing Fuzzy Semantic Granules from Natural Language Texts. A Computational Semiotics Approach to Understanding Word Meanings. In M. H. Hamza, editor, *Artificial Intelligence and Soft*

Computing, Proceedings of the IASTED International Conference, Ana-heim/Calgary/Zürich, pages 475–479. IASTED/Acta Press, 1999.

[28] B. B. Rieger. Perception Based Processing of NL Texts. Discourse Understanding as Visualized Meaning Constitution in SCIP Systems. In A. Lotfi, B. John, and J. Garibaldi, editors, *Recent Advances in Soft Computing (RASC-2002 Proceedings), Nottingham (Nottingham Trent UP)*, pages 506–511, 2002.

[29] B. B. Rieger and C. Thiopoulos. Situations, Topoi, and Dispositions: On the Phenomenological Modeling of Meaning. In J. Retti and K. Leidlmair, editors, *5th Austrian Artificial Intelligence Conference, ÖGAI '89, Innsbruck, KI-Informatik-Fachberichte 208*, pages 365–375, Berlin, 1989. Springer.

[30] G. Salton and M. J. McGill. *Introduction to Modern Information Retrieval.* McGraw Hill, New York, 1983.

[31] C. J. van Rijsbergen. *Information Retrieval.* Butterworths, London, Boston, 1975.

[32] V. N. Vapnik. *Statistical Learning Theory.* Wiley & Sons, New York, 1998.

Compositionality in Quantitative Semantics. A Theoretical Perspective on Text Mining

Alexander Mehler

Bielefeld University
Alexander.Mehler@uni-bielefeld.de

Summary. This chapter introduces a variant of the principle of compositionality in quantitative text semantics as an alternative to the bag-of-features approach. The variant includes effects of context-sensitive interpretation as well as processes of meaning constitution and change in the sense of usage-based semantics. Its starting point is a combination of semantic space modeling and text structure analysis. The principle is implemented by means of a hierarchical constraint satisfaction process which utilizes the notion of hierarchical text structure superimposed by graph-inducing coherence relations. The major contribution of the chapter is a conceptualization and formalization of the principle of compositionality in terms of semantic spaces which tackles some well known deficits of existing approaches. In particular this relates to the missing linguistic interpretability of statistical meaning representations.

1 Introduction

Theory and practice of *automatic text analysis* hinge upon computable representations of natural language texts. This relates to computational linguistics as well as to applications such as mining, summarizing, extracting, categorizing, filtering, routing and tracking texts. Moreover, recent approaches to simulating processes of text comprehension and quantifying coherence relations also prove their relevance in cognitive linguistics [6, 12, 34, 40]. Generally speaking, *text representation models* specify the kind of mathematical entity (e.g. sets, graphs, feature distributions or stochastic processes) used to represent texts and their constituents. One of the most prominent representation models in this area is the *bag-of-features* approach [37] which represents texts as *sets* of weighted features, that is, as fuzzy sets of mostly lexical items whose membership values depend on local (i.e. text-specific) and global (i.e. corpus-specific) weighting parameters. As input to the *Vector Space* (VS) model, these sets are represented as vectors whose dimensions are defined by the lexical features taken into consideration. From its beginning this model proved to be successful in representing texts in the area of information retrieval [37].

A. Mehler: *Compositionality in Quantitative Semantics. A Theoretical Perspective on Text Mining*, StudFuzz **209**, 139–167 (2007)
www.springerlink.com

Table 1. Prevalent deficits of the VS model, semantic spaces and related approaches.

1. **Missing typing as a result of a limitation to similarity relations**

   The VS model and related approaches do not classify similarity relations so that in these frameworks similarity judgements are underspecified.[1] Asserting the similarity of two texts according to these models does not clarify the proper source of similarity: *Are they similar because they deal, for example, with the same topic, instantiate the same text type or are written by the same author?*

2. **Insensitivity to indirect similarity relations**

   According to the VS model, texts are similar to the extent they share approximately equally weighted words. Consequently, this model judges texts to be *dissimilar* which do not share such words, even if they deal with the same or related topics and, thus, are similar not by virtue of their expression plane, but of their content plane.

3. **Unstructured representations as a result of an insensitivity to sign structure**

   Similarity judgements based on the VS model or related approaches are insensitive to the (constituency and dependency) structure of signs. Texts, for example, sharing a subset of lexical items may be judged to be similar even if they are completely differently structured because of instantiating, for example, highly divergent genres or text types. Consequently, meaning representations within these frameworks are unstructured in the sense that signs of whatever linguistic stratum are always mapped onto single feature vectors – irrespective of their internal structure.

4. **Missing structure-sensitive similarity measures**

   A direct consequence of the latter insensitivity is the focus on (dis-)similarity measures (e.g. the cosine measure) of feature vectors leaving out structure sensitive similarity measures (e.g. the tree edit distance).

5. **Missing iterative computability**

   From the point of view of linguistic dynamics, a deficit of some approaches to semantic spaces is that they do not allow to iteratively compute output spaces for incrementally presented input texts. That is, a semantic space is computed for the input corpus as a whole irrespective of any order of its element texts.

Nevertheless, it faces a lot of problems especially when used as a representation model in computational linguistics. We concentrate on five of them – see Table (1).

A candidate for solving the first problem is given by *semantic spaces* which are utilized in cognitive linguistics [7, 12, 17], computational linguistics [20, 33, 38] and information retrieval [4]. Semantic spaces start from a geometric interpretation of *usage-based structural semantics*. In the case of words this means, for example, that the more similar their usage, the less

---

[1]In the case of lexical items, sense relations (e.g. synonymy or co-hyponymy) can be distinguished. In the case of textual units, coherence relations such as rhetorical relations (e.g. elaboration, contrast, circumstance – cf. Mann & Thompson [18]) are candidates of typing.

distant their feature vectors in semantic space. Rieger [29, 30] reconstructs *lexical meanings* as the output of a two-stage process of syntagmatic and paradigmatic learning. In this model, paradigmatic similarity is a function of the words' syntagmatic regularities which, in turn, are explored based on their lexical contexts. Consequently, Riegerian semantics suspends the requirement that words co-occur in order to be judged as paradigmatically similar. Thus, Rieger's semantics *is sensitive to indirect similarity relations.* This model of meaning constitution – developed in the framework of *Fuzzy Linguistics* (FL) [29, 30, 31, 32, 35] – is, to my knowledge, the first formally elaborated approach to learning indirect meaning relations of signs. Because of the semiotic universality of syntagmatics and paradigmatics, FL gives a perspective on machine learning in still unexplored areas.

Latent Semantic Analysis (LSA) [17] is an alternative model of semantic spaces which originates from efforts to explore indirect similarities in information retrieval [4] according to the following requirements:

- Words should be interrelated even if they do not or only rarely co-occur, but tend to be used in similar contexts.
- Texts should be interrelated even if they do not share any or only few lexical items, but whose lexical constituents are similar in the sense of the latter claim.

In a series of experiments it could be shown that LSA meets these requirements and thus is sensitive to indirect similarity relations on the level of words and texts. Comparable to FL, LSA models syntagmatic and paradigmatic learning. But other than FL, LSA amalgamates both processes in a single step based on singular value decomposition [20]. Consequently, we prefer FL as it is more explicit in terms of cognitive modeling.

Both approaches, FL and LSA, also suspend the notion of a semantic atom. That is, other than early efforts in feature semantics [8], they do not rely on predefined semantic dimensions, but utilize lexical units observed in the input corpora in order to span the output spaces in which the focal signs are interrelated. Nevertheless, a shortcoming of both approaches is that they are insensitive to typed relations as well as to text structure. Thus, they do not solve the problem of missing typing and structural insensitivity (see Table 1). LSA, for example, maps texts – irrespective of their structure and length – as well as words onto single meaning points. Moreover, it concentrates on unsystematic sense relations. Thus, neighborhood relations of meaning points are *not* categorized according to sense relations.

A crucial step towards *structure-sensitive semantic spaces* was done by Kintsch [13] who proposes an extension of LSA in order to map sense relations of predicate-argument structures by means of a function which is sensitive to this class of linguistic structures.[2] More specifically, Kintsch focuses on the

---

[2]See Ruge [36] for an alternative model of predicate-argument structures based on semantic spaces.

context variability of words, that is, on their tendency to modify their meaning and, thus, to adopt text specific interpretations subject to the context of their usage. In the discussion of the validity of the *principle of compositionality* (CP) this phenomenon is analyzed by example of NN and AN constructions (modeled as predicate-argument structures of the general form $P(A)$).[3] The decisive point is that as far as these contexts are not enumerable, context variability argues the validity of the CP.[4]

Generally speaking, in order to model the similarity relations of predicate-argument structures $P(A)$ as representations of simple assertions, Kintsch proposes a *predication operation*

$$\pi_k \colon S^{k+2} \to S \tag{1}$$

where S is the set of meaning points. This operation computes the meaning point $\|P(A)\| \in S$ as the weighted mean of the meaning points $\|A\|, \|P\| \in S$ and all k nearest neighbors of $\|P\|$ also nearest to $\|A\|$. In this model, the meaning representation of predicate-argument structures $P(A)$ is a function of the meaning representation of A and of P, *but only according to the relevance of P to interpret A*. Recursively applying this operation, Kintsch arrives at mapping correlates of SVO structures:

$$(P(A_1))(A_2) \tag{2}$$

This operation does not strictly follow the CP: Although the meaning of $P(A)$ is a function of the meanings of its parts and the way they are combined, the contribution of P to $P(A)$ varies nontrivially with its argument A:

- Firstly, this relates to the selection of a subset X of nearest neighbors of P in semantic space. As long as the underlying universe U of meaning points is fixed, this selection is covered by the CP. But if U changes in the course of text comprehension, that is, if new meaning points are generated as intermediary representations of the text-specific interpretation of A, U cannot be enumerated in advance. In this case, the semantic contribution of P has to be *computed "at runtime" subject to its usage context*.
- Secondly, this context is not bound to A, but includes selectional effects of the preceding co-text, whose text-specific interpretations have to be reflected too. In cognitive linguistics, this phenomenon is called *text priming* [39]. In contrast to word priming, it relates to the effect that the preceding text of a lexical item enforces context adequate interpretations while inhibiting inadequate ones.
- Thirdly, P and A do not have stable meanings allowing for deterministic, functional selections of context adequate readings, but change in the course of changing usage habits.

---

[3]Cf. Osherson & Smith [24] and Kamp & Partee [11], but also Lahav [16] who rates this variability as an argument against the compositionality of natural languages.

[4]See Janssen [10] and Partee [25] for a systematic introduction to the CP.

This chapter proposes a variant of the CP – called *Latent Compositionality Principle* (LCP) – which reflects these considerations on the level of texts.[5] It serves as the kernel of a quantitative semantics modeling sense relations not only of lexical items, but of units belonging to whatever stratum of textual resolution. The variant includes processes of text priming, perspective interpretation, meaning constitution and change. Its algorithmic specification is based on the framework of fuzzy constraint satisfaction processes superimposed by a hierarchical order. This model departs from approaches to coherence as an "unordered" process of *parallel constraint satisfaction* [40]. As will be shown, turning away from the paradigm of parallel processing allows tackling a central deficit of semantic spaces (see Table 1): their *insensitivity to structure*.

The chapter is organized as follows: Section (2) introduces the LCP. Section (3) presents formal preliminaries and Section (4) introduces an algorithmic specification of the LCP in terms of *hierarchical constraint satisfaction processes* whose numerical specification is proposed in Section (5). Section (6) compares this model with the VS model, FL and LSA. Finally, Section (7) gives a conclusion and prospects future work.

2 Latent Compositionality

Starting from the relational concept of meaning as defined in situation semantics [2], the informal specification of the LCP introduces two extensions: First, *coherence relations* are additionally referred to as determinants of interpretation. Second, the LCP refers to *usage regularities* according to which the interpretation of lexical items can change their usage conditions and thus their interpretation in subsequent communication situations. Consequently, the LCP introduces a kind of dynamics which relates to *learning* linguistic knowledge (e.g. routinization, schematization etc.) and which is left out in the classical reading of the CP:

> *The meaning of a linguistic item x is a relation over*
> - *its usage-regularities,*
> - *its usage contexts as systems of syntactic dependency as well as cohesion and coherence relations to which it participates,*
> - *the meanings of its components,*
> - *the way they are combined and*
> - *described situations.*
>
> *The interpretation of x in a given context is the situation it describes subject to concrete values of the latter parameters.*

Although this version of the LCP accounts for different parameters of text meaning constitution, it is nevertheless underspecified in the sense that it

---

[5]Note that Kintsch [see 13] does not generalize his model to texts as a whole.

contains several innumerable parameters. Usage regularities, for example, are dynamic entities which cannot be enumerated as lexicon entries. In order to tackle their dynamics, the LCP needs to be redefined by including *procedural models of cognitive processes* which allow computing parameter values subject to the operative contexts:

> *The meaning of a linguistic unit x is a procedure P generating its interpretation based on its usage regularities, the contexts of its usage, the meanings of its components and the way they are combined. In order to guarantee interpretability of the measurements performed by P, it is required that not only its input and output have modeling function w.r.t cognitive entities, but also P w.r.t cognitive processes.*

This variant of the LCP asks for procedures which do not only produce "good results" in terms of machine learning, but model socio-cognitive processes and their lexico-grammatical manifestations. In the following sections, this criterion is met by means of the *Construction-Integration* (CI) *theory* of Kintsch [12]. That is, text interpretation is conceived as a process of alternating construction and integration processes starting with elementary text components and finally integrating – if successful – the input text. In this model, construction leads to the rather unconstrained generation of possibly incoherent candidate interpretations which are selected in subsequent integration processes in order to derive the most coherent interpretation.

At present, the LCP lacks specifications with respect to a quantitative semantics. Comparable to the CP [10], it still contains undefined parameters. This relates to the notion of *context, meaning, linguistic items*, their *components*, the way they are *combined* and the *usage regularities* of elementary items. In the following sections, the LCP is specified with respect to the structural meaning aspect and the linguistic context (i.e. cotext) of text components down to the level of lexical items. Further, usage regularities are referred to in terms of Miller and Charles [23]. In other words, we follow the weak contextual hypothesis according to which the similarity of the contextual representations of words contributes to their semantic similarity. As demonstrated in [21], we extend this hypothesis to the level of texts by saying that the contextual similarities of the lexical constituents of two text segments contribute to their semantic similarity. Subsequently, the LCP is specified in a way which accounts for the fact that the connotations of complex signs do not only systematically depend on the sense relations of their parts, but systematically vary with their textual context.

The LCP builds on situation semantics and, thus, distinguishes between *meaning* and *interpretation*: Linguistic items are interpreted subject to their contextual embedding. This interpretation includes the items' linguistic meanings as additional parameters which are determined irrespective of the items' contextual embedding. In this sense we distinguish between the linguistic meaning of a text and its discourse specific interpretation.

3 Preliminaries

In this section, formal preliminaries are introduced. First we introduce the set
of types and tokens:

Definition 1. Let $C = \{x_1, \ldots, x_n\}$ be a text corpus, $\mathbb{S}$ a segmentation mapping each text $x \in C$ onto an ordered rooted tree $\mathbb{S}(x) = (S(x), E, x, O_1, O_2)$
as a model of its kernel hierarchical structure in the sense of an *ordered hierarchy of content objects* [28] and $\mathbb{L}: T(C) \to L(C)$ a lemmatization mapping
each token $\boldsymbol{a} \in T(C)$ onto its type $a \in L(C)$; $T(C) \subset S(C)$ is the set of tokens
and $L(C)$ the set of types of corpus C. O_1 is an order relation mapping the
syntagmatic order of all immediate constituents of any segment of x. That is,
$O_1(y_i, y_j)$ iff $y_i, y_j \in S(x)$ are immediate constituents of the same $z \in S(x)$
according to $\mathbb{S}$ so that y_i precedes y_j in z. O_2 is the linear order relation
induced by the postorder traversal of $\mathbb{S}(x)$.
  We define $S(x)$, $x \in S(x)$, as the set of all segments of x according to $\mathbb{S}$ and
$S(C) = \cup_{x \in C} S(x)$. Further, $T(x) \subset S(x)$ is the set of all tokens of x according
to $\mathbb{S}$ and $T(C) = \cup_{x \in C} T(x)$. Next, $L(x) = \{a \mid \exists \boldsymbol{a} \in T(x) \colon \boldsymbol{a} \models_T a\}$ is the set
of all types classifying at least one token in $T(x)$. Thus, $L(C) = \cup_{x \in C} L(x)$.
We write S, T and L instead of $S(C)$, $T(C)$ and $L(C)$ if the corpus C is
known from the context. $\mathbb{L}$ induces a *type-token classification* $(T, L, \models_T)$ where
$\models_T \subseteq T \times L$ and $\boldsymbol{a} \models_T a$ iff $\boldsymbol{a} \in T$ is a token according to $\mathbb{S}$ instantiating the
type $a \in L$ according to $\mathbb{L}$ – the notation $\boldsymbol{a} \models_T a$ is borrowed from [3].

  Now we give an abstract definition of semantic spaces which grasps the
varying space models of FL [33], LSA [17] and derivations thereof:

Definition 2. Let a corpus C, a segmentation $\mathbb{S}$ and a lemmatization $\mathbb{L}$ be
given according to definition (1). Further, let $\mathbb{X}$ be an uncountable set, e.g.
$\mathbb{X} = \mathbb{R}^n$ for some $n > 0, n \in \mathbb{N}$, and $(\mathbb{X}, d)$ be a metric space. A *semantic space*
is a quintuple $(L, S, \alpha, \beta, (\mathbb{X}, d))$ where $\alpha \colon L \to \mathbb{X}$ is a function mapping types
$a \in L$ onto representations of the contexts of their tokens $\boldsymbol{a} \in L$ in segments
$x \in S$. Further, $\beta \colon S \to \mathbb{X}$ is a function mapping segments $x \in S$ onto $\mathbb{X}$ by
operating on the context representations of their components according to $\mathbb{S}$
down to the level of tokens $\boldsymbol{a} \in T(x)$ as instances of types $a \in L(x)$.

  Note that the uncountability of $\mathbb{X}$ (in contrast to L and S)[6] is required
to guarantee an infinite inventory for representing the meanings of newly
invented or composed signs as well as to account for meaning change.
  Definition (2) does not specify semantic spaces by defining the functions
α, β, but requires that candidate models of semantic spaces, instantiating this

---

[6]That is, for any $u \in L \cup S$ there is always an element $\mathbf{x} \in \mathbb{X}$ so that either
$\alpha(u) = \mathbf{x}$ or $\beta(u) = \mathbf{x}$, but not vice versa. In terms of glossematics, the metric space
$(\mathbb{X}, d)$ is, so to say, a representation of those part of the "meaning substance" which
is articulated by sign systems in order to shape their system of structural meanings.

definition, map the two-level process of meaning constitution, i.e. of consecutive syntagmatic and paradigmatic learning [29]. This is reflected by the metric d operating on context representations of signs in L and S, respectively. That is, learning semantic similarities of linguistic items is required to operate on measurements of their (lexical constituents') usage regularities (in the case of complex text segments). Further, because of decoupling of $\mathbb{X}$, L and S, def. (2) allows different signs to have the same representation in $(\mathbb{X}, d)$. This is needed to distinguish, for example, signs which are used equally.

In the following section, a *Fuzzy Constraint Satisfaction Problem* (FCSP) is defined in terms of *semantic spaces*. Normally, FCSPs are defined in terms of *fuzzy sets* [41]. But this contrast is only a question of notational variants since meaning points in semantic space are easily translated into fuzzy sets [29]: Let $(L, S, \alpha, \beta, (\mathbb{X}, d))$ be a semantic space. For any sign $v \in U = L \cup S$ we define a fuzzy set $\mathbb{F}(v) \in \mathbb{F}(U)$ with the membership function[7]

$$\forall u \in U : \mu_{\mathbb{F}(v)}(u) = \begin{cases} d(\hat{\alpha}(v), \hat{\alpha}(u)) & : u, v \in L \\ d(\hat{\alpha}(v), \hat{\beta}(u)) & : u \in L, v \in S \\ d(\hat{\beta}(v), \hat{\alpha}(u)) & : u \in S, v \in L \\ d(\hat{\beta}(v), \hat{\beta}(u)) & : u, v \in S \end{cases} \tag{3}$$

where $\hat{\alpha}$ and $\hat{\beta}$ are standardizations of α and β onto the unit interval.

Analogously, any similarity or distance measure operating on S can be transformed into an operation on fuzzy sets [1].

4 Hierarchical Constraint Satisfaction

Hierarchical constraint satisfaction problems (HCSP) are now introduced as specifications of the LCP in terms of quantitative text semantics. The basic idea is that coherence relations span networks of interpretational constraints whose evaluation order is constrained by the focal text's kernel hierarchical structure. This hierarchy is assumed in accordance with the notion of *logical document structure* [27] and related models. Thus, we suppose texts to have a hierarchical structure – henceforth called *integration hierarchy* – which is superimposed by constraints in the form of cohesion and coherence relations.[8]

The concept of integration hierarchy reflects the idea that texts are interpreted bottom-up from left to right starting with elementary text components thereby integrating more and more complex components up to the level of texts. It is assumed that this process of constraint satisfaction tries to maximize the degree to which all operative constraints are fulfilled subject to their preference order. Since constraints are seen to be based on coherence

---

[7]$\mathbb{F}(U)$ ist the set of all fuzzy sets over U.

[8]Cohesion relations are treated as special cases of coherence relations according to [22].

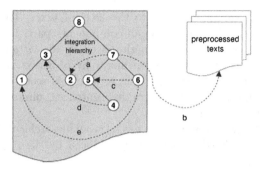

Fig. 1. An example of an integration hierarchy augmented by inter- (a, b), intra-(c) and transrelational (d, e) coherence relations from the perspective of the focal text component 7. Numbers indicate the *postorder* in which the kernel hierarchy is traversed (see Section 4.1) and, thus, the components' interpretations are computed.

relations as informational uncertain interpretation restrictions two cases have to be considered:

- *prioritized constraints* according to preference relations regarding the order of their fulfillment;
- *fuzzy constraints* which may be only partially fulfilled because of a non-specific or dissonant information base.

As will be shown, both types of constraints are easily mapped in the framework of fuzzy sets [5] and its adoption in terms of semantic spaces.

Integration hierarchies superimpose an order on the interpretation of text components. This process evolves in a continuum spanned by two extreme cases. On the one hand, it may be strictly compositional – *without being restricted by any coherence relation* interfering with the kernel hierarchical text structure. On the other hand, it may evolve according to the context principle [10] – *in strict contradiction to the order induced by the integration hierarchy.* The exact course of interpretation depends on the operative coherence relations. In this sense, the LCP covers a continuum which includes strict compositionality (as a characteristic of logical semantics) and strict contextuality (as a characteristic of connectionist approaches in the sense of the principle of *parallel constraint satisfaction*).

Based on these preliminaries, the building blocks of HCSPs can now be introduced with a focus on monological texts:

4.1 Integration Hierarchy

In the present approach, text components interact along syntactic dependency and textual coherence relations. As an output of this interaction, the components' discourse specific, perspective interpretations participate in the process of text interpretation which – if successful – leads to the integration of more and more complex components up to the level of texts. This process does not run in parallel in the sense that all components participate on an equal footing. Rather, it is seen to be superimposed by an *integration hierarchy*. For simplicity reasons, the integration hierarchy of a text is identified with its *logical document structure* [28]. That is, a text is seen to be subdivided, amongst others, into sections, paragraphs, sentences, phrases and words by the operative segmentation $\mathbb{S}$.[9]

According to this concept a text x has a kernel hierarchical structure which is modeled by the *ordered rooted tree* $\triangle_x$ induced by the segmentation $\mathbb{S}$ of definition (1):

$$\triangle_x = (V, E, x, O_1, O_2) =_{\text{def}} \mathbb{S}(x) \tag{4}$$

whose root x is the text to be integrated and whose leafs are elementary text components in the sense of $\mathbb{S}$. For any $y \in V$, $(y, y_{i_1}), \ldots, (y, y_{i_n}) \in E$ and $O_1(y_{i_1}, y_{i_2}), \ldots, O_1(y_{i_{n-1}}, y_{i_n})$ we abbreviate

$$y = I(y_{i_1}, \ldots, y_{i_n}) \tag{5}$$

Moreover, we distinguish between the syntactic integration operation I and its semantic counterpart $\mathbf{I}$. Thus, using the homomorphism $\|\|$ for interrelating syntactic and semantic structure, we write:

$$\begin{aligned} \|y\| &= \|I(y_{i_1}, \ldots, y_{i_n})\| \\ &= \|I\|(\|y_{i_1}\|, \ldots, \|y_{i_n}\|) \\ &= \mathbf{I}(\|y_{i_1}\|, \ldots, \|y_{i_n}\|) \end{aligned} \tag{6}$$

In Section (5.1), the semantics of $\mathbf{I}$ is defined with the help of aggregation operations on semantic spaces.

Next, the notion of an integration hierarchy $\triangle_x = (S(x), E, x, O_1, O_2)$ is recursively referred to all segments of x. That is, for any $y \in S(x)$ we define:

$$\triangle_y = (S(y), E', y, O_1, O_2) \tag{7}$$

where $E' = \{(y_i, y_j) \in E \mid y_i, y_j \in S(y)\}$.

As explained subsequently, the integration hierarchy of a text is transcended by coherence relations constituting a graph-like texture by embedding its components into contexts as a source of their perspective interpretation.

4.2 Constraints

The basic building blocks of constraint-satisfaction problems are constraints. In HCSPs these constraints are defined by analogy with the distinction of relations and their formats in the relational calculus:

---

[9]A more elaborate approach is to utilize the RST-tree of a text according to the *Rhetorical Structure Theory* (RST) [18].

- *Constraints:* In terms of HCSPs, text components are *variables* whose values range over their possible interpretations. Following this terminology, coherence relations are constraints restricting the interpretations of their arguments. Let $D_{i1}, \ldots, D_{ik}$ be the domains of the variables $x_{i1}, \ldots, x_{ik}$. Then a coherence relation[10] of these components can be defined as a fuzzy constraint $R_i(x_{i1}, \ldots, x_{ik})$ with the membership function

$$\mu_{R_i} : D_{i1} \times \ldots \times D_{ik} \to [0,1] \tag{8}$$

  specifying to which degree compound labels $(d_{i1}, \ldots, d_{ik})$, $d_{ij} \in D_{ij}$, satisfy R_i. This is an example of an undirected coherence relation. In the case of directed relations the values of the dependent variables are functions of the values of their independent counterparts. This is symbolized as

$$R_i(x_{ik} \mid x_{i1}, \ldots, x_{ik-1}) \, . \tag{9}$$

  In this paper, only *directed coherence relations* with at most one dependent variable are considered. Further, we restrict the set of independent variables of a coherence relation $R_i(x_{ik} \mid x_{i1}, \ldots, x_{ik-1})$ to those segments of the focal text x which precede x_{ik} according to the postorder O_2 of $\triangle_x$ (see Equation 4). The only relaxation of this restriction is that the independent variables precede the (uniquely defined) mother node of x_{ik} in $\triangle_x$ according to the postorder of $\triangle_x$. In Figure (1), segment 1 may be dependent, for example, on segment 2, but not on segment 4. Finally, we suppose that constraints are typed based on an appropriate type system of coherence relations as introduced in linguistics (cf. Halliday & Hasan [9]). This allows to classify their semantics by means of constraint schemata as introduced in Section (4.3). *Note that for simplicity reasons we suppose that no two equally typed coherence relations connect exactly the same text components.*

- *Inter-, intra- and transrelational constraints:* From the point of view of a given text component y, coherence relations are classified as follows:
  - If $R_i(x_{i_k} \mid x_{i_1}, \ldots, x_{i_{k-1}})$ is a coherence relation exclusively of components of y, R_i is called *intrarelational* w.r.t y.
  - If $R_i(y \mid x_{i_1}, \ldots, x_{i_k})$ is a coherence relation connecting y exclusively with components outside of it, R_i is called *interrelational* w.r.t y.
  - If $R_i(x_{i_k} \mid x_{i_1}, \ldots, x_{i_{k-1}})$ relates x_{i_k} as a component of y with components outside of y, R_i is called *transrelational* w.r.t y.

  Component 7 in Figure (1) exemplifies these three classes.

- *Contexts:* A coherence relation is a discourse specific instance of a certain coherence type which imposes a constraint on the interpretation of the relation's arguments according to the semantics of that type. Consequently,

---

[10]The term *relation* as part of the compound *coherence relation* is not used in its mathematical sense – i.e. as a subset of a certain Cartesian product –, but in its linguistic sense as denoting certain dependencies of discourse components [15].

the *context* of a component y in a text x equals the system of all co-herence relations in which it participates in x. This context imposes a *complex* constraint on the perspective, i.e. context-sensitive interpretation of y which is required to fulfill as many of the constraints involved as possible. Moreover, the higher the priority of a constraint the higher the degree to which it is demanded to be fulfilled.[11] More specifically, the context of a component y of a text x is a directed hypergraph

$$\rhd_{x,y} = (V, E) \tag{10}$$

whose edge set E collects all coherence relations $R_i(y \mid x_{i1}, \ldots, x_{ik})$ to which y participates as the dependent variable. E collects these coherence relations in terms of directed hyperedges $(\{R_i, \{x_{i1}, \{\ldots, \{x_{ik}\}\}\}\}, \{y\}) \in E$. For simplicity reasons, we write $R_i(y \mid x_{i1}, \ldots, x_{ik}) \in E$ to denote hyperedges of this kind.[12]

- *Texture:* This allows defining the *texture* of a text x with integration hierarchy $\triangle_x = (V, E, x, O_1, O_2)$ as a hypergraph whose set of hyperedges equals the union of E and the contexts of all of its components.

4.3 Constraint Schemata

The fulfillment of a constraint is bound to the semantics of the corresponding type. Anaphoric relations are, for example, evaluated differently from rhetorical relations whose evaluation departs from lexical cohesion relations. On the other hand, anaphoric relations $R(x_i \mid x_j)$, $S(x_m \mid x_n)$ are expected to be evaluated in a uniform manner. In order to specify the semantics of constraints in accordance with the type system of coherence relations, *constraint schemata* are introduced by analogy with the distinction of relation schemata and their instances in the relational calculus. These schemata are abstract specifications of how linguistic items can be embedded into discourse and how they have to be interpreted according to this embedding. More specifically, constraint schemata vary according to (i) their signatures – i.e. the non-necessarily fixed number and roles of their instances' obligatory/optional arguments –, (ii) the patterns of linguistic units by which these roles are preferably manifested and (iii) the way in which their instances (i.e. coherence relations) restrict the interpretation of their respective arguments (i.e. text components). A constraint schema $\mathbf{R}$ can thus be represented as a triple

$$\mathbf{R} = (\{A_0\} \cup (A_i)_{i \in I \setminus \{0\}}, \{G_0\} \cup (G_i)_{i \in I \setminus \{0\}}, \otimes_{\mathbf{R}}), \tag{11}$$

---

[11]Compared to the complexity of discourse processing this is a limited concept of context which, by analogy with RST, restricts the context of a text segment to its *cotext*. But insofar as components of the situative, cognitive or social context of interpretation can be textually manifested, this specification is easily extensible.

[12]Note that we include a coherence relation label R_i into the sequence of input nodes in order to allow different relations to operate on the same set of nodes.

where $\{A_0\} \cup (A_i)_{i \in I \setminus \{0\}}$ is a family of attributes with domains $D(A_0) = D(A_i) = \mathbb{P}(\mathbb{X})$, $\{G_0\} \cup (G_i)_{i \in I \setminus \{0\}}$ is a family of stochastic grammars where G_i models the set of linguistic manifestations of the ith attribute A_i and $\otimes_{\mathbf{R}} : D(A_1) \times \ldots \times D(A_n) \to D(A_0)$, $n = |I|$, is a construction operation (defined in Section 5.2) mapping the n-fold Cartesian product of the power set $\mathbb{P}(\mathbb{X})$ of the universe of $(\mathbb{X}, d)$ onto $\mathbb{P}(\mathbb{X})$.

In order to exemplify constraint schemata we consider predicate-argument structures, anaphoric reference, rhetorical relations and lexical cohesion:

- In the simplest case, predicate-argument structures are manifested by nominal arguments and verbal predicates. Modeling more complex manifestations (e.g. nominal phrases as manifestations of arguments) already involves a complexity up to the level of sentence grammars. In the following sections, we suppose that sentential text components are represented as predicate argument or as SVO structures, respectively.
- In the simplest case, the constraint schema of anaphoric relations comprises the role of an anaphora A_1 and of a corresponding antecedent A_2 where the former is preferably manifested by pronouns and the latter by nouns. In this case, the grammars G_1, G_2 contain probabilistic lexicalization rules.
- With respect to lexical cohesion, two types have to be distinguished:
  - In the case of a sense relation, e.g. hyponymy, holding between two lexical items of a text, the role of the hyperonym and of the hyponym can be distinguished.
  - In the case of a collocational dependency of unsystematic lexical cohesion [9] of more than two words (entering into a lexical chain) the number of attributes is not restricted but their roles are all the same.

  In both of these cases, the grammars contain lexicalization rules.[13]
- In contrast to this, complex grammars are needed in order to map rhetorical relations. An elaboration, for example, operates on a nucleus in the role of an elaborated and a satellite in the role of an elaborating text span. Both spans are manifested at least by components of the phrase level, but also by higher-level units up to the level of texts. In computational linguistics exists a range of rhetorical text grammars [19] which may be used as a starting point to build the corresponding grammars G_i.

If R is a coherence relation of type $\mathbf{R}$, we write

$$R \models_{\mathbf{C}} \mathbf{R} \tag{12}$$

where $\mathbb{C}$ is the set of constraint schemata under consideration.

---

[13]More complex grammars have to be considered in the case of lexical chains where, for example, Markovian constraints on the choice of items have to be considered as a function of chain length.

4.4 Prioritization

In HCSPs, constraints are evaluated in a definite order. Anaphora have to be resolved, for example, before they can participate in other coherence relations. In order to map this sort of prioritization, a function $r_C \colon \{\mathbf{R}_i \,|\, i \in \mathcal{I}\} \to \mathbb{N}_0$ is introduced which maps each constraint schema $\mathbf{R}$ onto its rank $r_C(\mathbf{R}) = j$ where the lower j, the higher the rank of $\mathbf{R}$. We suppose that r_C is an injective, total function on $\mathbb{C}$. This allows for any text component y of a text x to define a linear order relation $\sqsubseteq_{x,y} \subseteq \mathbb{LR}_x(y)^2$ over the set $\mathbb{LR}_x(y)$ of all coherence relations to which y participates as the dependent variable, i.e. $\mathbb{LR}_x(y) = E$ for $\rhd_{x,y} = (V, E)$ (see Section 4.2). We first decompose $\mathbb{LR}_x(y)$ into two sets: the set $\mathbb{L}_x(y)$ of coherence relations interrelating y solely with components preceding it in the sense of O_2 and the set $\mathbb{R}_x(y)$ of relations which contain at least one independent variable preceding z, but not y (see Section 4.2):

Definition 3. *Let z be the mother constituent of y in text $x \in \mathbb{C}$ in the sense of $\triangle_x = (S(x), E, x, O_1, O_2)$, i.e. $(z, y) \in E$. Further, let $\rhd_{x,y} = (V', E')$ be the context of y in x. Then we define: $\mathbb{L}_x(y) = \{R(y \,|\, y_{i_1}, \dots, y_{i_k}) \in E' \,|\, \forall l \in \{1, \dots, k\} \colon O(y_{i_l}, y)\}$; $\mathbb{R}_x(y) = E' \setminus \mathbb{L}_x(y)$.*

Definition 4. *Let x, y, z be given according to definition (3). Further, let $A = \{y_{i_1}, \dots, y_{i_k}\}$, $B = \{y_{j_1}, \dots, y_{j_l}\}$, $R \models_C \mathbf{R}$, $S \models_C \mathbf{S}$ and $R \in X$ an abbreviation of $R(y \,|\, y_{i_1}, \dots, y_{i_k}) \in X$ for $X \in \{\mathbb{L}_x(y), \mathbb{R}_x(y)\}$. Then:*[14]

- $\sqsubseteq_{1_{x,y}} = \{(R(y \,|\, y_{i_1}, \dots, y_{i_k}), S(y \,|\, y_{j_1}, \dots, y_{j_l})) \,|\, R \in \mathbb{L}_x(y) \wedge S \in \mathbb{L}_x(y) \wedge (r_C(\mathbf{R}) < r_C(\mathbf{S}) \vee (\mathbf{R} = \mathbf{S} \wedge (\inf_{O_2}(A \triangle B) \in A)))\}$.
- $\sqsubseteq_{2_{x,y}} = \{(R(y \,|\, y_{i_1}, \dots, y_{i_k}), S(y \,|\, y_{j_1}, \dots, y_{j_l})) \,|\, R \in \mathbb{R}_x(y) \wedge S \in \mathbb{R}_x(y) \wedge (\sup_{O_2}(A \triangle B) \in B \vee (A = B \wedge r_C(\mathbf{R}) < r_C(\mathbf{S})))\}$.
- $\sqsubseteq_{3_{x,y}} = \{(R(y \,|\, \dots), S(y \,|\, \dots)) \,|\, R \in \mathbb{L}_x(y) \wedge S \in \mathbb{R}_x(y)\}$.

Finally, we set $\sqsubseteq_{x,y} = \sqsubseteq_{1_{x,y}} \cup \sqsubseteq_{2_{x,y}} \cup \sqsubseteq_{3_{x,y}}$.

$\sqsubseteq$ induces a linear order $(R_{o_1}, \dots, R_{o_n})$ of all interrelational coherence relations $R_{o_l} \in \mathbb{LR}_x(y)$ to which y participates as the dependent variable in x. This allows mapping these relations onto indices indicating their processing order by means of a function $r \colon \mathbb{LR}_x(y) \to \mathbb{N}$ with $r(R_{o_i}) = i$, $i \in \{1, \dots, n\}$.

4.5 Semantic Flexibility

The semantics of coherence relations is modeled by means of *construction operations*.[15] As generic specifications of how coherence relations restrict the interpretation of their arguments, they operate on sets of meaning points in semantic space. The inclusion of these operations into constraint schemata

---

[14]Note that $A \triangle B$ is the symmetric difference of the sets A and B.

[15]Construction and integration operations roughly correspond to meaning calibration in the sense of Kamp & Partee [11].

aims at accounting for coherence not only in terms of *integration* but also of *construction* in the sense of Kintsch's CI model.

Construction operations may *generate* alternative interpretations of text segments which are not required to belong to the interpretation domains of these segments at the beginning of the CI process. Rather, these alternatives may be generated subject to the operative constraints *at runtime*. This extension, which departs from Thagard's [40] model of coherence as a *set-theoretical separation problem*, is indispensable when trying to tackle the flexibility and, thus, "uncountability" of meaning [13] according to which words calibrate their meanings to fit their variable usage contexts without requiring that these variants are enumerated in the lexicon. Obviously, this uncountability holds all the more for complex signs above the level of words.

In order to substantiate the notion of construction operations algorithmically, *structure-sensitive semantic spaces* are needed. That is, a data structure which allows to model the language systematic meaning as well as the context-sensitive interpretation of texts and their components. Other than semantic spaces which map words and texts onto single, unstructured meaning points (see Table 1), the mathematical apparatus of hypergraphs will be utilized in order to build structure-sensitive semantic spaces which also take their structure into account. Thus, the classical model of semantic spaces is extended in a way which maps signs onto *structured*, but not simply distributed meaning representations. This allows to conceive construction operations on structure-sensitive semantic spaces as models of cognitive processes operating on sign meanings which, in turn, are modeled by means of semantic spaces.

4.6 Usage-Based Semantics

According to Miller & Charles [23] the semantic similarity of words is conditioned by the similarity of their contextual representations. This similarity results from processes of text comprehension which incrementally revise the words' meanings subject to their varying contextual embeddings. We use the following notation in order to represent this variability: Let $\|a\|^{C_i}$ be the meaning representation of lexeme a as a result of processing corpus $C_i = \{x_1, \ldots, x_i\}$. The incremental revision of $\|a\|^{C_i}$ as a result of processing the sequence of texts $x_{i+1}, \ldots, x_{i+n}$ (subject to their contexts $\triangleright_{x_{i+1}}, \ldots, \triangleright_{x_{i+n}}$, spanned by interrelational coherence relations of the texts as a whole) is noted as

$$\|a\|^{C_i} \xrightarrow{x_{i+1}} \|a\|^{C_{i+1}} \xrightarrow{x_{i+2}} \ldots \xrightarrow{x_{i+n-1}} \|a\|^{C_{i+n-1}} \xrightarrow{x_{i+n}} \|a\|^{C_{i+n}} \tag{13}$$

This process of incrementally learning the meanings of lexical items denotes an important difference compared to the CP which, due to its focus on synchrony, abstracts from meaning constitution and change. An implementation of this incremental learning process would obviously solve the problem of missing iterative computability enumerated in Table (1), but is out of the scope of the present chapter.[16]

4.7 Structure-Sensitive Semantic Spaces

Starting from a corpus C_t, a semantic space $(L, S, \alpha, \beta, (\mathbb{X}, d))$ according to definition (2) and an arbitrary component y of a focal text $x \in C$, *Structure-Sensitive Semantic Spaces* are stepwise introduced by means of triples[17]

$$(X, G_1, G_2) \tag{14}$$

in which X is a set of meaning points and G_1, G_2 are hypergraphs representing y's integration hierarchy $\triangle_y$ and context $\triangleright_{x,y}$ in x, respectively. The definition of these parameters varies with elementary (i.e. lexical) and complex text components as well as their meanings and interpretations, respectively:

- *The meaning $\|\boldsymbol{y}\|^C$ of an elementary text component $\boldsymbol{y} \models_T y$ (i.e. the meaning of a token as a leaf of x's integration hierarchy) w.r.t corpus C is modeled irrespective of its contextual embedding in x as a triple*

$$\|\boldsymbol{y}\|^C = (\{\alpha(y)\}, G_\emptyset(\boldsymbol{y}), G_\emptyset(\boldsymbol{y})) , \tag{15}$$

  where $G_\emptyset(\boldsymbol{y}) = (\{\boldsymbol{y}\}, \emptyset)$ is a hypergraph with an empty set of edges. It indicates that the meaning of tokens is neither sensitive to their integration hierarchy (which is empty by definition) nor to their context in x.
- *The interpretation $\|\boldsymbol{y}\|^{C,x}$ of a token $\boldsymbol{y} \models_T y$ is a triple*

$$\|\boldsymbol{y}\|^{C,x} = (\mathbb{X}(\boldsymbol{y}), G_\emptyset(\boldsymbol{y}), \triangleright_{x,y}) \tag{16}$$

in which $\mathbb{X}(\boldsymbol{y})$ is a subset of $\mathbb{X}$ (i.e. the universe of the metric space of the operative semantic space) and $\triangleright_{x,y} = (V, E)$ is the context of y in x. $\mathbb{X}(\boldsymbol{y})$ is recursively computed: Suppose that all coherence relations $R_i(\boldsymbol{y} \mid \ldots) \in \mathbb{L}_x(\boldsymbol{y})$ to which $\boldsymbol{y}$ participates in x are indexed according to their rank as defined in Section (4.4). That is, $\mathrm{r}(R_i) = i$, $R_i \in \mathbb{L}_x(\boldsymbol{y})$, and $\perp_{\mathbb{L}_x(\boldsymbol{y})} = \mathrm{r}(\sup_{\sqsubseteq_{x,y}} \mathbb{L}_x(\boldsymbol{y}))$ is the index of the coherence relation $R \in \mathbb{L}_x(\boldsymbol{y})$ of lowest rank to which $\boldsymbol{y}$ participates. Then we define:

$$\|\boldsymbol{y}\|_0^{C,x} = \|\boldsymbol{y}\|^C$$

$$\ldots$$

$$\begin{aligned}
\|\boldsymbol{y}\|_i^{C,x} &= \|R_i(\boldsymbol{y} \mid x_{i_1}, \ldots, x_{i_k})\|_i^{C,x} \\
&= \|R_i\|_{i-1}^{C,x}(\|\boldsymbol{y}\|_{i-1}^{C,x} \mid \|x_{i_1}\|_{i-1}^{C,x}, \ldots, \|x_{i_k}\|_{i-1}^{C,x}) \\
&= \mathbf{R}(\|\boldsymbol{y}\|_{i-1}^{C,x} \mid \|x_{i_1}\|^{C,x}, \ldots, \|x_{i_k}\|^{C,x}) \\
&= (\mathbb{X}_i(\boldsymbol{y}), G_\emptyset(\boldsymbol{y}), (V_{i-1} \cup \cup_{l=1}^k \{x_{i_l}\}, E_{i-1} \cup \{R_i\}))
\end{aligned} \tag{17}$$

---

[16]See Perlovsky [26] for an approach to solve the iterative learning of dynamic category systems in the area of text categorization.

[17]For simplicity reasons, we will write C instead of C_t.

where $R_i \models_C \mathbf{R}$, $\mathbb{X}_i(\boldsymbol{y}) = \otimes_{\mathbf{R}}(\mathbb{X}_{i-1}(\boldsymbol{y}) \,|\, \mathbb{X}(x_{i_1}), \ldots, \mathbb{X}(x_{i_k}))$, $\mathbb{X}_0(\boldsymbol{y}) = \{\alpha(\boldsymbol{y})\}$ (since $\|\boldsymbol{y}\|_0^{C,x} = \|\boldsymbol{y}\|^C$) and $\|x_{i_l}\|^{C,x} = (\mathbb{X}(x_{i_l}), \triangle_{x_{i_l}}, \rhd_{x,x_{i_l}})$ for each $l \in \{1, \ldots, k\}$. Note that constraint schemata model the supposedly fixed semantics of their instance relations so that $\|R_i\|_{i-1}^{C,x} = \mathbf{R}$. Note further that $\|x_{i_l}\|_{i-1}^{C,x} = \|x_{i_l}\|^{C,x}$ since, by definition (see Equation 9), x_{i_l} is an independent variable of R_i which, according to the postorder of x, is preordered w.r.t $\boldsymbol{y}$, i.e. the interpretation of x_{i_l} is known and unaffected by R_i when interpreting $\boldsymbol{y}$ in any of the steps $0, \ldots, i, \ldots, \perp_{\mathbb{L}_x(\boldsymbol{y})}$. The construction operation $\otimes_{\mathbf{R}}$ as part of the construction schema $\mathbf{R}$ is defined in Section (4.5). This allows to set

$$
\begin{aligned}
\|\boldsymbol{y}\|_{\perp_{\mathbb{L}_x(\boldsymbol{y})}}^{C,x} &= (\mathbb{X}_{\perp_{\mathbb{L}_x(\boldsymbol{y})}}(\boldsymbol{y}), G_{\emptyset}(\boldsymbol{y}), \\
&\quad (V_{\perp_{\mathbb{L}_x(\boldsymbol{y})}-1} \cup \cup_{l=1}^{k}\{x_{\perp_{\mathbb{L}_x(\boldsymbol{y})_l}}\}, E_{\perp_{\mathbb{L}_x(\boldsymbol{y})}-1} \cup \{R_{\perp_{\mathbb{L}_x(\boldsymbol{y})}}\}))\\
&= (\mathbb{X}_{\perp_{\mathbb{L}_x(\boldsymbol{y})}}(\boldsymbol{y}), G_{\emptyset}(\boldsymbol{y}), (V_{\perp_{\mathbb{L}_x(\boldsymbol{y})}}, \mathbb{L}_x(\boldsymbol{y})))
\end{aligned}
\tag{18}
$$

It remains to process those constraints operating on $\boldsymbol{y}$ as a dependent variable which belong to $\mathbb{R}_x(\boldsymbol{y})$. We suppose these constraints to be processed (by analogy with the elements of $\mathbb{L}_x(\boldsymbol{y})$) according to their linear order induced by $\sqsubseteq_{x,\boldsymbol{y}}$ and $\mathrm{r}\colon \mathbb{L}\mathbb{R}_x(\boldsymbol{y}) \to \mathbb{N}$, respectively (see Section 4.4).[18] That is, $\mathrm{r}(R_i) = i$, $R_i \in \mathbb{R}_x(\boldsymbol{y})$, and $\perp_{\mathbb{R}_x(\boldsymbol{y})} = \mathrm{r}(\sup_{\sqsubseteq_{x,\boldsymbol{y}}} \mathbb{R}_x(\boldsymbol{y}))$ is the index of the coherence relation $R \in \mathbb{R}_x(\boldsymbol{y})$ of lowest rank to which $\boldsymbol{y}$ participates. Next we set:

$$
\begin{aligned}
\|\boldsymbol{y}\|_i^{C,x} &= \|R_i(\boldsymbol{y} \,|\, x_{i_1}, \ldots, x_{i_k})\|_i^{C,x}\\
&= \|R_i\|_{i-1}^{C,x}(\|\boldsymbol{y}\|_{i-1}^{C,x} \,|\, \|x_{i_1}\|_{i-1}^{C,x}, \ldots, \|x_{i_k}\|_{i-1}^{C,x})\\
&= \mathbf{R}(\|\boldsymbol{y}\|_{i-1}^{C,x} \,|\, \|x_{i_1}\|^{C,x}, \ldots, \|x_{i_k}\|^{C,x})\\
&= (\mathbb{X}_i(\boldsymbol{y}), G_{\emptyset}(\boldsymbol{y}), (V_{i-1} \cup \cup_{l=1}^{k}\{x_{i_l}\}, E_{i-1} \cup \{R_i\}))
\end{aligned}
$$

$$
\cdots
$$

$$
\|\boldsymbol{y}\|_{\perp_{\mathbb{R}_x(\boldsymbol{y})}}^{C,x} = (\mathbb{X}_{\perp_{\mathbb{R}_x(\boldsymbol{y})}}(\boldsymbol{y}), G_{\emptyset}(\boldsymbol{y}), \rhd_{x,\boldsymbol{y}})
\tag{19}
$$

Note that, per definition, $\|\boldsymbol{y}\|_{\top_{\mathbb{R}_x(\boldsymbol{y})}-1}^{C,x} = \|\boldsymbol{y}\|_{\perp_{\mathbb{L}_x(\boldsymbol{y})}}^{C,x}$ where $\top_{\mathbb{R}_x(\boldsymbol{y})} = \mathrm{r}(\inf_{\sqsubseteq_{x,\boldsymbol{y}}} \mathbb{R}_x(\boldsymbol{y}))$ is the index of the topmost ranked coherence relation within $\mathbb{R}_x(\boldsymbol{y})$. Finally, we set:

$$
\|\boldsymbol{y}\|^{C,x} = \|\boldsymbol{y}\|_{\perp_{\mathbb{R}_x(\boldsymbol{y})}}^{C,x}
\tag{20}
$$

- *The meaning of a complex component $y = I(y_{j_1}, \ldots, y_{j_k})$ is defined irrespective of its contextual embedding in x as*

---

[18] Obviously, a more appropriate model is to cyclically process these constraints so that all interrelational coherence relations of the immediate constituents of a component to be integrated are repeatedly processed until their interpretations converge.

$$\|y\|^C = \|I(y_{j_1}, \ldots, y_{j_k})\|^C$$
$$= \|I\|^C(\|y_{j_1}\|^C, \ldots, \|y_{j_k}\|^C)$$
$$= \mathbf{I}(\|y_{j_1}\|^C, \ldots, \|y_{j_k}\|^C)$$
$$= (\mathbf{X}(y), \triangle_y, G_\emptyset(y)) \tag{21}$$

where $\mathbf{X}(y) \subseteq \mathbf{X}$ is computed with the help of an aggregation function (defined in Section 5.1):

$$\mathbf{X}(y) = \bigoplus_{l \in \{1,\ldots,k\}, \|y_{j_l}\|^C = (\mathbf{X}(y_{j_l}), \triangle_{y_{j_l}}, G_\emptyset(y_{j_l}))} \mathbf{X}(y_{j_l}) \tag{22}$$

That is $\mathbf{X}(y)$ is a subset of meaning points computed by means of the aggregation operation $\oplus$ operating on the meaning representations of y's immediate constituents according to $\triangle_y$. Since (21) defines the *meaning* of y, it disregards its context. Thus, the third parameter of $\|y\|^C$ is $G_\emptyset(y)$.

- *The interpretation of a complex component $y = I(y_{j_1}, \ldots, y_{j_k})$ of x is defined as*

$$\|y\|^{C,x} = (\mathbf{X}(y), \triangle_y, \rhd_{x,y}) \tag{23}$$

Other than before, both parameters $\rhd_{x,y}$ *and* $\triangle_y$ restrict the interpretation of y in x. By analogy with equation (17) we suppose that all coherence relations $R_i(y \mid \ldots)$ to which $y \in \mathbb{L}_x(y)$ participates are indexed according to their rank. That is, $\mathrm{r}(R_i) = i$, $R_i \in \mathbb{L}_x(y)$, and $\bot_{\mathbb{L}_x(y)} = \mathrm{r}(\sup_{\sqsubseteq_{x,y}} \mathbb{L}_x(y))$. Now we define:[19]

$$\|y\|_0^{C,x} = \|I(y_{j_1}, \ldots, y_{j_k})\|^{C,x}$$
$$= \|I\|^{C,x}(\|y_{j_1}\|^{C,x}, \ldots, \|y_{j_k}\|^{C,x})$$
$$= \mathbf{I}(\|y_{j_1}\|^{C,x}, \ldots, \|y_{j_k}\|^{C,x})$$
$$= (\oplus_{l \in \{1,\ldots,k\}, \|y_{j_l}\|^{C,x} = (\mathbf{X}(y_{j_l}), \triangle_{y_{j_l}}, \rhd_{x,y_{j_l}})} \mathbf{X}(y_{j_l}), \triangle_y, G_\emptyset(y))$$

$$\cdots$$

$$\|y\|_i^{C,x} = \|R_i(y \mid x_{i_1}, \ldots, x_{i_k})\|_i^{C,x}$$
$$= \|R_i\|_{i-1}^{C,x}(\|y\|_{i-1}^{C,x} \mid \|x_{i_1}\|_{i-1}^{C,x}, \ldots, \|x_{i_k}\|_{i-1}^{C,x})$$
$$= \mathbf{R}(\|y\|_{i-1}^{C,x} \mid \|x_{i_1}\|^{C,x}, \ldots, \|x_{i_k}\|^{C,x})$$
$$= (\mathbf{X}_i(y), \triangle_y, (V_{i-1} \cup \{x_{i_1}, \ldots, x_{i_k}\}, E_{i-1} \cup \{R_i\})) \tag{24}$$

where $\mathbf{X}_i(y) = \otimes_{\mathbf{R}}(\mathbf{X}_{i-1}(y) \mid \mathbf{X}(x_{i_1}), \ldots, \mathbf{X}(x_{i_k}))$, $\|x_{i_l}\|^x = (\mathbf{X}(x_{i_l}), \triangle_{x_{i_l}}, \rhd_{x,x_{i_l}})$ for each $l \in \{1, \ldots, k\}$. This allows to set

$$\|y\|_{\bot_{\mathbb{L}_x(y)}}^{C,x} = (\mathbf{X}_{\bot_{\mathbb{L}_x(y)}}(y), \triangle_y,$$
$$(V_{\bot_{\mathbb{L}_x(y)}-1} \cup \cup_{l=1}^k \{x_{\bot_{\mathbb{L}_x(y)_l}}\}, E_{\bot_{\mathbb{L}_x(y)}-1} \cup \{R_{\bot_{\mathbb{L}_x(y)}}\}))$$
$$= (\mathbf{X}_{\bot_{\mathbb{L}_x(y)}}(y), \triangle_y, (V_{\bot_{\mathbb{L}_x(y)}}, \mathbb{L}_x(y))) \tag{25}$$

---

[19]Note that $\|y\|_0^x$ depends on the interpretations and not on the meanings of y's constituents.

Note that other than in (18), the integration hierarchy $\triangle_y$ is referred to as the second parameter. Since $\triangle_y$ is evaluated before any constraint in $\rhd_{x,y}$ is processed, we can, by analogy with (20), write (thereby skipping the details of intermediary equations for reasons of brevity):

$$\|y\|^{C,x} = \|y\|_{\perp_{\mathbb{R}_x(y)}}^{C,x} = (\mathbb{X}_{\perp_{\mathbb{R}_x(y)}}(y), \triangle_y, \rhd_{x,y}) \tag{26}$$

The computation of $\|y\|^{C,x}$ now depends on the interpretation of its constituents according to their embedding into x which because of transrelational dependencies may cross the borders of y. In addition, $\|y\|^{C,x}$ reflects all interrelational coherence relations defining y's context in x.

The CI process exemplified by $\|y\|^{C,x}$ works as follows: First, an intermediate representation $\|y\|_0^{C,x}$ is computed by means of the aggregation operation $\oplus$ operating on the interpretations of y's immediate constituents. Thereafter $\|y\|_0^{C,x}$ is object of one construction operation after another modeling the semantics of the coherence relations to which y participates. In other words: The interpretation of y may – in spite of its hierarchical organization – transcend its integration hierarchy because of valuating transrelational dependencies w.r.t y and thus depart from the classical principle of compositionality.

Equations (16-25) implement a CI model with a kernel hierarchical order superimposed by graph-inducing coherence relations. As a result of evaluating such an HCSP, the interpretation of a text component y – whether elementary or complex – can deviate from its meaning. Thus, y's interpretation does not only depend on its constituents and the way they are combined, but also on its embedding into the focal text. The equivalent of this process in terms of semantic spaces is a construction operation whose application may result in a shift, spreading (extension) or contraction (reduction) of the subspace representing y's text-specific interpretation. The degree to which y's meaning and text-specific interpretation deviate, is a function of y's context within the focal text: The more this embedding deviates from y's or its constituents' usage regularities the greater the deviation.

5 Quantitative Semantics

Construction and integration are procedural building blocks of HCSPs – they model possibly cooperative or competing processes of text interpretation which are now defined in terms of operations on fuzzy sets.

5.1 Integration

The primary order of text interpretation is – due to the integration hierarchy – *bottom-up*: Starting with the leafs of $\triangle_x$, a successful interpretation process finally reaches the root of $\triangle_x$. In the case of complex text components y,

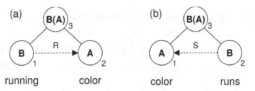

Fig. 2. In (a), constraint R is processed only when interpreting A since $R \in \mathbb{R}_x(B)$ (x denotes the focal text). That is, R is processed in step 2 so that B may have different interpretations in step 1 and 2 (indicated by the postorder indices of the constituents). In (b), $S \in \mathbb{L}_x(B)$ is processed as part of interpreting B in step 2.

this process either reflects coherence constraints or not. In the former case, the meaning representations of y's immediate constituents are input to integration (cf. Equation 22). In the latter case, the interpretations of these constituents are used instead (cf. Equation 25). Equation (22) and (25) refer to the operation $\oplus$. This operation is defined in terms of fuzzy theory with the help of a weighted mean of membership functions and, thus, as an element of a class of aggregation operations $h: [0,1]^n \rightarrow [0,1]$ (cf. Klir & Folger [14]):

$$h_p(a_1, \ldots, a_n; w_1, \ldots, w_n) = \left(\sum_{i=1}^{n} w_i a_i^p \right)^{1/p}, p \in \mathbb{N}, p \geq 2, \sum_{i=1}^{n} w_i = 1 \quad (27)$$

This allows to define the membership function of the aggregation $X = \oplus_{i=1}^{n} X_i$ of n fuzzy sets $X_i \in \mathbb{F}(U)$, $i = 1, \ldots, n$, as

$$\forall x \in U : \mu_X(x) = h_p(\mu_{X_1}(x), \mu_{X_2}(x), \ldots, \mu_{X_n}(x), w_1, \ldots, w_n) \quad (28)$$

In terms of semantic spaces and their constitutive vectors, aggregation functions are redefined as follows:

$$\mathbf{h}_p(\mathbf{a}_1, \ldots, \mathbf{a}_n, \mathbf{w}) = \left(\left(\sum_{j=1}^{m} w_i a_{ij}^p \right)^{1/p}, \ldots, \left(\sum_{j=1}^{m} w_i a_{ij}^p \right)^{1/p} \right)^T \quad (29)$$

where $\mathbf{a}_i \in \mathbb{X} \subseteq \mathbb{R}^m$, $i \in \{1, \ldots, n\}$; $\mathbf{w} = (w_1, \ldots, w_n)^T$ is a weighting vector.

In [13], weights are defined as functions of the distances of the k nearest meaning points of the points assigned to the predicate B and its argument A, respectively; – k is a parameter of the model. The corresponding distances are computed with the help of the cosine measure. They are used to initially bias links in a connectionist network whose weights are computed by means of a spreading activation process. We use a non-connectionist model instead to represent, for example, the interpretation of $B(A)$ as a function of two subsets of meaning points, the one interpreting A and the other B subject to the coherence relation $P(B|A)$, $P \models_c \mathbf{P}$, calibrating B dependent on A – $\mathbf{P}$ denotes the constraint schema of predicate-argument structures. This is demonstrated in Figure (2) and specified in more detail in Section (5.4). The construction schema involved in this example is described in Section (5.2). The integration process takes the output of this construction process as input to generate a set of meaning points interpreting $B(A)$: Let $\{\mathbb{X}_i \subset \mathbb{X} \mid i \in \mathcal{I}\}$ be a family of sets of meaning points. The aggregation function $\oplus \colon \mathbb{P}(\mathbb{P}(\mathbb{X})) \to \mathbb{P}(\mathbb{X})$ is defined as follows:

$$\overset{p}{\underset{i \in \mathcal{I}}{\bigoplus}} \mathbb{X}_i = \overset{p}{\bigoplus}(\{\mathbb{X}_i \subset \mathbb{X} \mid i \in \mathcal{I}\}) = \overset{p}{\bigoplus}(\mathbb{X}_1, \ldots, \mathbb{X}_{|\mathcal{I}|})$$

$$= \{\mathbf{h}_p(\mathbf{x}_{j_1}, \ldots, \mathbf{x}_{j_k}, \mathbf{w}_j) \mid \{\mathbf{x}_{j_1}, \ldots, \mathbf{x}_{j_k}\} \in \mathrm{cl}(\mathbb{X}_1 \cup \ldots \cup \mathbb{X}_{|\mathcal{I}|})\} \quad (30)$$

where $p \in \mathbb{N}$, $p \geq 2$, and

$$w_{j_l} = \frac{\max_d -d(\mathbf{m}, \mathbf{x}_{j_l})}{\sum_{n=1}^{k} \max_d -d(\mathbf{m}, \mathbf{x}_{j_n})} \in [0, 1] \quad (31)$$

is the weight of the j_l-th argument of the vector $\mathbf{w}_j = (w_{j_1}, \ldots, w_{j_k})^T$ and

$$\mathbf{m} = \arg \min_{\mathbf{x}_{j_m} \in \{\mathbf{x}_{j_1}, \ldots, \mathbf{x}_{j_k}\}} \frac{1}{k} \sum_{n=1}^{k} d(\mathbf{x}_{j_m}, \mathbf{x}_{j_n}) \quad (32)$$

is the median of $\mathbf{x}_{j_1}, \ldots, \mathbf{x}_{j_k}$ and d is a distance measure, e.g. the Mahalanobis distance, with the maximum value $\max_d$. Finally, $\mathrm{cl}(\mathbb{X}_1 \cup \ldots \cup \mathbb{X}_{|\mathcal{I}|})$ is a partitive clustering of the set of meaning points $\mathbb{X}_1 \cup \ldots \cup \mathbb{X}_{|\mathcal{I}|}$. We utilize hierarchical agglomerative clustering based on average linkage with subsequent partitioning. This partitioning refers to a lower bound [33]

$$\theta = \bar{\eta} + \frac{1}{2}\sigma \quad (33)$$

where $\bar{\eta}$ is the mean and σ the standard deviation of the absolute value of the differences of the similarity levels of consecutive agglomeration steps. This gives a threshold for selecting an agglomeration step for dendrogram partitioning whose similarity distance to the preceding step is greater than θ. We use the first step exceeding θ.

Equations (30-31) define an aggregation function on sets of meaning points which maps each alternative interpretation of the focal component to a single meaning point.

5.2 Construction

Construction is performed subject to coherence relations. It generates/selects interpretation candidates of text segments. As elementary segments do not have an integration hierarchy, their meaning representations are input to construction processes whose semantics is separately defined for each constraint type. They make use of a set of functions for generating/selecting meaning points defined as follows: Let $(L, S, \alpha, \beta, (\mathbb{X}, d))$ be a semantic space, $A, B \subset \mathbb{X}$, $\bar{A} = (\mathbf{a}_1, \ldots, \mathbf{a}_k)$ a sequence of points $\mathbf{a}_i \in \mathbb{X}$ and $\mathbf{a}, \mathbf{b} \in \mathbb{X}$. Then:

$$
\begin{aligned}
n_{x,\rho}(\mathbf{b} \,|\, \mathbf{a}) &= \{\beta(y) \in \mathbb{X} \,|\, y \in S(x) \wedge d(\mathbf{b}, \beta(y)) \leq d(\mathbf{a}, \mathbf{b}) \wedge \\
&\quad d(\mathbf{a}, \beta(y)) < \min(\rho, d(\mathbf{a}, \mathbf{b}))\} \\
n_{x,\rho}(\mathbf{b} \,|\, \neg\mathbf{a}) &= \{\beta(y) \in \mathbb{X} \,|\, d(\mathbf{a}, \beta(y)) \geq d(\mathbf{a}, \mathbf{b}) \wedge d(\mathbf{b}, \beta(y)) < \rho\} \\
N_{x,\rho}(B \,|\, A) &= n_{x,\rho}(\pi(B) \,|\, \pi(A)) \\
N_{x,\rho}(B \,|\, \neg A) &= n_{x,\rho}(\pi(B) \,|\, \neg\pi(A)) \\
\bar{N}_{x,\rho}(B \,|\, \bar{A}) &= n_{x,\rho}(\pi(B) \,|\, \kappa(\bar{A})) \\
\bar{N}_{x,\rho}(B \,|\, \neg\bar{A}) &= n_{x,\rho}(\pi(B) \,|\, \neg\kappa(\bar{A}))
\end{aligned}
\tag{34}
$$

where $\rho = \bar{\eta} - \frac{1}{2}\sigma$ ($\bar{\eta}$ now denotes the mean distance of meaning points and σ its standard deviation). Further, $\pi : \mathbb{P}(\mathbb{X}) \to \mathbb{X}$ with

$$
\pi(\{\mathbf{a}_1, \ldots \mathbf{a}_n\}) = \mathbf{h}_p(\mathbf{a}_1, \ldots, \mathbf{a}_n, (\frac{1}{n}, \ldots, \frac{1}{n})^T)
\tag{35}
$$

and $\kappa : \mathbb{P}(\mathbb{X}) \to \mathbb{X}$ is a function, which in contrast to π, reflects the syntagmatic order of its argument points:

$$
\kappa((\mathbf{a}_i \,|\, i \in \mathcal{I})) = \mathbf{h}_p(\mathbf{a}_1, \ldots, \mathbf{a}_n, \mathbf{w})
\tag{36}
$$

where $\mathbf{w} = (w_1, \ldots, w_n)^T$ reflects the syntagmatic order of the segments whose meaning points are collected by $\{\mathbf{a}_i \,|\, i \in \mathcal{I}\}$. Such a weighting scheme has been defined in [21]. κ defines a sequence sensitive weighted mean of vectors. It is used to map lexical chains as cotexts of text interpretation.

Equation (34) is now used to define construction operations as part of constraint schemata. We concentrate on the following constraint types:

- *Anaphoric reference:* Let $\mathbf{A}$ be the type of anaphoric relation and $A(b \,|\, a)$ an instance of this type with b in the role of an anaphora w.r.t a. Let further $\mathbb{X}_a, \mathbb{X}_b$ be the interpretations of a and b according to equation (20) or (26), respectively. Then, $\otimes_{\mathbf{A}}$ is defined as

$$
\otimes_{\mathbf{A}}(\mathbb{X}_b \,|\, \mathbb{X}_a) = \mathbb{X}_a
\tag{37}
$$

- *Sense relations:* We consider four types of sense relations:
  - *Synonymy:* Let $\mathbf{S}_1$ be the type of synonymy and $S(b \,|\, a)$ one of its instances with b to be interpreted as a synonym of a. Let further $\mathbb{X}_a, \mathbb{X}_b$ be the interpretations of a and b according to (20) or (26). Then:

$$
\otimes_{\mathbf{S}_1}(\mathbb{X}_b \,|\, \mathbb{X}_a) = \mathbb{X}_a
\tag{38}
$$

  - *Partial synonymy:* Let $\mathbf{S}_2$ be the type of partial synonymy and $S(b \,|\, a)$ an instance of this type with b to be interpreted as a partial synonym of a. Let further $\mathbb{X}_a, \mathbb{X}_b$ be the interpretations of a and b according to (20) or (26), respectively. Then:

$$
\otimes_{\mathbf{S}_2}(\mathbb{X}_b \,|\, \mathbb{X}_a) = N_{x,\rho}(\mathbb{X}_b \,|\, \mathbb{X}_a)
\tag{39}
$$

– *Hyponymy:* Let $\mathbf{H}_1$ be the type of hyponymy and $H(b\,|\,a)$ one of its instances with b in the role of the hyponym to be interpreted w.r.t its hyperonym a. Let further $\mathbb{X}_a, \mathbb{X}_b$ be the interpretations of the components a and b according to (20) or (26), respectively. Then:

$$\otimes_{\mathbf{H}_1}(\mathbb{X}_b\,|\,\mathbb{X}_a) = N_{x,\rho}(\mathbb{X}_b\,|\,\mathbb{X}_a) \cup N_{x,\rho}(\mathbb{X}_a\,|\,\mathbb{X}_b) \qquad (40)$$

– *Hyperonymy:* Let $\mathbf{H}_2$ be the type of hyperonymy and $H(b\,|\,a_1,\ldots,a_n)$ an instance of this type with b in the role of the hyperonym w.r.t its hyponyms $a_1,\ldots,a_n$. Let further $\mathbb{X}_{a_1},\ldots,\mathbb{X}_{a_n}$, $\mathbb{X}_b$ be the interpretations of the components $a_1,\ldots,a_n$ and b according to (20) or (26), respectively. Then:

$$\otimes_{\mathbf{H}_2}(\mathbb{X}_b\,|\,\mathbb{X}_{a_1},\ldots,\mathbb{X}_{a_n}) = N_{x,\rho}(\mathbb{X}_b\,|\,\cup_{i=1}^n \mathbb{X}_{a_i}) \qquad (41)$$

• *Predicate-argument structures:* Let $\mathbf{P}$ be the type of predicate-argument relation and $P(b\,|\,a)$ an instance of this type with b in the role of the predicate. Let further $\mathbb{X}_a, \mathbb{X}_b$ be the interpretations of a and b according to equation (20) or (26), respectively. Then:

$$\otimes_{\mathbf{P}}(\mathbb{X}_b\,|\,\mathbb{X}_a) = N_{x,\rho}(\mathbb{X}_b\,|\,\mathbb{X}_a) \qquad (42)$$

• *Lexical chains:* If a token a continues a lexical chain $c = (a_1,\ldots,a_m)$, a is interpreted in the context of $\|c\|^{C,x}$. More specifically, let $\mathbf{L}$ be the type of unsystematic lexical cohesion and $L(a\,|\,a_1,\ldots,a_m)$ an instance of this type with c in the role of the chain. Let further $\mathbb{X}_a, \mathbb{X}_{a_1},\ldots,\mathbb{X}_{a_m}$ be the interpretations of $a, a_1,\ldots,a_m$, according to equation (20) or (26), respectively (note that since lexical chains are not necessarily constituents of the integration hierarchy of a text, but superimpose it, the chain is mapped by means of a hyperedge). Then we define:

$$\otimes_{\mathbf{L}}(\mathbb{X}_a\,|\,\mathbb{X}_{a_1},\ldots,\mathbb{X}_{a_n}) = \bar{N}_{x,\rho}(\mathbb{X}_a\,|\,(\pi(\mathbb{X}_{a_1}),\ldots,\pi(\mathbb{X}_{a_n}))) \qquad (43)$$

• *Rhetorical relations:* We consider two types of rhetorical relations:
    – Let $\mathbf{E}$ be the type of *elaboration* and $E(a\,|\,b)$ one of its instances with b in the role of the elaborating segment. Let further $\mathbb{X}_a, \mathbb{X}_b$ be the interpretations of a and b according to equation (20) or (26). Then:

$$\otimes_{\mathbf{E}}(\mathbb{X}_a\,|\,\mathbb{X}_b) = N_{x,\rho}(\mathbb{X}_a\,|\,\mathbb{X}_b) \qquad (44)$$

    – Let $\mathbf{C}$ be the type of *contrast* and $C(a\,|\,b)$ one of its instances with b in the role of the contrasting span. Let further $\mathbb{X}_a, \mathbb{X}_b$ be the interpretations of a and b according to (20) or (26), respectively. Then:

$$\otimes_{\mathbf{C}}(\mathbb{X}_a\,|\,\mathbb{X}_b) = N_{x,\rho}(\mathbb{X}_a\,|\,\neg\mathbb{X}_b) \qquad (45)$$

5.3 Consonant vs. Dissonant Constraints

Coherence relations operating on the same component may cooperate or compete to varying degrees. An ambiguous segment may, for example, be interrelated with components of the same text (e.g. elaborations) supporting its dissonant readings. *How are such relations handled in the present framework?* This can be answered as follows:

- *Consonant relations* narrow down a region within the focal space as small as possible to locate the discourse specific interpretation of the focal segment (possibly down to a single meaning point).
- *Dissonant relations* have the opposite effect: processing them enlarges the region of candidate interpretations, possibly by splitting it into discontinuous regions as representations of competing interpretations. This is mainly due to the integration step and clustering performed within it (see Equation 30).

Obviously, relations resulting in many, large regions of the semantic space as representations of candidate interpretations reflect an insufficient, incomplete solution of the focal HCSP.

5.4 Optimization

HCSPs induce an optimization problem as follows: To solve an HCSP induced by a text x means to integrate its segments' interpretations along x's integration hierarchy bottom-up from left to right as far as possible by reflecting as many coherence relations as possible according to their descending degree of priority. More specifically, suppose that $(a_1, \ldots, a_n), a_i \in S(x), i \in \{0, \ldots, n\}$, is the postorder of x's integration hierarchy $\triangle_x$. This postorder uniquely determines the order of all construction and integration steps. It allows to assign any component $y \in S(x)$ a number $f(y) = i$ iff $y = a_i$, $i = 1, \ldots n$, indicating at which of the n different steps of the interpretation process (induced by x) y is processed. More specifically, for each segment $y \in S(x)$ (note that $S(x)$ includes x) two sets of constraints are distinguished:

1. All constraints $R \in \mathbb{L}_x(y)$ are processed in step $f(y) = i$ either according to equation (18) or (25) (depending on whether y is simple or complex).
2. For all constraints $R(y \mid y_{i_1}, \ldots, y_{i_k}) \in \mathbb{R}_x(y)$ we identify segment $a_i = \sup_{O_2}\{y_{i_1}, \ldots, y_{i_k}\}$ and, thus, $i = f(a_i)$ as the step at which R is processed either according to (19) or (26) after all constraints in $\mathbb{L}_x(a_i)$ have been processed.

This process narrows down a subset of meaning points as an interpretation of each segment to be integrated which becomes the smaller the more specific and less dissonant its interrelational coherence relations and the less dissonant its constituents' interpretations are. Now we can, finally, define the

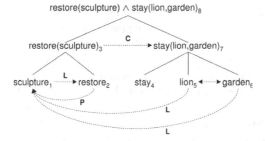

Fig. 3. Outline of the texture of the text sample.

interpretation of x with immediate constituents $y_1, \ldots, y_k$ (for the case that x does not participate in any interrelational coherence relation) as:

$$\|x\|^{C,x} = \mathbf{I}\left(\|y_1\|^{C,x}, \ldots, \|y_k\|^{C,x}\right) \tag{46}$$

It is important to note that x's interpretation is reduced to an integration of the interpretations of its parts, where these interpretations may result from construction processes transcending the borders of x by means of transrelational coherence relations. This goes beyond the CP.

6 A Dry Run

In this section we outline the processing of the following text sample in the framework of the VS model, FL, LSA and the present LCP approach – this example is motivated by the fact that constructions as, for example, *stone lion* play a prominent role in the discussion of the CP [see 11]:

All sculptures are restored. Only the lions stay in the garden.

- The vector space model represents this text as a set of words by filtering out function and other words. The remaining words (say *garden, lion, restore, sculpture, stay*) are used to build a vector of weighted descriptors in order to locate the text sample within the vector space.
- LSA extends this approach. Instead of directly referring to weighted counts of word occurrences, it utilizes factor analytic representations. That is, the sample is represented by means of those factors which predominate in locating the representations of the input words *garden, lion, restore, sculpture, stay* within the semantic space. Thus, we may speak of a *bag of feature vector*-based representation which still ignores the structure of the text sample and its coherence relations.
- [20] describes a variant based on FL which derives a representation of input texts based on their integration hierarchy. This variant is in accordance with equation (21) although it maps texts on single meaning points and still disregards any coherence relations.

The present framework partly overcomes this structure insensitivity. What we expect is that the text as a whole connotes words like *sculptural art*, *park*, *collection [of sculptures]* or even *castle* and *baroque*, but not *veldt*, *elephant* (because of *lion*) or *zoo* (because of *lion* and *[zoological] garden*). In order to motivate such an analysis, suppose the sample is structured as outlined in Figure (3). That is, *sculpture* is reinterpreted in the context of *to restore* which, before, is reinterpreted in the role of a predicate. We assume that the meanings of these words have been learned by means of a corpus documenting their various readings so that *to restore* can be interpreted in the sense of *to restaurate*, while the interpretation of *sculpture* enforces its reading as a concrete piece of work and inhibits its generic reading in the sense of *statuary*. Next, *restore(sculpture)* is integrated so that the second sentence can be processed. We assume *lion* and *garden* to be mutually dependent and being interpreted in the context of *sculpture* which inhibits the *animal* reading of *lion* as well as the *zoo[logical garden]* reading. This allows integrating *stay(lion,garden)* and using it in the role of a contrasting segment in order to calibrate the interpretation of *restore(sculpture)* before the whole text is integrated. According to (45) this means that *restore(sculpture)* is interpreted in a way which further inhibits the *lion* and *garden* reading, while enforcing the *sculpture* and *restauration* reading.

This analysis presupposes the present order of sentences; it does not allow reinterpreting *lion* if the order of the sentences is reversed. Moreover, the example presupposes that all coherence relations as well as the integration hierarchy have been identified before. In this sense, the present approach is – in spite of its formal character – only a preliminary study which outlines a way to replace the bag-of-features approach.

7 Conclusion

In this chapter, a principle of latent compositionality was introduced into quantitative semantics. It was implemented by means of an HCSP which assumes a kernel hierarchical structure superimposed by graph-inducing coherence relations as the fundamental text representation model. This implementation utilized an extended model of semantic spaces which is sensitive to text structure and thus leaves behind the bag-of-feature approach. The major task of this model is to better account for complex units and, thus, to better reconstruct their connotations in a framework of a higher level of cognitive plausibility. The evaluation of this model will be a major task of future work. In the long run, it is seen as part of an algebraic reconstruction of statistical approaches to semantics which reach a level of stringency comparable to formal logical semantics, *but in combination with empirical falsifiability*.

References

[1] H. Bandemer and W. Näther. *Fuzzy Data Analysis*. Kluwer, Dordrecht, 1992.

[2] J. Barwise and J. Perry. *Situations and Attitudes*. MIT Press, Cambridge, 1983.

[3] J. Barwise and J. Seligman. *Information Flow. The Logic of Distributed Systems*. University Press, Cambridge, 1997.

[4] S. Deerwester, S. T. Dumais, G. W. Furnas, T. K. Landauer, and R. Harshmann. Indexing by Latent Semantic Analysis. *Journal of the American Society for Information Science*, 41(6):391–407, 1990.

[5] D. Dubois, H. Fargier, and H. Prade. Propagation and Satisfaction of Flexible Constraints. In R. R. Yager and L. A. Zadeh, editors, *Fuzzy Sets, Neural Networks, and Soft Computing*, pages 166–186. van Nostrand Reinhold, New York, 1994.

[6] P. Foltz, W. Kintsch, and T. Landauer. The Measurement of Textual Coherence with Latent Semantic Analysis. *Discourse Processes*, 25(2&3):285–307, 1998.

[7] P. W. Foltz. Latent Semantic Analysis for Text-Based Research. *Behavior Research Methods, Instruments & Computers*, 28(2):197–202, 1996.

[8] A. J. Greimas. *Strukturale Semantik. Methodologische Untersuchungen*. Viehweg, Braunschweig, 1971.

[9] M. A. K. Halliday and R. Hasan. *Cohesion in English*. Longman, London, 1976.

[10] T. M. V. Janssen. Compositionality (with an Appendix by Barbara H. Partee). In J. van Benthem and A. ter Meulen, editors, *Handbook of Logic and Language*, pages 417–473. Elsevier, Amsterdam, 1997.

[11] H. Kamp and B. Partee. Prototype Theory and Compositionality. *Cognition*, 57(2):129–191, 1995.

[12] W. Kintsch. *Comprehension. A Paradigm for Cognition*. Cambridge University Press, Cambridge, 1998.

[13] W. Kintsch. Predication. *Cognitive Science*, 25:173–202, 2001.

[14] G. J. Klir and T. A. Folger. *Fuzzy Sets, Uncertainty, and Information*. Prentice Hall, Englewood, 1988.

[15] A. Knott and T. Sanders. The Classification of Coherence Relations and their Linguistic Markers: An Exploration of Two Languages. *Journal of Pragmatics*, 30:135–175, 1998.

[16] R. Lahav. Against Compositionality: The Case of Adjectives. *Philosophical Studies*, 57(3):261–279, 1989.

[17] T. K. Landauer and S. T. Dumais. A Solution to Plato's Problem: The Latent Semantic Analysis Theory of Acquisition, Induction, and Representation of Knowledge. *Psychological Review*, 104(2):211–240, 1997.

[18] W. C. Mann and S. A. Thompson. Rhetorical Structure Theory: Toward a Functional Theory of Text Organization. *Text*, 8:243–281, 1988.

[19] D. Marcu. *The Theory and Practice of Discourse Parsing and Summarization*. MIT Press, Cambridge, Massachusetts, 2000.

[20] A. Mehler. *Textbedeutung. Zur prozeduralen Analyse und Repräsentation struktureller Ähnlichkeiten von Texten*. Peter Lang, Frankfurt a. M., 2001.

[21] A. Mehler. Hierarchical Orderings of Textual Units. In *Proceedings of the 19th International Conference on Computational Linguistics, COLING'02, Taipei*, pages 646–652, San Francisco, 2002. Morgan Kaufmann.

[22] A. Mehler. Zur textlinguistischen Fundierung der Text- und Korpuskonversion. *Sprache und Datenverarbeitung*, 1:29–53, 2005.

[23] G. A. Miller and W. G. Charles. Contextual Correlates of Semantic Similarity. *Language and Cognitive Processes*, 6(1):1–28, 1991.

[24] D. N. Osherson and E. E. Smith. On the Adequacy of Prototype Theory as a Theory of Concepts. *Cognition*, 9(1):35–58, 1981.

[25] B. H. Partee. Compositionality. In F. Landman and F. Veltman, editors, *Varieties of Formal Semantics. Proceedings of the fourth Amsterdam Colloquium, September 1982*, pages 281–311, Dordrecht, 1984. Foris.

[26] L. Perlovsky. Neural Networks, Fuzzy Models and Dynamic Logic. In this volume.

[27] R. Power, D. Scott, and N. Bouayad-Agha. Document Structure. *Computational Linguistics*, 29(2):211–260, 2003.

[28] A. Renear, E. Mylonas, and D. Durand. Refining our Notion of What Text Really Is: The Problem of Overlapping Hierarchies. In N. Ide and S. Hockey, editors, *Research in Humanities Computing*, pages 263–280. Oxford University Press, Oxford, 1996.

[29] B. B. Rieger. Feasible Fuzzy Semantics. In K. Heggstad, editor, *COLING-78 7th International Conference on Computational Linguistics*, pages 41–43. ICCL, Bergen, 1978.

[30] B. B. Rieger. Fuzzy Word Meaning Analysis and Representation in Linguistic Semantics. In *Proceedings of the 8th International Conference on Computational Linguistics (COLING '80), Tokyo*, pages 76–84, 1980.

[31] B. B. Rieger. Feasible Fuzzy Semantics. On Some Problems of How to Handle Word Meaning Empirically. In H. Eikmeyer and H. Rieser, editors, *Words, Worlds, and Contexts. New Approaches in Word Semantics (Research in Text Theory 6)*, pages 193–209. de Gruyter, Berlin/New York, 1981.

[32] B. B. Rieger. Semantic Relevance and Aspect Dependency in a Given Subject Domain. In D. E. Walker, editor, *COLING '84 - Proceedings 10th International Conference on Computational Linguistics*, pages 298–301, Stanford, 1984. ACL.

[33] B. B. Rieger. *Unscharfe Semantik: Die empirische Analyse, quantitative Beschreibung, formale Repräsentation und prozedurale Modellierung vager Wortbedeutungen in Texten*. Peter Lang, Frankfurt a. M., 1989.

[34] B. B. Rieger. Situation Semantics and Computational Linguistics: Towards Informational Ecology. In K. Kornwachs and K. Jacoby, editors,

Information. New Questions to a Multidisciplinary Concept, pages 285–315. Akademie-Verlag, Berlin, 1995.

[35] B. B. Rieger. Computing Granular Word Meanings. A Fuzzy Linguistic Approach in Computational Semiotics. In P. Wang, editor, *Computing with Words*, pages 147–208. Wiley, New York, 2001.

[36] G. Ruge. *Wortbedeutung und Termassoziation. Methoden zur automatischen semantischen Klassifikation*. Olms, Hildesheim, 1995.

[37] G. Salton. *Automatic Text Processing: The Transformation, Analysis, and Retrieval of Information by Computer*. Addison-Wesley, Reading, Massachusetts, 1989.

[38] H. Schütze. Automatic Word Sense Discrimination. *Computational Linguistics*, 24(1):97–123, 1998.

[39] A. J. C. Sharkey and N. E. Sharkey. Weak Contextual Constraints in Text and Word Priming. *Journal of Memory and Language*, 31(4):543–572, 1992.

[40] P. Thagard. *Coherence in Thought and Action*. MIT Press, Cambridge, 2000.

[41] L. A. Zadeh. Toward a Theory of Fuzzy Information Granulation and its Centrality in Human Reasoning and Fuzzy Logic. *Fuzzy Sets and Systems*, 90:111–127, 1997.

Part III

Quantitative Linguistic Modeling

A Structuralist Framework for Quantitative Linguistics

Stefan Bordag and Gerhard Heyer

Leipzig University
{sbordag,heyer}@informatik.uni-leipzig.de

Summary. Recent advances in the quantitative analysis of natural language call for a theoretical framework that explains, how these advances are possible. This helps to unify different approaches and algorithms in quantitative linguistics. We consider the linguistic tradition of structuralism as a basis for such a framework. In what follows, we focus on syntagmatic and paradigmatic relations and attempt to describe them in a coherent way. We present an abstract version of a (neo-)structuralist language model and show how already known algorithms fit into it. We also show how new algorithms can be derived from it. As has already been predicted by linguists like Firth and Harris, it is possible to construct a computational model of language based on linguistic structuralism and statistical mathematics. The model we propose specifically helps to explain fully unsupervised algorithms for natural language processing which are based on well known methods like co-occurrence measures and clustering.

1 Introduction

One of the fundamental questions in quantitative linguistics is whether it is possible to compute semantic relations between words using statistical methods only. If the answer is yes then other questions immediately arise, like *how* it is possible and, most importantly, *why* it is possible. In this paper we attempt to fuse a linguistic theory, namely structuralism, with a mathematical view of language, cf. Manning & Schütze [29] and Jurafsky & Martin [23], in order to provide a framework on which further research can be based. The language model that will be described and represented by this framework is always derived from a large sample of natural language text (see www.wortschatz.uni-leipzig.de), and never from a hypothetical language. The observed relations and other structural elements should always be rooted in the data and then be made explicit by means of an algorithm that operates on the particular natural language sample. In general, we also presume the distributional hypothesis [18] which has been successfully used within the proposed framework in order to retrieve semantic relations, see [3] and [2]. In

S. Bordag and G. Heyer: *A Structuralist Framework for Quantitative Linguistics*, StudFuzz **209**, 171–189 (2007)
www.springerlink.com   © Springer-Verlag Berlin Heidelberg 2007

fact, the ideas in this paper can be viewed as an attempt to continue Harris' work. For a similar work which suggests another root see Rieger [35]. The global context introduced in the present chapter corresponds to his syntagmatic alpha-abstraction whereas the similarity operation corresponds to his paradigmatic delta-abstraction (or semantic topology). It is important to note that the approach taken here is to view human cognition and language structure as being separable from each other – in contrast to what is proposed in Rieger's approach to *Semiotic Cognitive Information Processing* (SCIP) systems [36]. Nevertheless, it is only through Rieger's work on syntagmatic and paradigmatic learning that these two layers of abstraction can be referred to as the basis of the algorithms modeled here. We restrict ourselves to describe linguistic structures within an assumedly structured sample of text data. As such, the proposed framework might be considered as agreeing with Rieger's approach to SCIP systems [37] with respect to one stratum, namely the language itself.

When dealing with the problem of designing algorithms for computing semantic relations, the first thing to do is to define which relations are going to be calculated. While some semantic relations, like cohyponymy, seem to be more viable than others (e.g. synonymy), it is the salience of the relation and the theoretical foundation of the method to calculate it that is of greatest influence. If the relation is made up arbitrarily, and there are no objective means to differentiate between words using this relation, then, of course, there will never be an algorithm able to calculate them better than a random baseline-algorithm. On the other hand, if a relation is of large impact on language use, (in particular on the co- or non-co-occurrence of words) and if there is a host of hints on how to recognize such words, then algorithms can be constructed that incorporate as many of these hints as possible in order to make reliable predictions. Nevertheless, most of these algorithms will never make 100% correct predictions, which means that other means for improving the results will be necessary. (Without incorporating any further sources of knowledge, the overall effectiveness of the approach can be improved, for example, by combining several algorithms to solve a problem and by making use of a broadened classificator-bagging. Since linguistic relations are often mutually exclusive (e.g a word's synonym never is its antonym at the same time) meta-rules can also effectively help to combine different algorithms and allow the overall results to become more accurate.)

In the following sections, we describe in formal detail some structuralist notions for the analysis of language from which efficient algorithms for the calculation of semantic relations in language can be derived. First, we consider the notion of language levels. Then try to reconstruct de Saussure's notion of *syntagmatic* and *paradigmatic* relations. Finally we discuss how the analysis of syntagmatic and paradigmatic relations might serve as a basis for deriving semantic relations.

2 Language Levels

Language is not a random concatenation of words. Following the famous Swiss linguist Ferdinand de Saussure (who has studied and taught linguistics at the University of Leipzig), meaning (and other notions of linguistic description) can be defined solely by reference to the structural relations existing amongst the words of a language [11]. The reference to the structural relations can be seen as relations between words and therefore this, and especially the distinction between syntagmatic and paradigmatic relations, constitute the basis for the hypothesis that semantic relations can be computed.

One of the first and thus far, unshaken findings in linguistics is that language splits up into several language levels, commonly named phonology, morphology, syntax, and semantics. On each level, there is always the same pair of principles working together which allows combining simple, or atomic, units (like morphs) into complex units (like word forms or phrases), which then represent the simple units for the next level. These two principles are

1. *composition*, and
2. *abstraction* through equivalence classes.

Composition is used to place atoms into a stream of atoms according to some rules, which then represent a complex unit. Abstraction, on the other hand, or, selection in traditional structuralism, allows classifying sets of atoms into equivalence classes of atoms that all have something in common but are somehow distinguishable from all other atoms. These findings seem trivial, but as will be shown later, constitute the cornerstone of a theoretical framework, in which automatic calculation of semantic relations can be explained.

To substantiate the notion of language levels, we define a certain level L_l with $l = \{phones, morphs, wordforms, sentences, texts, \ldots\}$ of a language L to consist of two sets $L_l = (A_l, C_l)$ with A_l being the set of all possible atoms on this level and C_l being the subset of all possible combinations of atoms. Since order matters and there are rules of composition, C_l has to be seen as a set of sets of tuples of various lengths: $C = \{C_{l,1}, C_{l,2}, \ldots, C_{l,n}\}$ (in $C_{l,n}$, l is the level and n is the length of the tuples in this set) with $c_k \in C_{l,n}$ and for $k = 2$ for example $c_2 = \langle a_i, a_j \rangle$. The rules of composition can (and will) be disregarded for certain analyses and algorithms in which case C_l is a set of sets instead of tuples.

It is possible for an atom of a level L_i to also be an atom on level L_{i+n} though it still would have an own representation both on L_i and on L_{i+n}. For example, the atom 'a' as phone on the lowest phonological level can also be an atom on the level of morphology at the same time, and even on the sentences level, as is the case of the interjection 'A!'. However, a rule of composition on a level L_i can modify the rules of composition on a level L_{i-n}. For example, if the word 'yesterday' occurs when constructing a sentence, it is likely that instead of the word 'do' the word 'did' will occur in that sentence.

In order to elucidate the definitions given so far, let us consider some simplified examples. The language to be described is the language L denoting written English. The language levels are

$$l = \{\text{letters}, \text{morphs}, \text{wordforms}, \text{sentences}, \text{texts}\} \tag{1}$$

Notice that it is assumed that for each level there exists a set of permissible complex units (that is a subset of the power set of its constituting atoms, i.e. the set of their possible combinations). In the following sections, we always assume that for each level, the set of permitted complex units can be determined empirically. The framework itself does not provide us with those sets, but defines the principles of how the construction processes may take place, or can be inferred, based on empirical evidence. First, we define the lowest level L_{letters} of L:

$$L = (A_{\text{letters}}, C_{\text{letters}}) \tag{2}$$

where $A_{\text{letters}} = \{a, b, \ldots, z\}$ is the alphabet of the language L and C_{letters} is defined as follows: It is possible to concatenate (as a special case of composition) letters in order to produce strings of the form $\{a, b, \ldots, z\}$ with the length n. For example for $n = 3$:

$$C_{l,3} = \{(a, a, a), (a, a, b), \ldots, (z, z, z)\}. \tag{3}$$

C_{letters} can then be defined as a subset of all possible strings of arbitrary length: $C_{\text{letters}} \subseteq \bigcup_{n>0}\{a_i \,|\, a_i \in A_{\text{letters}}\}$. Thus, strings like 'aa' or 'baaa' or 'aaab' are complex units on the level of letters, though not all combinations of letters will really be observed in a text, thus subset relation.

As the next level we define the level of morphs L_{morphs} of L:

$$L_{\text{morphs}} = (A_{\text{morphs}}, C_{\text{morphs}}) \tag{4}$$

In general, the atoms of a given level are a subset of the complex units of the lower levels. In this case, however, there is only one lower level, so the definition is simple: $A_{\text{morphs}} \subseteq C_{\text{letters}}$. Introducing morphs as surface units of what actually should be the morpheme level allows the methods introduced below to find classifications and relations between morphs. This eventually allows various morphs to be viewed as morphemes. A morpheme would then be a collection of morphs that were found to belong to the same class or share the same attribute. Once such relations are defined and available it is possible to properly reintroduce this level as a morpheme level, instead of a morph level. The same applies to all other levels, such as a word level based on initially a word forms level.

The subset relation describes that the string 'boat' is a morph whereas the string 'aaab' is not a morph (since it has not been observed in the data), both having a length of 4 letters $n = 4$. Not all observed complex units on the level of letters are really morphs, thus it must be a subset relation.

The complex units C_{morphs} of this level can again be viewed as a subset of all possible compositions of the atoms of this level of all possible lengths (of the atoms): $C_{\text{morphs}} \subseteq \bigcup_{n>0} \{a_i \mid a_i \in A_{\text{morphs}}\}$ and is the set of morph combinations. For example, 'boat' is a morphological unit of length 1, 'boatboat' (by doubling the morph 'boat') a morphological unit of length 2, as well as 'boats' (by taking the morph 'boat' and concatenating it with the morph 's'), also a morphological unit of length 2.

The third level of word forms, then, $L_{\text{wordforms}}$ follows in the same way and thus is abbreviated. However, these next levels are more important than the others since these are the ones which are usually most easily observed:

$$L_{\text{wordforms}} = (A_{\text{wordforms}}, C_{\text{wordforms}}) \tag{5}$$

The atoms of this level are the word forms as they occur in written English texts. They are a subset of all possible combinations of morphs of all possible lengths, which excludes 'boatboat' but includes 'boat' and 'boats':

$$A_{\text{wordforms}} \subseteq C_{\text{morphs}} \tag{6}$$

The complex units of this level can be concatenated using additional characters like space, comma and so on as part of the alphabet. The atoms and the complex units on this level are usually (not for example for Chinese or Korean) easily observable since words are separated by spaces and sentences by points, exclamation marks etc.

The fourth level, the level of sentences, $L_{\text{sentences}}$, again follows from its predecessor, the level of word forms:

$$L_{\text{sentences}} = (A_{\text{sentences}}, C_{\text{sentences}}) \tag{7}$$

The atoms on this level are all sentences as they have been encountered in written English texts. They are a subset of the set of all possible combinations of word forms to sentences of all possible lengths:

$$A_{\text{sentences}} \subseteq C_{\text{wordforms}} \tag{8}$$

The complex units of this level can be concatenated using additional characters like a point, an exclamation mark, etc. some of which have already been introduced at the previous level.

In the above example we have described the relations holding between language levels without considering rules of composition, or the meaning of the obtained expression. Rules of composition are based on compliance on the *syntagmatic* level, whereas the meaning of expressions is related to *paradigmatic* relations as described in the following section.

3 Paradigmatic and Syntagmatic Relations

So far, the construction of complex units has been introduced as a subset of all possible combinations of atoms. In general, however, this is not enough to describe the mechanisms of a language.

One of the most important distinctions made by de Saussure is the dichotomy between *syntagmatic* and *paradigmatic* relations. Syntagmatic or paradigmatic relations in a language system relate two atoms that must belong to the same level. Two atoms stand in a syntagmatic relation, only if they are *composed*, or appear together in some expression. On the other hand, two atoms are in a paradigmatic relation, only if they appear in *similar contexts*, or are interchangeable to some extent. Syntagmatic and paradigmatic relations constitute fundamental semantic relationships: Two atoms that appear together comply in function and meaning. An example would be 'torch' and 'shines'. Two atoms that appear in similar contexts have a similar grammar and meaning. Here, simple examples are 'torch' and 'sun'.

Examples of syntagmatic relations on the level of word forms typically include dependencies between nouns and verbs, compounds of nouns and nouns, and head-modifier constructions based on adjectives and nouns, or between nouns and nouns. Syntagmatic relations are often responsible for changes of composition rules on a lower level. Paradigmatic relations vary depending on the measure of similarity presumed. On the word form level, paradigmatic relations range from semantic fields to well defined logical relations such as hyponymy, co-hyponymy, hyperonymy, synonymy and antonymy. It should be noted, however, that the notions of syntagmatic and paradigmatic relations do not pertain only to the level of word forms. By involving the notion of different language levels, it is one of the intentions of the proposed framework to generalize the notion of syntagmatic and paradigmatic relations and to apply it also to language levels other than word forms. In this way, algorithms that have been developed for one language level may be more easily transferred to other levels.

The distinction between syntagmatic and paradigmatic relations follows from the main assumption of structuralism that the value of a certain language element, i.e. atom, can exist only due to the existence of another language elements of the same language. The basic relations between the various language elements are equalities and inequalities – or better, similarities and dissimilarities. Dissimilarities in a language are expressed by opposition or contrast of elements.

"Nicht dass eines anders ist als das andere ist wesentlich, sondern dass es neben allen anderen und ihnen gegenüber steht. Und der ganze Mechanismus der Sprache [...] beruht auf Gegenüberstellungen dieser Art [...]." [11, p. 145].

In order to approach the main aim of this paper, we now define for each level the notion of local context $K_c(a_i)$:

Definition 1. The *local context* $K_c(a_i)$ of a given atom l on a given level $a_i \in A_l$ is the set of all atoms a with which the atom a_i occurs together in a complex unit $c_n \in C$:

$$K_c(a_i) = \{a \mid a \in c_n \wedge a_i \in c_n \wedge c_n \in C \wedge a \neq a_i\}$$

Since an atom occurs n times in a statistical observation, there will be a maximum of n possible contexts of a_i. It is important that, for example on the sentence level two sentences which differ only in the word order, represent the same context because the composition rules, in this case the syntax, and with them, of course, the order of atoms could be disregarded. There are various possibilities of how to instantiate this notion of local context and all of these possibilities might be used with different aims in mind. It is now possible to formalize the notion of a syntagmatic relation $SYN(a_i, a_j)$ between two atoms using the above definition of a local context.

Definition 2. The *syntagmatic relation* $SYN(a_i, a_j)$, which is symmetrical, between two atoms $a_i, a_j \in A_l$ of the same level l, holds true if and only if there exists a local context for one of the atoms in which the other appears:

$$SYN(a_i, a_j) \Longleftrightarrow \exists (K_c(a_i))[a_j \in K_c(a_i)]$$

On a level like the sentence level, where atoms are words, it is possible for a word to co-occur with any other word. Assuming that the number of different sentences, and along with this the cardinality of C in general, is enumerably infinite, this would result in a syntagmatic relation between any two words, or possible atoms in general. To accommodate this objection, we might recall a comment of Wittgenstein in his Tractatus Logico Philosophicus:

"In order to recognize the symbol in the sign we must consider its *significant use*." [45, p. 3.326].

Assuming that the 'significant use' of an atom will be reflected in terms of frequency, there will be an expectation value X that states the absolute number of joint occurrences that must have been observed in relation to the number of all local contexts. To define this value, parameters like frequency of atoms, their Zipfian distribution, the size of the text corpus etc. can be taken into account, see for example [8, 14] or [40]. In effect, for any atom a_i that is part of local contexts of another atom a (with $a_i, a \in A_l$) it can be decided whether or not a_i is a *significant* constituent of these contexts based on the number of such contexts, $SIG(a_i, K_c(a))$:

$$SIG(a_i, K_c(a)) \Longleftrightarrow |\{K_c(a_i)|a_i \in K_c(a)\}| > X \tag{9}$$

Given this relation that an atom is a *significant* constituent of the context of another atom, we can now define the *statistical* syntagmatic relation.

Definition 3. The *statistical syntagmatic* relation $SYNS(a_i, a)$ between two atoms $a_i, a \in A_l$ holds if and only if the atom a_i is a *significant* constituent of the contexts of an atom a:

$$SYNS(a_i, a) \Longleftrightarrow SIG(a_i, K_c(a))$$

This relation corresponds to Rieger's α-abstraction [35] which is assumed between a and a_i if their occurrences correlate. While for the local context of two atoms it was sufficient to consider specific single instances of the complex units of a level, for paradigmatic relations it is necessary to compare the contexts of atoms of one particular level. For this purpose, we introduce the notion of a *global context*.

Definition 4. The *global context* $K_G(a)$ of an atom a of a given level l is the set of all atoms a_i of the same level with which the atom a stands in statistical syntagmatic relation $SYNS(a_i, a)$:

$$K_G(a) = \{a_i \mid SYNS(a_i, a)\}$$

Since the goal was to be able to compare the contexts of a given atom with the contexts of another atom, and the contexts of an atom have been summarized by the notion of the global context, we assume a comparison operator $SIM(K_G(a_i), K_G(a_j))$ that compares global contexts and yields the value 1 in all those cases where the similarity is "similar enough", and 0 in all other cases. In what follows, for simplicity's sake we write $SIM(K_G(a_i), K_G(a_j))$ in this case, the similarity operation on two contexts, yields the value 1. This similarity operation results in a semantic space [35] and, thus, corresponds to Rieger's δ-abstraction.

Examples of comparison operator instances include known similarity measures like the Tanimoto measure, the cosine (global contexts can be interpreted as vectors), Euclidian distance, and many more. The question as to which one is the "best", cannot be answered easily. On the one hand, this is due to evaluation difficulties, on the other hand it is also important to ask for the desired results i.e which paradigmatic relations are to be calculated by some instance of a similarity measure. Details of this discussion are not crucial for the model introduced here, and can be left open for further research. It should be noted, however, that there is no uniform usage of the notion of similarity in the literature. At many instances 'word similarity' is used to refer to a comparison of the global contexts of a given word, whereas sometimes it is also – misleadingly – used to refer to the co-occurrence measure (Def. 3) of statistical syntagmatic relations, as similar words are returned by such a computation [7, 10, 41, 43].

In order to define paradigmatic relations completely, we further need to introduce the notion of linguistic categories. However, as a preliminary step, two atoms a_i and a_j of the same level ($a_i, a_j \in A_l$) can be considered as standing in *paradigmatic relation* $PARA(a_i, a_j)$, only if their global contexts are similar to each other:

$$PARA(a_i, a_j) \Longleftrightarrow SIM(K_G(a_i), K_G(a_j)) \tag{10}$$

When considering antonyms, it might look like a mistake at this point to stipulate that a paradigmatic relation may only hold if the contexts of two atoms are similar to each other, since the meanings of two antonyms should be opposite. In this case however, opposition in meaning is expressed by one or just a few opposite values of features that basically are the same for the antonymous expressions, and is not based on a significant difference in contexts. For example, the words "dim" and "bright" are opposites, but still will be found in the same contexts.

This, and the complete definition of a paradigmatic relation below, also shows that paradigmatic relations can be derived from syntagmatic relations. From the point of view of quantitative linguistics this is substantial, as it justifies our focus on statistical measures of co-occurrence and similarity of contexts.

Another justification for this focus derives from the fact that using instantiations of these definitions it was possible to create working algorithms. One example is the algorithm for unsupervised morpheme boundary detection [4]. It computes a global context for each word based on sentence cooccurrences (local contexts). Then it compares the surface structure of the input word with the most similar words from the global context. This enables the algorithm to correctly detect morpheme boundaries even within irregular word formations.

4 Linguistic Categories

One of the most fundamental distinctions in linguistics, in addition to the definition of composition and abstraction, is the notion of linguistic *categories*, i.e. the distinction of word classes on the word level, vocals and consonants on the phoneme level etc. Traditionally, categories are sets of distribution classes, and as such have been described and validated by subjecting samples of a natural language to so-called substitution tests [16, 15]. The exact degree of differentiation of linguistic categories can vary greatly depending on the language level and the purpose of the categories. On the level of word forms, a basic distinction is between four main categories of words: nouns (N), verbs (V), adjectives (A) and functional words (S). Though rudimentary, this distinction already allows for plenty of possibilities to distinguish between paradigmatic and syntagmatic relations. For example [26] and [27] used categories on the word form level in syntactic functions like 'adj-of', 'subj-of', etc. in order to distinguish between semantic relations. Paradigmatic relations are always distribution classes with respect to one category. Hence, on the level of words, relations between a noun and a noun (NN) can be syntagmatic and paradigmatic, while relations between a noun and a verb (NV), or between an adjective and a noun (AN) can be only syntagmatic.

Any category classification $CLASS(a_i)$ can be introduced as a function that maps any atom a_i of a given level l into a set of values. For the word level, the set of values commonly used is a tag label from a set of labels like $\{A, N, V, S\}$. In general, and for any level, the set of values can simply be a set of numbers:

$$CLASS(a_i) \longrightarrow \{1, 2, 3, 4, \ldots\} \tag{11}$$

Classes and similarities in their function can indeed be found on all levels. On the phoneme level it is the classification of vowels and consonants, as it has been subject to several refinements, see [22, 42] and others. On the morpheme level there are well-known classifications of morphemes, the derivational, inflectional and root morphemes. On the word level there are several classifications of lexical categories, for example [39], whereas on the phrase level concatenations of word forms are classified into noun phrases, verb phrases etc. Finally, there are also classifications on the sentence level, the simplest one being between questions, assertions and exclamations, and more advanced ones such as introduced by RST [30], with nucleus sentences, explanatory sentences etc. The higher the level, the more complicated are the interactions between the elements of category classes and the underlying classes.

Once categories are available, an important precondition for automatically computing syntagmatic and paradigmatic relations is available. The global context of an atom (e. g. a word) – itself being a set of atoms whose contexts are similar to the context of the given atom – will usually contain a variety of category classes. Filtering this set with reference to the category of a particular atom in focus will divide the set into syntagmatic and paradigmatic relations. This way, the final definition of a paradigmatic relation between two atoms can now be expressed as holding only if the two atoms belong to the same level, have similar contexts (this includes also those atoms which are not in the global context but still have similar contexts), and belong to the same category class.

Definition 5. An atom a_i stands in *paradigmatic relation* $PARA(a_i, a)$ with the atom a_j, $a_i, a_j \in A_l$, if and only if their global contexts $K_G(a_i)$ and $K_G(a_j)$ are similar to each other, and the atom a_i belongs to the same category class as the atom a_j:

$$PARA(a_i, a_j) \Longleftrightarrow [SIM(K_G(a_i), K_G(a_j)) \wedge (CLASS(a_i) = CLASS(a_j))]$$

To give some examples from two levels, the word level and the phrase level, consider the contextual sentence "The X shines.". On the word level, word forms like "lamp", "sun", "surface" would fit in here, whereas on the phrase level phrases like "rising morning sun", or "due to a loose contact dangerous desk lamp" might fit. All of these atoms have something in common in that all of them "shine". The implications and relation of this finding to paradigmatic relations will be detailed out in the next paragraphs.

First, however, the problem of how a category class might be obtained automatically has to be discussed. For simplicity's sake, only word classes are focused on as it seems to contradict one of the initial aims of this paper, i.e. to describe a model that is at any point fully computable. The problem of obtaining word classes has up to now been considered unsolvable without supervision (as well as any other category classification on other levels). There are several well-known ways to obtain word classes, mostly based on Hidden Markov Models and supervised learning, which all require some sort of a human-tagged training set, see [5, 6, 9].

A fully unsupervised word class tagger would need a different approach. First of all, it is important to state that for our purposes it is not necessary that the $CLASS(a_i)$ – function uses a mapping onto a set of tags like $\{A, N, V, S\}$, because a set of numbers would suffice equally well. As is obvious from Def. 5, it is only important to know whether the category of two atoms is equal or not. From the implicit definition of the category class of an atom a_i it follows that all atoms of the same class share similarities in how they are used in complex forms, and at the same time contrast against other classes. This can be considered a clustering task where the units to be clustered are atoms, and the features of these atoms are their global contexts (e.g. words from the word level).

A first proof-of-concept implementation of this idea indicates that clustering of word forms according to their significant left and right neighbors (as special cases of global contexts) clusters together primarily words belonging to the same syntactic category. Similar algorithms can be specified for other levels, and the output of an algorithm of one level can be incorporated into the algorithms of other levels in order to improve their performance. This work will be continued and evaluated.

5 Semantic Compliance

After exploring the atoms and their relationships, we now have a look into the mechanics of complex units that are built from atoms. First of all, we assume that all observed complex units are meaningful. Note that in contrast to a truth-functional semantics we do not try to compute whether a sentence is true or false. Rather we try to compute whether it is meaningful or meaningless with respect to a body of knowledge represented by a large sample of texts.

The combinations of complex units from atoms on a certain level make use of implicit rules that obey what might be called *compliances* and depend on semantic relations between the atoms on a particular level. Some compliances (sometimes also called agreements) might involve higher levels – e.g. whether we combine some morphemes into 'done' or 'did' on the level of words depends on the value that the feature TIME has been assigned in the sentence in which the lemma 'to do' is currently used.

In order to explore the possibilities of how semantic relations can be computed, we make a digression and look at the way morpho-syntactic categories like PERSON, GENDER, or NUMBER have been used to explain syntagmatic relations in syntax. The main function of these categories is to explain morpho-

Table 1. Present tense, singular.

PERSON	PRONOUN	VERB
1	ich	geh+e
2	du	geh+st
3	er	geh+t

syntactic compliance on a morphological and sentence/phrase level. In fact, de Saussure describes syntagmatic and paradigmatic relations by way of an example from morphology [11]. Secondary categories like PERSON, GENDER, or NUMBER follow a syntagmatic pattern, which usually combines into tables. Thus, in the following example from German, the pronouns *ich, du, er* are syntagmatically related to the endings of a verb in present tense singular with respect to the secondary category PERSON – see table (1).

Obviously, the paradigm expresses a compliance condition such that a first person pronoun always needs to be complemented by a first person verb-ending, and vice versa; other combinations are not permitted.

Formalizing this relation, we might introduce an attribute PERS applying to pronouns (for simplicity's sake, nouns (N)), and verbs (V), taking as values the natural numbers 1, 2, and 3:

$$\text{PERS}(N) = \{1, 2, 3\}$$
$$\text{PERS}(V) = \{1, 2, 3\}$$

The compliance condition amounts to stipulating that in any meaningful expression in a natural language always:

$$\text{PERS}(N) = \text{PERS}(V).$$

In effect it serves the function of restricting the possible values that a verb ending succeeding a pronoun can take, and will result in a statistically significant syntagmatic relation between the congruent pairs of a verb form and its related pronoun.

The idea now is to generalize the notion of morpho-syntactic categorial compliance sketched above, to introduce an analogous notion of semantic categories, and to exploit this analogy in order to compute semantic categorial compliance. The morpho-syntactic categories result in compliances of atoms in complex units on the word level, which are given through the sentence level. In the same manner, the semantic categories are compliances of atoms on the sentence level, which are given through the next higher level, for example, the text level. In effect, a globally ambiguous word used in a text will usually be used in a meaning which complies with the meaning of the text. This explains the possibility to disambiguate words automatically by using context extracted from the surrounding text, as is evidenced by the extensive literature on this topic [25, 38] or [1]. It also means that in a text only words

Table 2. An example of a semantic compliance table.

Description	Morpho-syntactic category PERSON		Semantic category PROPERTY	
atom	you	read	sun	lamp
syntactic category	N	V	N	N
Morpho-syntactic/ semantic category value	2	2	*shines*	*shines*

that comply with other words in the same text will occur, since otherwise the text would become meaningless. This might be expressed by using a notion of coherence, where, the more compliance there is, the more coherence can be found within a text. However more compliance also implies that less new information is conveyed by a text of the same length. Thus, conveyance of new information must always go along with sufficient coherence in order to be understandable.

Analogous to the morpho-syntactic paradigm tables, semantic compliance tables could now be constructed in the same way:

Similarly as in morphology, the exact name and value of the semantic categories may not be important. In order to automatically establish semantic relations, we only need to be able to distinguish between equalities and inequalities, i.e. compliance, with respect to some abstract categories, and this is only possible by using quantitative methods.

Besides obvious semantic categories like 'LOCATION' or 'PROPERTY', there are also less obvious ones like SEMANTIC ORIENTATION (positive or negative) of words. Hatzivassiloglou & McKeown [19] compute the positive or negative orientation of adjectives exploiting the fact that conjoined adjectives typically have the same orientation in a small training set. Turney [43] computes ORIENTATION by using the web search engine Altavista in order to obtain the occurrences of all words near positive or negative words (i.e. a training set as well). Though such algorithms use a manually created training set, and thus do not fit in the intended paradigm of fully unsupervised algorithms, they will become important once other algorithms become available that generate a small set of proposals for positive/negative algorithms.

It may be interesting to view the development of the notions in this framework from another perspective. Morpho-syntactic categories play a role in the construction of complex units on the morpheme level (thus producing word forms), and are influenced by the sentence level. Semantic categories play a role in the construction of complex units on the word level (thus producing sentences), and are influenced by the text level. To go a step further, it might be interesting to consider which factors influence the construction of complex units on the sentence level (thus producing texts), which is influenced by the level of text-collections. Much work on this issue has been done by Mehler [31], who also introduces the notion of coherence in this context.

6 Structuralist Relations

In addition to semantic categories we encounter a set of *abstract* semantic relations in any sufficiently large corpus of natural language that are not dependent upon the values of certain specific semantic categories. These abstract semantic relations are important for many applications of NLP, and have traditionally been manually encoded in lexical-semantic structures like WordNet [32], GermaNet [24] or [17], and EuroWordNet [44]. One important application of these resources is to infer other relations, cf. Richardson [34].

In WordNet, the collection of semantic relations includes:

- *Hyperonymy* and *hyponymy* (in WordNet called hyperonymy), also sometimes called the is-a-relation. It holds between two words whenever one has a more abstract meaning than the other.
- Two words are *meronyms* whenever one denotes something that is part of the other word's denotation. In WordNet, meronyms are split into several different types: part-of, member-of, substance. The differences between these types depend on the type of the denotation (countable, fluid, etc.).
- *Synonymy* means exchangeability within a sentence without changing the meaning of the sentence. In the logical tradition, true synonymy means exchangeability in all sentences of a language (i.e. the global context of a word). However, this is very rare since it works against the economy of language (to use as few different tokens as possible). Instead, synonymy here means exchangeability in local contexts.
- *Antonymy* means all kinds of oppositions.
- *Derivation* in WordNet covers all cases where one word is morphologically derived from another word; often this implies that the two words belong to different word classes. Derivation is usually viewed as a syntagmatic relation.

For some reason *cohyponymy* has not been included in either WordNet or GermaNet. Deriving this relation from the hyperonymy relation is possible, but sometimes yields errors because hyperonymy has not been coded consistently. In GermaNet the authors have tried to resolve that problem by introducing artificial nodes of non-lexicalized words/meanings. However, this technique has not been used consistently enough to prevent all errors. Furthermore, new errors occur when artificial nodes are created in order to derive cohyponymy relations between words.

Some of these relations are paradigmatic, whereas others are syntagmatic (like derivation). In general this depends on whether or not two words are of the same syntactic category.

Using the framework described above, it is now possible to explicate each of these relations by structuralist means. The following descriptions are not precise yet, but they give an indication of how to use the framework. At the same time they may serve as clues for the development of algorithms that automatically extract these relations from a large corpus as it is the subject of our research.

Hyperonymy is a paradigmatic relation. Two words belong to the same word class and their contexts are similar to each other. It holds then between two words, if their contexts are equal while the significance of the co-occurrence in the corpus is low. For example, sentences may include an abstract categorization of the named objects in order to improve comprehension by redundancy: *"My prairy dog, a rodent, is really cute!"*

Cohyponymy is a paradigmatic relation that holds between two words if they share a common hyperonym as well as a high co-occurrence significance and a high context similarity. This is because there is a large number of semantic categories on which two cohyponyms comply, and hence there will be many similar sentences that contain either the one or the other or both words: *"In our zoo we have elephants, buffalos and giraffes."*.

Synonymy is a paradigmatic relation between two words that holds if there is a significant number of sentences that are similar to each other but differ by the alternation of these two words.

Antonymy is a paradigmatic relation between two words that holds if they comply in all semantic categories, except in one or just a few in which they oppose. In some cases they co-occur in sentences along with negating tokens: "The water is *hot*, not *cold!*". In fact, [28] proposes an unsupervised algorithm that makes use of negating patterns like 'X not Y', 'either X or Y' where X and Y have automatically been determined as similar by means of co-occurrence measurements.

Derivation is a syntagmatic relation. When one word is a derivation of another, they are likely to co-occur but will not necessarily have similar contexts. In order to extract them it will be useful to step one level down and include the morpheme level: *"using"* → *"user"*.

7 Conclusion

We have tried to create a structuralist framework that enables computable definitions for syntagmatic and paradigmatic relations on all language levels. In particular, we have tried to show how this framework creates the possibility to compute semantic relations, and thus to treat semantic relations explicitly as something independent from human intuition. Though being independent from human intuition on the methodology level, each operation is founded within a linguistic theory and is straightforward to grasp, unlike another similar model, i.e. *Latent Semantic Analysis* [12, 13, 33]. The abstract framework is intended to incorporate some of the most important structuralist concepts. By providing precise mathematical models for each specific concept, the framework can be instantiated.

The main mathematical model we have used in our previous work [20, 21], is one for measuring statistical co-occurrences. Further research will continue, in particular towards the full instantiation of the framework accompanied by an evaluation of the resulting algorithms. Here, GermaNet and WordNet are going to be used to evaluate the results of the extraction algorithms for semantic relations. As has been mentioned above, an algorithm for the unsupervised extraction of word classes is crucial for implementing the full computability of the framework. Thus, research will also continue in this direction.

References

[1] S. Banarjee and T. Pedersen. An Adapted Lesk Algorithm for Word Sense Disambiguation Using WordNet. In *CICLing-2002*, 2002.

[2] C. Biemann, S. Bordag, G. Heyer, U. Quasthoff, and C. Wolff. Language-independent Methods for Compiling Monolingual Lexical Data. In *Proceedings of CICLing 2004*, pages 215–228. Springer, 2004.

[3] C. Biemann, S. Bordag, and U. Quasthoff. Automatic Acquisition of Paradigmatic Relations Using Iterated Co-occurrences. In *Proceedings of LREC 2004*, 2004.

[4] S. Bordag. Two-step Approach to Unsupervised Morpheme Segmentation. In *Proceedings of PASCAL 2006*, 2006.

[5] T. Brants. TnT – a statistical part-of-speech tagger. In *Proceedings of the Sixth Applied Natural Language Processing Conference ANLP-2000*, 2000.

[6] E. Brill. A Simple Rule-based Part-of-speech Tagger. In *Proceedings of ANLP*, pages 152–155, 1992.

[7] P. F. Brown, P. V. de Souza, R. L. Mercer, T. J. Watson, V. J. Della Pietra, and J. C. Lai. Class-based n-gram Models of Natural Language. *Computational Linguistics*, 18(4):467–479, 1992.

[8] K. W. Church and W. Gale. Poisson Mixtures. *Journal of Natural Language Engineering*, 1(2):163–190, 1995.

[9] D. Cutting, J. Kupiec, J. Pedersen, and P. Sibun. A Practical Part-of-speech Tagger. In *Proceedings of ANLP*, pages 133–140, 1992.

[10] I. Dagan, L. Lee, and F. C. N. Pereira. Similarity Based Models of Word Cooccurrence Probabilities. *Machine Learning*, 1-3(34):43–69, 1999.

[11] F. de Saussure. *Grundfragen der allgemeinen Sprachwissenschaft*. de Gruyter, c. bally and a. sechehaye (eds.), 3rd edition, 2001.

[12] S. Deerwester, S. T. Dumais, G. W. Furnas, T. K. Landauer, and R. Harshmann. Indexing by Latent Semantic Analysis. *Journal of the American Society for Information Science*, 41(6):391–407, 1990.

[13] S. T. Dumais. Latent Semantic Indexing (LSI): TREC-3 Report. In D. K. Harman, editor, *Overview of the Third Text Retrieval Conference (TREC-3)*, pages 219–230, Gaithersburg, 1995. National Institute of Standards and Technology.

[14] T. E. Dunning. Accurate Methods for the Statistics of Surprise and Coincidence. *Computational Linguistics*, 19(1):61–74, 1993.

[15] G. Grewendorf. Parametrisierung der Syntax. Zur kognitiven Revolution in der Linguistik. In L. Hoffmann, editor, *Deutsche Syntax. Ansichten und Einsichten*, pages 11–73. de Gruyter, 1993.

[16] G. Grewendorf, F. Hamm, and W. Sternefeld. *Sprachliches Wissen. Eine Einführung*. Suhrkamp, 1989.

[17] B. Hamp and H. Feldweg. GermaNet – a Lexical-Semantic Net for German. In *Proceedings of ACL Workshop Automatic Information Extraction and Building of Lexical Semantic Resources for NLP Applications*, 1997.

[18] Z. S. Harris. *Mathematical Structures of Language*. Wiley, 1968.

[19] V. Hatzivassiloglou and K. R. McKeown. Predicting the Semantic Orientation of Adjectives. In *Proceedings of ACL/EACL-97*, pages 174–181, 1997.

[20] G. Heyer, U. Quasthoff, T. Wittig, and C. Wolff. Learning Relations Using Collocations. In A. Maedche, S. Staab, C. Nedellec, and E. Hovy, editors, *Proceedings IJCAI Workshop on Ontology Learning*, 2001.

[21] G. Heyer, U. Quasthoff, and C. Wolff. Knowledge Extraction from Text: Using Filters on Collocation Sets. In *Proceedings of LREC-2002 and IICS 2002*, pages 153–162. Springer, Berlin/New York, 2002.

[22] R. Jakobson. Two Aspects of Language and Two Types of Aphasic Disturbances. In R. Jakobson, editor, *Selected Writings II. Word and Language*, pages 239–259. The Hague, 1956.

[23] D. Jurafsky and J. H. Martin. *Speech and Language Processing*. Prentice Hall, 2000.

[24] C. Kunze and A. Wagner. Integrating GermaNet into EuroWordNet, a Multilingual Lexical-semantic Database. *Sprache und Datenverarbeitung – International Journal for Language Data Processing*, 1999.

[25] M. Lesk. Automatic Sense Disambiguation Using Machine Readable Dictionaries: How to Tell a Pine Cone from an Ice Cream Cone. In *Proceedings of SIGDOC*, pages 24–26, 1986.

[26] D. Lin. Automatic Retrieval and Clustering of Similar Words. In *Proceedings of COLING/ACL-98*, pages 768–774, 1998.

[27] D. Lin. Extracting Collocations from Text Corpora. In *First Workshop on Computational Terminology*, 1998.

[28] D. Lin, S. Zhao, L. Qin, and M. Zhou. Identifying Synonyms among Distributionally Similar Words. In *Proceedings of the 18th International Joint Conference on Artificial Intelligence (IJCAI)*, 2003.

[29] C. Manning and H. Schütze. *Foundations of Statistical Natural Language Processing*. MIT Press, 1999.

[30] D. Marcu. *The Theory and Practice of Discourse Parsing and Summarization.* MIT Press, 2000.

[31] A. Mehler. Hierarchical Orderings of Textual Units. In *Proceedings of the 19th International Conference on Computational Linguistics, COLING'02, Taipei,* pages 646–652, San Francisco, 2002. Morgan Kaufmann.

[32] G. A. Miller. WordNet: An online lexical database. *International Journal of Lexicography,* 4(3):235–312, 1990.

[33] C. H. Papadimitriou, H. Tamaki, P. Raghavan, and S. Vempala. Latent Semantic Indexing: A Probabilistic Analysis. In *Proceedings of the Seventeenth ACM SIGACT-SIGMOD-SIGART Symposium on Principles of Database Systems, June 01-04, Seattle,* pages 159–168. ACM, 1998.

[34] S. D. Richardson. *Determining Similarity and Inferring Relations in a Lexical Knowledge Base.* PhD thesis, The City University of New York, 1997.

[35] B. B. Rieger. *Unscharfe Semantik. Die empirische Analyse, quantitative Beschreibung, formale Repräsentation und prozedurale Modellierung vager Wortbedeutungen in Texten.* Peter Lang, Bern/Frankfurt/New York, 1989.

[36] B. B. Rieger. Distributed Semantic Representations of Word Meanings. In J. D. Becker, I. Eisele, and F. W. Mündemann, editors, *Parallelism, Learning, Evolution. Proceedings of the Workshop on Evolutionary Models and Strategies and of the Workshop on Parallel Processing (WOPPLOT'89),* pages 243–273, Berlin/New York, 1991. Springer.

[37] B. B. Rieger. Situations, Language Games, and SCIPS. Modeling Semiotic Cognitive Information Processing Systems. In J. Albus, A. Meystel, D. Pospelov, and T. Reader, editors, *Architectures for Semiotic Modeling and Situation Analysis in Large Complex Systems. Proceedings of the ISIC-Workshop and the 10th International IEEE-Symposium on Intelligent Control,* pages 130–138, Bala Cynwyd, 1995. AdRem.

[38] M. Sanderson. Word Sense Disambiguation and Information Retrieval. In *Proceedings of the 17th ACM SIGIR Conference,* pages 142–151. ACM, 1996.

[39] A. Schiller, S. Teufel, and C. Thielen. Guidelines für das Taggen deutscher Textcorpora mit STTS. Technical report, IMS-CL, Universität Stuttgart and SfS, Universität Tübingen, 1995.

[40] P.-N. Tan, V. Kumar, and J. Srivastava. Selecting the Right Interestingness Measure for Association Patterns. In *Proceedings of ACM SIGKDD Conference on Knowledge Discovery and Data Mining,* pages 32–41. ACM, 2002.

[41] E. Terra and C. L. A. Clarke. Frequency Estimates for Statistical Word Similarity Measures. In *Proceedings of HLT-NAACL 2003,* pages 165–172, 2003.

[42] N. S. Trubetzkoy. *Grundzüge der Phonologie.* Travaux du Cercle Linguistique de Prague 7. Kraus, Nendeln, 1939.

[43] P. D. Turney. Thumbs Up or Thumbs Down? Semantic Orientation Applied to Unsupervised Classification of Reviews. In *Proceedings of ACL-02*, pages 417–424, 2002.

[44] P. Vossen. Introduction to EuroWordNet. *Special Issue on EuroWordNet of Computers and the Humanities*, 32(2-3):73–89, 1998.

[45] L. Wittgenstein. Tractatus Logico-Philosophicus. Frankfurt a. M., 2003.

Quantitative Analysis of Syntactic Structures in the Framework of Synergetic Linguistics

Reinhard Köhler

University of Trier
koehler@uni-trier.de

Summary. This paper describes a first attempt to set up and test a basic synergetic-linguistic model of a syntactic subsystem in analogy to the existing models of lexical and morphological subsystems, which have been tested successfully. The modelling is based on selected syntactic units, properties, and interrelations, which are integrated into a common model structure. The empirical testing is performed on data[1] from the "SUSANNE" corpus [15].

1 Introduction

A first, embryonic model of a syntactic subsystem of language in the framework of synergetic linguistics [9] is set up in analogy to the existing models of lexical [2, 3, 9] and morphological [10, 11, 12, 13, 14] subsystems. As basic units, *syntactic constructions* are selected, which are, for the purpose of the present study, operationalised on the basis of the *constituency relation*, i.e., we consider *constituent types*. The properties analysed are

- *Frequency* (of occurrence in the text corpus),
- *Length* (number of the terminal nodes [= words] which belong to the given constituent),
- *Complexity* (number of immediate constituents of the constituent under consideration),
- *Position* (in the mother constituent or in the sentence, counted from left to right),
- *Depth* of embedding (number of production steps from the start symbol),
- *Information* (in the sense of information theory, corresponding to the memory space needed for the temporary storage of the grammatical relations of the constituent)

---

[1] I would like to thank Claudia Prün and Sabine Weber for their help with the extraction of the data from the corpus.

R. Köhler: *Quantitative Analysis of Syntactic Structures in the Framework of Synergetic Linguistics*, StudFuzz **209**, 191–209 (2007)
www.springerlink.com
© Springer-Verlag Berlin Heidelberg 2007

- *Polyfunctionality* (number of different functions of the construction under consideration),
- *Synfunctionality* (number of different functions with which a given function shares a syntactic representation)

and the relevant inventories, viz.

- the inventory of *syntactic constructions* (constituent types),
- the inventory of *syntactic functions*,
- the inventory of *syntactic categories*,
- the inventory of *functional equivalents* (i.e., of constructions with a similar function to the one under consideration).

2 Frequency, Complexity, and Length

The first step on the way to a model in the framework of the synergetic approach consists in setting up axioms. From earlier works (cf., e.g., Köhler [9, 10] and Hoffmann & Krott [7]) we take, together with the general central axiom of self-organisation and self-regulation of language systems, the communication requirement (Com) with its two aspects of the coding (Cod) and the application requirement (Usg). Further language-external requirements the system must meet are introduced below. The next step includes the search for functional equivalents which can meet the requirements, and the determination on their effects on other system variables. The influences of Cod, of which we consider here only that part which is connected with *syntactic* coding means as a functional equivalent, directly affect the inventory size of syntactic constructions (in perfect analogy to the lexical subsystem where lexicon size is affected by Cod). In a similar analogy to the situation in the lexicon, Usg represents the communicative relevance of an expression in the inventory and results in a corresponding frequency of application of the given construction (compare figure 1).

Before entering the next phase – the empirical testing of the hypotheses set up above – we introduce another axiom, viz. the requirement of *optimal coding* (OC), as known from earlier models, with two of its aspects: the requirement of minimising production effort (minP), and the requirement of maximisation of compactness (maxC). "Production effort" refers to the physical effort which is associated with the articulation while uttering an expression. In the case of syntactic constructions, this effort is determined by the number of terminal nodes (words) – even if the words are of different lengths[2] – and here is called

---

[2]The actual mean effort connected with the utterance of a syntactic construction is indirectly given by the number of its words and on the basis of the word length distribution (in syllables) and the syllable length distribution (in sounds). One has also to keep in mind, however, the influence of the Menzerath-Altmann law, which is, for the sake of simplicity, neglected here.

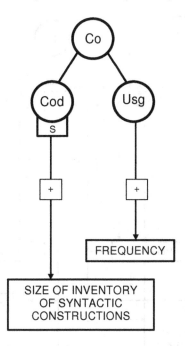

Fig. 1. The language-constituting requirement Cod (only syntactic coding means are considered) and the language-forming requirement Usg with two of their depending variables.

length of a syntactic construction. As in the case of lexical units, minP affects the relation between frequency and length, in that maximal economisation is realised when the most frequent constructions are the shortest ones (cf. figure 2). As a consequence, an optimised distribution of the frequency classes and a corresponding rank-frequency distribution can be expected, in a form similar – though probably not identical – to Zipf-Mandelbrot's law. There is, undoubtedly, an effect of shortening syntactic constructions in dependence on their frequency; however, this interrelation should be explained in the first place by the preferential application of shorter constructions over longer ones.

According to the data from the SUSANNE corpus, these distributions display, in fact, the expected forms (cf. figure 3). The well-known Waring distribution could be fitted to the empirical frequency spectrum (fitting with the Altmann Fitter 2.0 (1997) yielding the parameter estimations $b = .66990$ and $n = .47167$; the result of the Chi-square test was $\chi^2 = 81.0102$ with 85 degrees of freedom and a probability of $P[\chi^2] = .6024$), which is extremely good.

The requirement maxC is, among others, a consequence of the need for minimisation of production effort at the mother constituent level. This requirement can be met at the sentence level e.g. by an additional attribute instead of

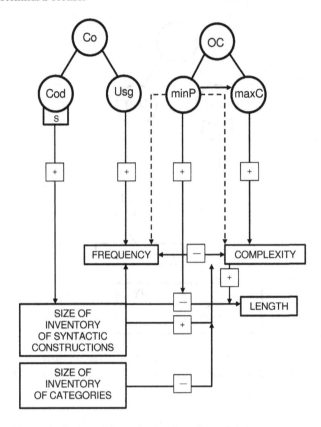

Fig. 2. The interrelation of complexity/length and frequency as a consequence of the requirement of optimal coding. The dotted lines represent the effect of minP as an order parameter for the distributions of the frequency and complexity classes.

a subordinate clause[3], with the effect that this economisation at the sentence level is achieved at the expense of an increased complexity. Length (measured in words), on the other hand, is stochastically proportional to complexity: The more immediate constituents a construction contains, the more terminal nodes it will consist of.

The *average* complexity of the syntactic constructions finally depends on the number of the necessary constructions in the inventory and on the number of elementary syntactic categories. This dependence results from simple com-

---

[3] An example: $_S[_{NP}[$Die Hörer$]$ konnten nichts verstehen, weil sie wieder einmal nicht vorbereitet waren$]] \rightarrow {}_S[_{NP}[$Die wieder einmal nicht vorbereiteten Hörer$]$ konnten nichts verstehen$]$. The first sentence has a length of 12 words, the second one only 9. On the other hand, the subject of the first sentence has only two immediate constituents and a length of 2, the subject of the second three immediate constituents and a length of 6.

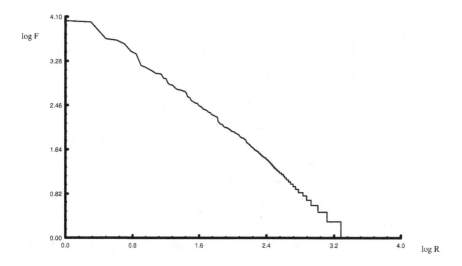

Fig. 3. The rank-frequency distribution of the constituent type frequencies in the SUSANNE corpus (logarithmic axes).

binatorial analysis: Every construction consists of a linear sequence of daughter nodes (immediate constituents) and is determined by their categories and their order. On the basis of G categories, G^K different constructions with K nodes can be generated, of which only a part – the "grammatical" one – is actually formed, in analogy to the only partial use of the principally possible phoneme (sound) combinations in the formation of syllables or morphemes – to phonotactic restrictions. Figure 4 shows the complexity distribution of all 90,821 occurrences of constituents in the SUSANNE corpus.

The empirical test of the hypotheses on the interrelation between frequency and complexity and complexity and length is shown in figures 5, 6, and 7.

Obviously, the expected tendencies are confirmed. Though there are not yet any theoretically corroborated hypotheses on the exact mathematical form of the interrelations (with the exception of the hypothesis on the dependence of length on complexity) and, therefore, no serious goodness-of-fit test can be performed, the general hypothesis on the existence of an inverse dependence is justified by the data.

The findings described above also possess a potentially important practical impact. Of the 4,621 different constituent types with their 90,821 occurrences, 2,710 types (58.6%) occur only once in the corpus, 615 types (32.3% of the rest, or 13.3% of the whole inventory) occur twice, 288 types (22.2% of the rest, or 6.2% of the inventory) thrice, 176 (17.5% of the rest, or 3.8% of the inventory) four times, etc. Less than 20% of the rules in the corresponding grammar can be applied more than four times, and less than 30% of the rules more than two times.

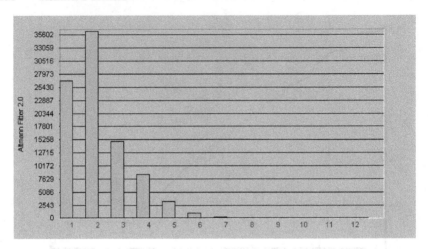

Fig. 4. The empirical frequency distribution of constituent complexity in the SU-SANNE corpus.

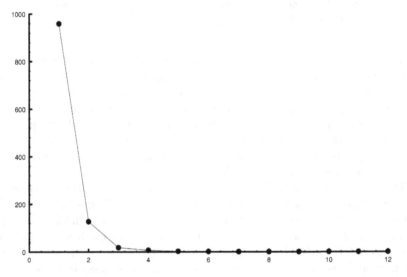

Fig. 5. The empirical dependence of the average constituent frequency as a function of constituent complexity. Fitting of the function $F = 858.83K^{-3.095}e^{.00727K}$, determination coefficient $D = .99$.

We can expect that investigations of other corpora and of languages other than English will yield comparable results. Similarly to how lexical frequency spectra are applied to problems of language learning and teaching, in compiling minimal vocabularies, in determining the text coverage of dictionaries, etc., the dependence described above could be taken into account, among others, when parsers are constructed (planning the degree of text coverage,

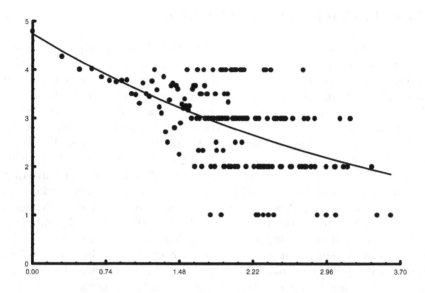

Fig. 6. The empirical dependence of the average constituent complexity as a function of constituent frequency (logarithmic x-axis). Fitting of the function $F = 4.789F^{-.1160}$, determination coefficient $D = .331$.

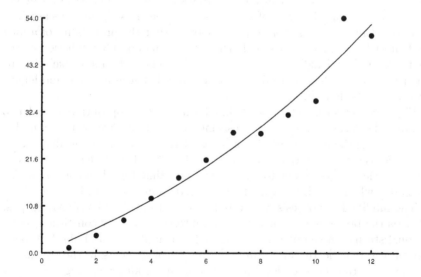

Fig. 7. The empirical dependence of complexity and length. Fitting of the function $L = 2.603K^{.963}e^{.0512K}$ determination coefficient $D = .960$.

estimation of the expenditure needed for setting up the rules, calculation of the degree of a text which can be automatically analysed, etc.).

3 Length, Complexity, and Position

A relation which has already been found by Otto Behaghel [1] is the *Gesetz der wachsenden Glieder* (the "law of growing members") as he called it: "Von zwei Gliedern von verschiedenem Umfang steht das umfangreichere nach" (Of two members of different sizes the larger one comes latest). Behaghel tested this empirical generalisation on data from German, Latin and Greek. Since then, word order variation has been considered, in the first place, from the point of view of typology. In linguistics, theme-rheme division and topicalisation as a function of syntactic coding by means of word order have to be mentioned and, in contrast, Givón's discourse pragmatic "the most important first" principle. An interesting and plausible hypothesis by John Hawkins [4], which can explain the preference "long after short" as observed by Behaghel and confirmed again by Hawkins on data from German, English, and Hungarian, is based on psycho-linguistic assumptions on mechanisms and border conditions of human language processing. Hawkins' Early Immediate Constituent (EIC) principle explains the data by the fact that fewer nodes of the syntactic structure have to be kept in memory during analysis if long constituents are placed after shorter ones which are grammatically of the same status.

Hoffmann [5] performed several empirical tests of consequences of this hypothesis, since the data collected by Hawkins were evaluated in an inadequate way (no statistical test of significance has been used; Hawkins assessed the data intuitively). Furthermore, she showed that the probability of a long constituent being placed after a shorter one is a monotonous function of the difference of their lengths: the greater the difference, the more likely is this constituent order. Recent studies on the relation between position and length of words can be found in [17] and [16].

Figure 8 shows this interrelation in a form which is appropriate in order to integrate the corresponding hypothesis into a comprehensive syntactic model. Instead of length (in number of words) as in Hawkins [4], complexity is considered as the relevant quantity, because the hypothesis refers to nodes in the syntactic structure, not to words. The fact that the phenomenon is also observable when length in words is considered seems to be an indirect effect.

The modified hypothesis was tested on data from the SUSANNE corpus. Whereas the previous investigations took into account only constituent pairs of 'equal status', in the present study, length, complexity, and absolute position data were collected and evaluated for all constituents in the corpus, in two ways: at the sentence level and recursively at all levels. Figures 9 and 10 show examples of the empirically observed interrelations. As can be seen, the hypothesis was extremely clearly confirmed; hence, a significance test is unnecessary, the more because a theoretical reason to postulate a specific function, which could be fitted to the data and be tested in this way, is not yet available. A corresponding formula will also have to take into account and combine, besides Hawkins' considerations, other interrelations and fac-

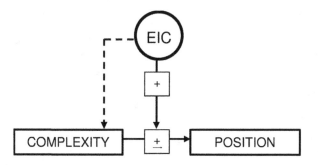

Fig. 8. Hawkins' EIC principle (modified: complexity instead of length).

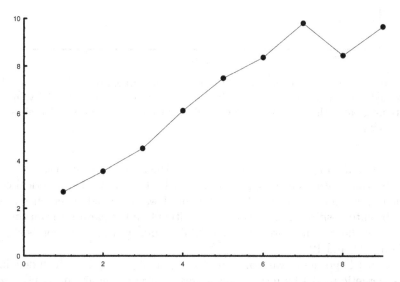

Fig. 9. The empirical dependence of the average constituent length (in number of words) and position in the mother constituent. The values of positions greater than 9 have not been taken into account because of their small frequency (< 10).

tors such as Givón's principle ("the most important first") and quantitative iconicity (e.g., Hai-man's "the more important the more linguistic material").

4 Position and Depth of Embedding

As another hypothesis for which data can easily be collected, a consequence of Yngve's Depth Saving principle [18] is integrated into the model. If in fact right branching structures are preferred due to memory efficiency in language

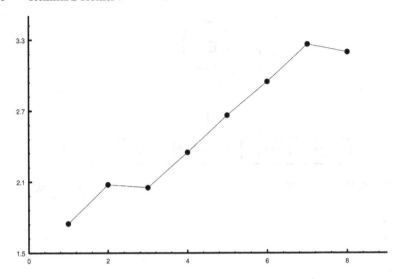

Fig. 10. The empirical dependence of the average constituent complexity (in number of immediate constituents) and position in the mother constituent. The values of positions greater than 8 have not been taken into account because of their small frequency (< 10).

processing, *all* constituents should show, on the average, an increasing depth of embedding with increasing position. In order to test this hypothetical consequence, depth of embedding[4] (depth value 1 was assigned at sentence level) and absolute position (in the mother constituent and, separately, from the beginning of the sentence) were evaluated. The empirical interrelation is shown in figures 11 and 12.

As this hypothesis was corroborated by a first test, it is included into the synergetic model by introducing a further system requirement, viz. *right branching preference* (RB), which controls the influence of constituent position on depth. Additionally, another axiom is set up which represents the necessary *limitation of the increase of depth* (LD) – an order parameter of the distribution of the variable depth. The three requirements EIC, RB, and LD can be subsumed under the general requirement of minimisation of memory effort (minM). Here, we also have to take into account that the requirement of maximal compactness (maxC) has an effect opposite to the depth limitation requirement, because more compactness is achieved by embedding of constituents, as discussed above. Figure 13 shows a model section which corresponds to the relations described above.

---

[4]Note that our operationalisation of depth differs from Yngve's.

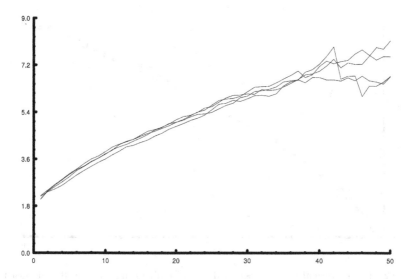

Fig. 11. The empirical dependence of depth of embedding on constituent position (measured in running words from the beginning of the sentence) for the four text types included in the SUSANNE corpus. Positions above 50 are not represented in the graph because of their small frequencies.

5 Position and Information

Another consideration is possible with respect to the position of the constituents: With increasing position, an increasing number of constituents has to be stored during the processing by the hearer or reader. Therefore, an increasing memory effort goes along with constituent position. If we assume that the language processor's memory has to store not only the nodes as representatives of the constituents but also the structural (grammatical) information which results from the analysis [8], it would be advantageous if the amount of structural information to be stored would grow slower than proportionally to the number of nodes. This would be possible if the number of allowed constituent types and functions decreases with increasing position. The more alternatives are permissible at a given position, the more storage space is needed for the representation of the constituent type of function actually found at that position. If this consideration is correct and if the self-organising mechanisms of the natural languages adjust their grammars (i.e., as a result of economical language behaviour) in this way (in order to allow for a higher complexity of constituents to be analysed by the processing device), then the logarithm of the number of alternative constituent types and functions which can actually be found at a given position in a text corpus decreases with the position. The logarithm is taken as a measure for the memory space needed. Figure 14 shows a model structure with this hypothesis integrated; the cor-

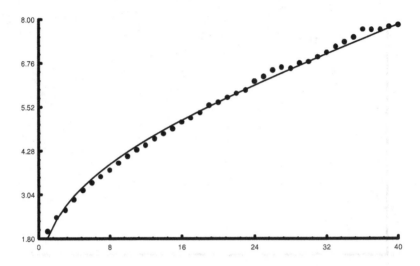

Fig. 12. The empirical dependence of depth of embedding on constituent position (measured in running words from the beginning of the sentence) for the entire SUSANNE corpus. Positions above 40 are not represented in the graph because of their small frequencies. Fitting of the function $T = 1.8188P^{3.51}e^{.00423P}$, coefficient of determination $D = .996$.

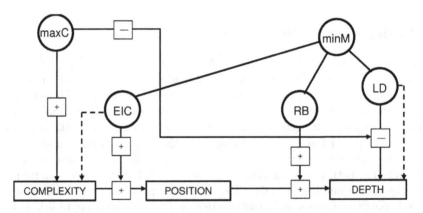

Fig. 13. Model section containing the quantities complexity, position, and depth with the relevant requirements.

responding requirement of minimisation of structural information (minS) is also a special aspect of the general requirement minM.

For the empirical test again two data sets were considered, one containing all constituents at sentence level, another all constituents at all levels. For every position in the mother constituent the number of different immediate follower constituent types was counted. At the sentence level, with increasing position, a nearly linear decrease of the corresponding information (the logarithm of the number of alternative followers) was found – beginning from a peak at position three, the follower of the favourite position of the finite verb (which is, according to the analysis in the SUSANNE corpus, an imme- diate constituent of the sentence). The curves of the functional dependence of structural information in two text types is represented in figure 15.

In the SUSANNE corpus, constituents are annotated with respect to their type and, in most cases, also to their grammatical function [15]. Therefore, it was also possible to measure the amount of information which is yielded from the number of alternative functions at a given position. An example of the results is shown in figures 16 and 17 – again an almost linear curve.

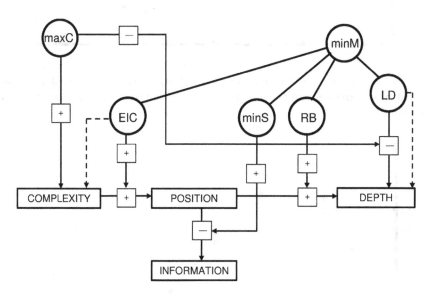

Fig. 14. With increasing position, the amount of structural information which must be stored increases.

At present, specific hypotheses providing the exact mathematical form of the functional dependencies and their interrelations have not yet been devel- oped. Such hypotheses must be set up on the basis of theoretical deduction from assumptions on the human language processing mechanism and take into account linguistic laws such as the Menzerath-Altmann [8], where this law is derived from a simple assumption about properties of the language processing device).

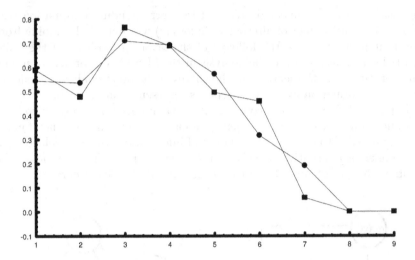

Fig. 15. Logarithm of the number of alternatively possible constituent types in dependence on the position (separately calculated for two of the four text types in the corpus).

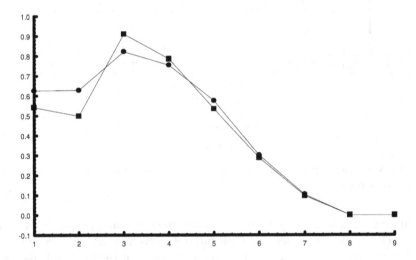

Fig. 16. Logarithm of the number of alternatively possible constituent functions in dependence on the position (separately calculated for two of the four text types in the corpus).

6 Other Quantities and Requirements

As in earlier models of linguistic subsystems, a requirement of minimisation of inventory size (minI) is postulated. In a syntactic subsystem, at least the

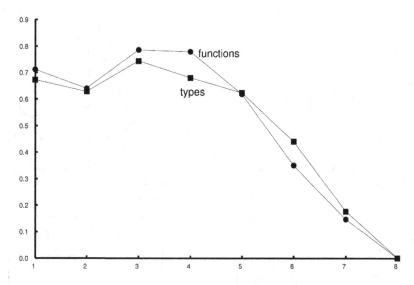

Fig. 17. Logarithm of the number of alternatively possible constituent types and functions in dependence on the position (separately calculated for an individual text).

following interrelations between inventories and other system variables must be investigated:

- An increase in the size of the inventory of syntactic constructions has, as discussed above, an increasing effect on the mean complexity of the constructions, whereas mean complexity is smaller the larger the inventory of categories. The smaller the inventory of categories the greater the functional load (or, multifunctionality). The requirement minI has a decreasing effect on all inventories, among others on the mean number of functional equivalents associated to a construction.
- The frequency distributions within the inventories are controlled by order parameters (figure 18).
- Theoretical and empirical analyses of the probability distributions of the quantities under consideration (frequency, length, complexity, position, depth, and information) and the analysis of the rank-frequency distributions of functional equivalents and of multifunctionality of a given construction will be given in a separate publication, because the empirical frequency distributions of the system variables and the distributions of functional diversification (syn-functionality) and of multifunctionality display an extremely heterogeneous behaviour and a large number of mathematical models is needed to cover the phenomena observed.

The model developed in the present paper is, as emphasised above, only a first attempt to analyse a part of the syntactic subsystem of language in the

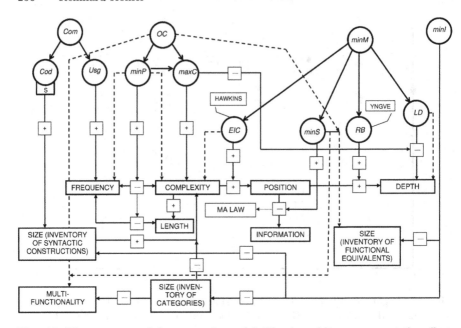

Fig. 18. The structure of the syntactic model. The dotted lines represent the effect of order parameters on distributions of system variables.

framework of synergetic linguistics and remains, at the moment, incomplete in several respects. In the first place, a theoretical derivation of the mathematical form of some of the functional interrelations and the frequency distributions is missing. Besides extensions of the model by further units and properties, a broader empirical basis and studies of data from other languages than English is needed. A particularly interesting question is the relation between the model structure described in this draft analysis and the Menzerath-Altmann law.

References

[1] O. Behaghel. Von deutscher Wortstellung. *Zeitschrift für Deutschkunde*, 44:81–89, 1930.

[2] K. Gieseking. Untersuchungen zur Synergetik der englischen Lexik. In R. Köhler, editor, *Korpuslinguistische Untersuchungen zur quantitativen und systemtheoretischen Linguistik*. Gardez! Verlag, Sankt Augustin, 2003.

[3] R. Hammerl. *Untersuchungen zur Struktur der Lexik: Aufbau eines Basismodells*. Wissenschaftlicher Verlag, Trier, 1991.

[4] J. Hawkins. *A Performance Theory of Order and Constituency*. Cambridge University Press, Cambridge, 1994.

[5] C. Hoffmann. Quantitativ-funktionalanalytische Untersuchungen zur Wortstellungsvariation. Master's thesis, Trier, 1996.

[6] C. Hoffmann. Word Order and the Principle of "Early Immediate Constituents" (EIC). *Journal of Quantitative Linguistics*, 6(2):108–116, 1998.

[7] C. Hoffmann and A. Krott. Einführung in die synergetische Linguistik. In R. Köhler, editor, *Korpuslinguistische Untersuchungen zur quantitativen und systemtheoretischen Linguistik*. Gardez! Verlag, Sankt Augustin, 2003.

[8] R. Köhler. Zur Interpretation des Menzerathschen Gesetzes. In *Glottometrika*, volume 6, pages 177–183. Brockmeyer, Bochum, 1984.

[9] R. Köhler. *Zur linguistischen Synergetik. Struktur und Dynamik der Lexik*. Brockmeyer, Bochum, 1986.

[10] R. Köhler. Elemente der synergetischen Linguistik. In *Glottometrika*, volume 12, pages 179–187. Brockmeyer, Bochum, 1990.

[11] R. Köhler. Synergetik und sprachliche Dynamik. In W. A. Koch, editor, *Natürlichkeit der Sprache und Kultur*, pages 96–112. Brockmeyer, Bochum, 1990.

[12] R. Köhler. Diversification of Coding Methods in Grammar. In U. Rothe, editor, *Diversification Processes in Language*, pages 47–55. Rottmann Medienverlag, Hagen, 1991.

[13] A. Krott. Some Remarks on the Relation between Word Length and Morpheme Length. *Journal of Quantitative Linguistics*, 3:29–37, 1996.

[14] A. Krott. Ein funktionalanalytisches Modell der Wortbildung. In R. Köhler, editor, *Korpuslinguistische Untersuchungen zur quantitativen und systemtheoretischen Linguistik*. Gardez! Verlag, Sankt Augustin, 2003.

[15] G. Sampson. *English for the Computer*. Clarendon Press, Oxford, 1995.

[16] L. Uhlíová. O vztahu mezi délkou slova a jeho polohou ve vìte. Slovo a slovesnost. (to appear).

[17] L. Uhlíová. Length vs. Order. Word Length and Clause Length from the Perspective of Word Order. *Journal of Quantitative Linguistics*, 4:266–275, 1997.

[18] V. Yngve. A Model and an Hypothesis for Language Structure. *Proceedings of the American Philosophical Society*, 104:444–466, 1960.

Appendix

The text corpus used is the SUSANNE corpus (cf. Sampson [15]) – a collection of 64 English texts with a total of 128,000 running words, which is available in a syntactically analysed and annotated form. As an example, the first sentence of text A01 from the SUSANNE corpus is shown below. The organisational and linguistic information for each word-form is given in six columns: The first column (reference field) gives a text and line code, the second (status

field) marks abbreviations, symbols, and misprints, the third gives the wordtag
according to the Lancaster tagset, the fourth the word form from the raw text,
the fifth the lemma, and the sixth the parse. In lines A01:0040j and A01:0050d
(example in figure 1), for example, the :o's mark the NP "the over-all ... of the
election" as logical direct object, the brackets with label Fr in lines A01:0060h
and A01:0060n mean that "in which ... was conducted" is a relative clause.

A01:0010a	- YB	\<minbrk\>	-	[Oh.Oh]
A01:0010b	- AT	The	the	[O[S[Nns:s.
A01:0010c	- NP1s	Fulton	Fulton	[Nns.
A01:0010d	- NNL1cb	County	county	.Nns]
A01:0010e	- JJ	Grand	grand	.
A01:0010f	- NN1c	Jury	jury	.Nns:s]
A01:0010g	- VVDv	said	say	[Vd.Vd]
A01:0010h	- NPD1	Friday	Friday	[Nns:t.Nns:t]
A01:0010i	- AT1	an	an	[Fn:o[Ns:s.
A01:0010j	- NN1n	investigation	investigation	.
A01:0020a	- IO	of	of	[Po.
A01:0020b	- NP1t	Atlanta	Atlanta	[Ns[G[Nns.Nns]
A01:0020c	- GG	+\<apos\>s	-	.G]
A01:0020d	- JJ	recent	recent	.
A01:0020e	- JJ	primary	primary	.
A01:0020f	- NN1n	election	election	.Ns]Po)Ns:s]
A01:0020g	- VVDv	produced	produce	[Vd.Vd]
A01:0020h	- YIL	\<ldquo\>	-	.
A01:0020i	- ATn	+no	no	[Ns:o.
A01:0020j	- NN1u	evidence	evidence	.
A01:0020k	- YIR	+\<rdquo\>	-	.
A01:0020m	- CST	that	that	[Fn.
A01:0030a	- DDy	any	any	[Np:s.
A01:0030b	- NN2	irregularities	irregularity	.Np:s]
A01:0030c	- VVDv	took	take	[Vd.Vd]
A01:0030d	- NNL1c	place	place	[Ns:o.Ns:o]Fn]Ns:o]Fn:o]S]
A01:0030e	- YF	+.	-	.O]
A01:0030f	- YB	\<minbrk\>	-	[Oh.Oh]
A01:0030g	- AT	The	the	[O[S[Ns:s.
A01:0030h	- NN1c	jury	jury	.Ns:s]
A01:0030i	- RRR	further	far	[R:c.R:c]
A01:0030j	- VVDv	said	say	[Vd.Vd]
A01:0030k	- II	in	in	[P:p.
A01:0030m	- NNT1c	term	term	[Np[Ns.

```
A01:0030n - YH       +<hyphen>          -            .
A01:0030p - NN1c     +end               end          .Ns]
A01:0040a - NN2      presentments       presentment  .Np]P:p]
A01:0040b - CST      that               that         [Fn:o.
A01:0040c - AT       the                the          [Nns:s101.
A01:0040d - NNL1c    City               city         .
A01:0040e - JB       Executive          executive    .
A01:0040f - NNJ1c    Committee          committee    .
A01:0040g - YC       +,                 -            .
A01:0040h - DDQr     which              which        [Fr[Dq:s101.Dq:s101]
A01:0040i  - VHD     had                have         [Vd.Vd]
A01:0040j  - JB      over<hyphen>all    overall      [Ns:o.
A01:0050a - NN1n     charge             charge       .
A01:0050b - IO       of                 of           [Po.
A01:0050c - AT       the                the          [Ns.
A01:0050d - NN1n     election           election     .Ns]Po]Ns:o]
A01:0050e - YC       +,                 -            .Fr]Nns:s101]
A01:0050f  - YIL     <ldquo>            -            .
A01:0050g - VVZv     +deserves          deserve      [Vz.Vz]
A01:0050h - AT       the                the          [N:o.
A01:0050i  - NN1u    praise             praise       [NN1n&.
A01:0050j  - CC      and                and          [NN2+.
A01:0050k - NN2      thanks             thank        .NN2+]NN1n&]
A01:0050m - IO       of                 of           [Po.
A01:0050n - AT       the                the          [Nns.
A01:0060a - NNL1c    City               city         .
A01:0060b - IO       of                 of           [Po.
A01:0060c - NP1t     Atlanta            Atlanta      [Nns.Nns]Po]Nns]Po]N:o]
A01:0060d - YIR      +<rdquo>           -            .
A01:0060e - IF       for                for          [P:r.
A01:0060f  - AT      the                the          [Ns:103.
A01:0060g - NN1c     manner             manner       .
A01:0060h - II       in                 in           [Fr[Pq:h.
A01:0060i  - DDQr    which              which        [Dq:103.Dq:103]Pq:h]
A01:0060j  - AT      the                the          [Ns:S.
A01:0060k - NN1n     election           election     .Ns:S]
A01:0060m - VBDZ     was                be           [Vsp.
A01:0060n - VVNv     conducted          conduct      .Vsp]Fr]Ns:103]P:r]Fn:o]S]
A01:0060p - YF       +.                 -            .O]
```

Latent Connotative Text Structure

Arne Ziegler[1] and Gabriel Altmann[2]

[1] Karl-Franzens Universität Graz
   arne.ziegler@uni-graz.at
[2] Ruhr-Universität Bochum
   RAM-Verlag@t-online.de

1 Introduction

A special kind of analysis of the denotative structure of a text can be performed by partitioning the text in denotative units called *hrebs* [18]. The hreb, which was called so in honor of its discoverer, L. Hřebíček, is the set of all entities of a text referring to the same object in reality or the same object in the text. Up to now, the *sentence hreb* has been defined [10, 11, 12, 13, 14] as the set of sentences with common reference, allowing the sentence to belong to different hrebs. The coherence of the text can be measured using the affiliation of sentences to hrebs.

In the present paper the *word hreb* is defined as the set of all words of the text which have the same denotation/reference. Simple techniques allow us to ascertain several linguistically relevant properties of the text, to represent them graphically and to evaluate them. The denotative analysis shows the denotative structuring of the text, the manifest view of the reality by means of the text, i.e. its manifold organization. For the analysis of the latent connotative structures which are not explicitly marked one needs somewhat finer methods which, fortunately, can be performed mechanically.

Hrebs allow us to analyze the text supra-semantically, to study the distribution of denotations both generally and sequentially, and the probabilistic coincidence analysis transforms the text in a graph whose different properties can be considered structural denotative properties of the text.

2 The Method of Denotative Analysis

In order to describe the denotative analysis we present the method in discrete steps.

A. Ziegler and G. Altmann: *Latent Connotative Text Structure*, StudFuzz **209**, 211–229 (2007)
www.springerlink.com   © Springer-Verlag Berlin Heidelberg 2007

Step 1: The Criteria

Set up the criteria for the establishment of hrebs. These may be different in different languages. For the establishment of hrebs, the authors used the following ones (for more detail see [18]):

(i) A word-like unit at a certain position of the text belongs to a unique hreb; in different places it can belong to different hrebs: *mein* (*"my"*) in *mein Vater* (*"my father"*) and in *mein Sohn* (*"my son"*) can belong to different hrebs, as is the case in the analyzed text example "Erlkönig" by Johann Wolfgang von Goethe [8].

(ii) Compounds can be decomposed if there are hrebs in the text to which the parts can be assigned. The same procedure is used with decomposable words (e.g. German verbs).

(iii) Conjugation and declination endings are ignored.

(iv) Synonyms belong to the same hreb, homonyms do not.

(v) Determined and indetermined nouns can belong to the same hreb or to different ones, e.g. in *Pilze sind giftig* (*"mushrooms are poisonous"*) and in *dieser Pilz nicht* (*"this mushroom not"*) the denotations are different. The problem can be eliminated if one disregards all articles and some kinds of pronouns or considers them as part of the noun, e.g. *der, dieser, jener, mancher, jeder* etc., so that they are not established as hrebs.

If one wants to establish *morpheme hrebs*, the procedure is identical but the text must be analyzed morphemically, thus even parts of a word could belong to different hrebs. A Hungarian finite verb can contain three hrebs: that of the activity, that of the person and that of the reference to the object, e.g. *látlak* ("I see you"). Even if the criteria are very clear, there is a possibility that ad hoc decisions must be made and "errors" can occur. This is the weakest point of the analysis, because a) it (still) cannot be performed mechanically, b) it must make do with compromises between controversial grammatical descriptions and c) it must frequently make intuitive decisions at the boundary of semantics and denotation. Thus the criteria are preliminary and should be modified. Nevertheless, the future aim of this research is the determination of a catalogue of acknowledged and binding criteria which enable us to furnish comparable boundary conditions under which the analyses – even under contrastive and comparative aspects – attain more objectivity.

Step II: Establishment of Hrebs

One marks the individual words of the text with position numbers, e.g. for the first two lines of "Erlkönig":

1   2   3   4   5   6   7   8
Wer reitet so spät durch Nacht und Wind

9 10 11   12   13   14   15
Es ist der Vater mit seinem Kind

Next, one establishes – analyzing word for word – the individual hrebs. Thus one finds (the number gives the position in text) e.g.

(reitet 2, reitet 203)

(so 3, so 38, so 165, so 180).

Table 1 explicitly lists all identified word hrebs of "Erlkönig". Each hreb was given a unique number for later reference, which essentially reflects its position in the list of hrebs ordered by the number of elements.

Table 1. Word hrebs in Goethes poem "Erlkönig".

2. wer 1, Vater 12, seinem 14, er 16, er 24, er 28, mein 32, Vater 42, du 43, mein 53, Vater 87, Vater 89, du 92, mein 103, Vater 139, Vater 141, du 144, mein 152, mein 154, ich 156, Vater 185, Vater 187, Vater 200, er 202, er 205, seinen 220	**1.** Kind 15, Knaben 19, ihn 26, ihn 30, Sohn 33, du 36, dein 39, Sohn 54, du 59, Kind 61, dir 72, mein 86, mein 88, mir 96, Kind 104, Knabe 113, du 114, dich 121, dich 136, mein 138, mein 140, Sohn 153, Sohn 155, dich 169, deine 172, du 177, mein 184, mein 186, mich 191, mir 195, Kind 211, Kind 223	**3.** Erlkönig 45, Erlenkönig 48, mir 65, ich 70, meine 80, Erlenkönig 95, mir 116, meine 118, meine 124, Erlkönigs 147, ich 167, mich 170, ich 182, er 190, Erlkönig 193
17. reitet 2, reitet 203	**31.** sicher 27	**41.** gar 66
8. so 3, so 37, so 165, so 180	**20.** hält 29, hält 206	**34.** bang 38
27. spät 4	**32.** warm 31	**42.** Spiele 68
28. durch 5	**21.** was 34, was 94	**43.** spiel 69
29. Nacht 6	**33.** birgst 35	**24.** manch 73, manch 83
5. und 7, und 51, und 90, und 130, und 132, und 134, und 142, und 175, und 217	**4.** der 11, den 18, dem 22, den 44, den 47, dem 78, der 109, den 127, dem 199, den 213	**11.** nicht 46, nicht 93, nicht 145, nicht 178
18. Wind 8, Wind 110	**35.** Gesicht 40	**45.** Blumen 75
9. es 9, es 55, es 158, es 160	**14.** siehst 41, siehst 143, seh 157	**46.** an 77
6. ist 10, ist 56, sind 76, sei 99, bist 176, war 224	**15.** schöne 67, schön 123 (adv.), schöne 173	**7.** mit 13, mit 49, mit 64, mit 71, mit 115, mit 215
44. bunte 74	**47.** Strand 79	**48.** Mutter 81
60. sollen 120	**36.** Kron 50	**49.** gülden 84
23. geh 63, gehn 117	**37.** Schweif 52	**50.** Gewand 85
12. hat 17, hat 82, hat 194	**22.** ein (Artikel) 57, ein 196	**51.** hörest 91

30. wohl 20	**38.** Nebelstreif 58	**52.** leise 97
10. in 21, in 105, in 207, in 219	**39.** liebes 60	**53.** verspricht 98
13. Arm 23, Armen 208, Armen 221	**40.** komm 62	**25.** ruhig 100, ruhig 102
54. bleibe 101	**85.** jetzt 188	**61.** warten 122
55. dürren 106	**86.** an (adv.) 192	**63.** nächtlichen 128
56. Blättern 107	**87.** Leids 197	**93.** Hof 214
57. säuselt 108	**88.** getan 198	**62.** führen 126
58. willst 111	**89.** grausets 201	**92.** erreicht 212
59. feiner 112	**90.** geschwind 204	**91.** ächzende 210
16. Töchter 119, Töchter 125, Töchter 148	**26.** das 209, das 222	**64.** Reihn 129
65. wiegen 130	**94.** Mühe 216	**19.** faßt 25, faßt 189
66. tanzen 133	**95.** Not 218	**70.** am 149
67. singen 135	**96.** tot 225	**71.** düstern 150
68. ein (adv.) 137	**69.** dort 146	**72.** Ort 151
73. genau 159	**74.** scheinen 161	**75.** die (Pl. fem.) 163
77. Weiden 164	**78.** grau 166	**79.** liebe 168
80. reizt 171	**81.** Gestalt 174	**82.** willig 179
83. brauch 181	**84.** Gewalt 183	

It has been shown that this order complies with a probability law. Starting from this table the text can be characterized in several ways [18].

Step III: Coincidence

If the words of the text are replaced by these numerical designations, then the text coded according to the membership of words in hrebs has the form as shown in table 2.

This presentation is more adequate for the analysis of the coincidence structure. This part of the analysis can be performed by means of a program. Concerned are exclusively *positional coincidences* in the framework of a higher unit – in this case the line – since it is this unit betraying the textual structure. There are manifold other associations like phonic ones in rhyme and alliteration, grammatical ones like government and congruency, semantic ones like synonymy, inclusion etc. Some of them are rule governed and thus strongly deterministic, and other kinds of associations could be taken into account. The positional coincidence regards merely the ad hoc common occurrence of two entities which is a purely text specific phenomenon. Consequently the probability of this coincidence can be computed. There are many other procedures taking as framework word pairs, sentences, distances between words or even the whole text [2, 3, 4, 7, 16, 17].

We proceed as follows: Let N be the number of lines (verses, sentences, chapters or other frame units) of the text, here $N = 32$. Let M be the number of lines containing hreb A, n be the number of lines containing hreb B, and

Table 2. Membership of words in hrebs (the second column is the continuation of the first).

2,17,8,27,28,29,5,18	58,59,1,1,7,3,23
9,6,4,2,7,2,1	3,16,60,1,61,15
2,12,4,1,30,10,4,13	3,16,62,4,63,64
2,19,1,31,2,20,1,32	5,65,5,66,5,67,1,68
2,1,21,33,1,8,34,1,35	1,2,1,2,5,14,2,11,69
14,2,2,4,3,11	3,16,70,71,72
4,3,7,36,5,37	2,1,2,1,2,14,9,73
2,1,9,6,22,38	9,74,75,76,77,8,78
1,39,1,40,23,7,3	3,79,1,3,80,1,15,81
41,15,42,43,3,7,1	5,6,1,11,82,8,83,3,84
24,44,45,6,46,4,47	1,2,1,2,85,19,3,1,86
3,48,12,24,49,50	3,12,1,22,87,88
1,2,1,2,5,51,2,11	4,2,89,2,17,90
21,3,1,52,53	2,20,10,13,26,91,1
6,25,54,25,2,1	92,4,93,7,94,5,95
10,55,56,57,4,18	10,2,13,26,1,6,96

x be the number of lines in which A and B co-occur (multiple occurrences of a hreb in a line is irrelevant). Then the probability that A and B co-occur at least in x lines is

$$P(X \geq x) = \sum_{j-x}^{\min(M,n)} \frac{\binom{M}{j}\binom{N-M}{n-j}}{\binom{N}{n}} \tag{1}$$

i.e. as the sum of individual hypergeometric probabilities. In very long texts one can pass to the Poisson distribution [1]. The smaller the probability of (1), the stronger the positional coincidence of the given two hrebs. In the following output (table 3), one sees the probabilities of coincidence of hreb 1 with all the other ones.

Table 3. Coincidences of hreb 1 with the other ones.

Hreb no.	1	22	0.42339	59	0.65625
with:		23	0.42339	60	0.65625
		25	0.65625	61	0.65625
2	0.10787	26	0.42339	65	0.65625
3	0.69768	30	0.65625	66	0.65625
4	0.99986	31	0.65625	67	0.65625
5	0.83778	32	0.65625	68	0.65625
6	0.30688	33	0.65625	69	0.65625
7	0.67013	34	0.65625	73	0.65625
8	0.89447	35	0.65625	79	0.65625
9	0.57328	38	0.65625	80	0.65625
10	0.57328	39	0.65625	81	0.65625
11	0.57328	40	0.65625	82	0.65625
12	0.73387	41	0.65625	83	0.65625
13	0.26815	42	0.65625	84	0.65625
14	0.73387	43	0.65625	85	0.65625
15	0.26815	51	0.65625	86	0.65625
16	0.96673	52	0.65625	87	0.65625
19	0.42339	53	0.65625	88	0.65625
20	0.42339	54	0.65625	91	0.65625
21	0.42339	58	0.65625	96	0.65625

In the following paragraphs we shall consider all possible hreb coincidences. If there is no coincidence, the hreb pairs are not mentioned as, e.g., (1,17), (1,18), (1,24) etc. in table 3.

Step IV: Graphic presentation

The output in table 3 is not lucid and not very informative. One gets a better view if the result is presented graphically. There are different ways to do this:

(a) One chooses by convention an ad hoc boundary α. This boundary need not agree with the usual significance levels in statistics. If the probability (1) of a coincidence is smaller than or equal to α, one joins the two given hrebs with an edge. In this way a graph with several components arises, as can be seen in figure 1 (components are labelled "K1", "K2" etc).

(b) One chooses a boundary α so that no vertex of the graph is isolated, all are joined with at least one other vertex.

(c) One chooses a boundary α so that one obtains one unique component. That is, all hrebs are joined with one another.

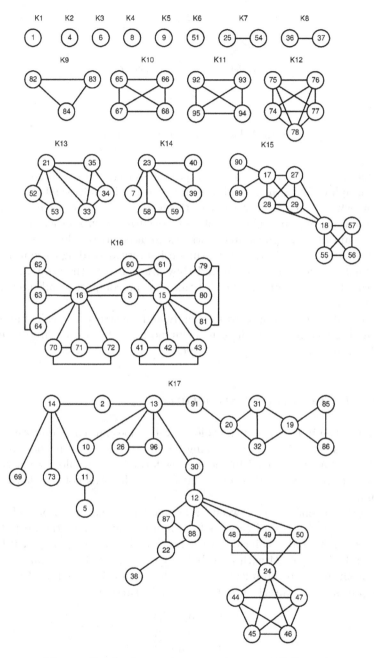

Fig. 1. Graph of coincidences in the poem "Erlkönig".

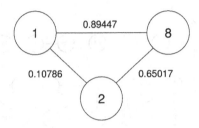

Fig. 2. Path lengths between hrebs.

If one chooses e.g. $\alpha = 0.1$ according to the steps (a), (b) or (c), one obtains the graph of "Erlkönig" as shown in figure 1. The individual components show what coincides with what in the text. If a hreb occurs many times, one obtains high probabilities of its co-occurrence with other ones, i.e. in the graph it can appear frequently as isolated. If two hrebs are seldom but always occur together, then the probability, according to equation (1) of their coincidence, is small and they are joined in the graph. A graph of this kind has many properties that can be computed and interpreted linguistically.

(d) And finally, one can proceed as usually and construct a so-called minimal spanning tree in which there is a unique route from each hreb to every other hreb, i.e. one constructs a tree.

3 Latent Connotative Structures

As can bee seen in the output in table 3, hreb 1 does not coincide with all the others. Hrebs 24, 27, 28 etc. are missing. Does this mean that it does not have any relationship with them? In those cases formula (1) would always yield 1 because one would sum from 0 to $\min(M, n)$ if this was not avoided by the program.

Note that connections between hrebs may span more than one edge, i.e. they do not need to be direct. Looking more accurately at the output one can detect that hreb 1 coincides with hreb 3 with a probability of 0.10787 and hreb 3 with hreb 24 with a probability of 0.69153. That means, there is a path from hreb 1 to hreb 24. But even direct coincidences have frequently higher values then indirect ones. Compare for example

$$(1,8): 0.89447$$
$$(1,2): 0.10787$$
$$(2,8): 0.65017$$

or in graphical form as shown in figure 2. In this case the path from hreb 1 to hreb 8 is shorter via hreb 2 ($0.10786 + 0.62017 = 0.72803$) than the direct path ($0.89447$). This means that in normal (not weighted) graphs with a given boundary α some semantic relations remain latent or disappear completely. Their existence can be made visible by means of weighted (fuzzy) graphs or one can take at the same time a new route to the morphemic level and first establish morpheme hrebs which reveal more of the association structure of the text than word hrebs. Here we remain at the word level and make the following assumptions:

(a) The manifest, direct connection (adjacency in the coincidence graph) of two hrebs means a *denotative connection*, whether it is ad hoc (text or author specific) or constant (as a fixed phrase in the language).
(b) If two hrebs are not adjacent to each other in the coincidence graph but there is a path between them, then they are said to be *(latently) connotatively connected.*
(c) If the weight of the edge between two adjacent vertices x and y in the coincidence graph has a higher value than the individual sums of weights from other paths, x to y (see figure 2), then they are said to be *denotatively and connotatively connected.*

The decision with all criteria depends on whether we accept "no adjacency" or "no adjacency above the criterion α". If one chooses a large α, then one can realize that all hrebs are at least denotatively connected. By criterion (c) we have to compare the weight of the direct adjacency with the shortest path between two vertices. The building of coincidences and adjacencies in the graph probably complies with Hřebíček's law of references [9, 10, 14]. There surely is a regularity within the growth of the number of connections connected with level α and the text length.

We consider the finding of the shortest path between two hrebs. The objective is merely to find the minimal paths in the coincidence graph (as presented in table 3). The strength of the coincidence computed by means of (1) is considered as the weight of the edge (or the distance) between the two adjacent vertices. The smaller this weight, the greater the association of the two hrebs. In order to find the shortest path between all pairs of edges in a graph, Floyd's algorithm [6] is used (see also [5, p. 152f.]).[1]

For the sake of illustration we show the shortest paths between hreb 1 and the other hrebs in table 4. In the same way we proceed to find the shortest paths between all pairs of hrebs of the text. The result cannot be shown here in its entirety because as it is some 20 pages long.

Comparing table 3 and table 4 one finds that values in table 3 are greater or equal to the values in table 4 and connections missing in table 3 are present. That means that even not coinciding hrebs or coinciding over the α boundary have a latent association that can be detected in different ways (for analogous problems cf. Mehler [15]).

---

[1] The freely available program used to calculate these structures and written in **Python** can be ordered from **himself@ralfjuengling.de**.

The hreb having shortest paths to all other hrebs is a *central hreb*. In order to ascertain centrality it is sufficient to compute for each hreb the sum of all its shortest paths to all other hrebs. These sums can be seen in table 5. If we order the sums of shortest paths and group them into classes in intervals of 5, we obtain the layers as shown in table 6.

Table 4. The shortest paths of hreb 1 to all other hrebs.

(1, 2)	0.107860	(1, 34)	0.485890	(1, 66)	0.478320
(1, 3)	0.341540	(1, 35)	0.485890	(1, 67)	0.478320
(1, 4)	0.510270	(1, 36)	0.478320	(1, 68)	0.478320
(1, 5)	0.259570	(1, 37)	0.478320	(1, 69)	0.293340
(1, 6)	0.306880	(1, 38)	0.485890	(1, 70)	0.508680
(1, 7)	0.453630	(1, 39)	0.485890	(1, 71)	0.508680
(1, 8)	0.460070	(1, 40)	0.485890	(1, 72)	0.508680
(1, 9)	0.360920	(1, 41)	0.361900	(1, 73)	0.293340
(1, 10)	0.200400	(1, 42)	0.361900	(1, 74)	0.485920
(1, 11)	0.234270	(1, 43)	0.361900	(1, 75)	0.485920
(1, 12)	0.387090	(1, 44)	0.494380	(1, 76)	0.485920
(1, 13)	0.199590	(1, 45)	0.494380	(1, 77)	0.485920
(1, 14)	0.199590	(1, 46)	0.494380	(1, 78)	0.485920
(1, 15)	0.268150	(1, 47)	0.494380	(1, 79)	0.361900
(1, 16)	0.414930	(1, 48)	0.480840	(1, 80)	0.361900
(1, 17)	0.319550	(1, 49)	0.480840	(1, 81)	0.361900
(1, 18)	0.387900	(1, 50)	0.480840	(1, 82)	0.359270
(1, 19)	0.319550	(1, 51)	0.359270	(1, 83)	0.359270
(1, 20)	0.319550	(1, 52)	0.485890	(1, 84)	0.359270
(1, 21)	0.423390	(1, 53)	0.485890	(1, 85)	0.382050
(1, 22)	0.423390	(1, 54)	0.494380	(1, 86)	0.382050
(1, 23)	0.423390	(1, 55)	0.325400	(1, 87)	0.480840
(1, 24)	0.543340	(1, 56)	0.325400	(1, 88)	0.480840
(1, 25)	0.494380	(1, 57)	0.325400	(1, 89)	0.382050
(1, 26)	0.205640	(1, 58)	0.485890	(1, 90)	0.382050
(1, 27)	0.382050	(1, 59)	0.485890	(1, 91)	0.268140
(1, 28)	0.382050	(1, 60)	0.361900	(1, 92)	0.478320
(1, 29)	0.382050	(1, 61)	0.361900	(1, 93)	0.478320
(1, 30)	0.293340	(1, 62)	0.508680	(1, 94)	0.478320
(1, 31)	0.382050	(1, 63)	0.508680	(1, 95)	0.478320
(1, 32)	0.382050	(1, 64)	0.508680		
(1, 33)	0.485890	(1, 65)	0.478320		

Table 5. Sums of shortest paths for all hrebs.

1:	38.727950	33:	49.967350	65:	56.797500
2:	35.048250	34:	49.967350	66:	56.797500
3:	45.370310	35:	49.967350	67:	56.797500
4:	43.860280	36:	47.084160	68:	56.797500
5:	37.453750	37:	47.084160	69:	45.559730
6:	43.049500	38:	48.658210	70:	58.909200
7:	46.633320	39:	53.502260	71:	58.909200
8:	41.682520	40:	53.502260	72:	58.909200
9:	43.848140	41:	52.379700	73:	42.575980
10:	37.145270	42:	52.379700	74:	47.918830
11:	37.093080	43:	52.379700	75:	47.918830
12:	43.991050	44:	48.950780	76:	47.918830
13:	37.108010	45:	48.950780	77:	47.918830
14:	36.884430	46:	48.950780	78:	47.918830
15:	46.704460	47:	48.950780	79:	55.204460
16:	50.409200	48:	48.615460	80:	55.204460
17:	44.103620	49:	48.615460	81:	55.204460
18:	42.775030	50:	48.615460	82:	38.576620
19:	50.026190	51:	48.843080	83:	38.576620
20:	47.161320	52:	55.491030	84:	38.576620
21:	52.713260	53:	55.491030	85:	52.932440
22:	48.713260	54:	60.330750	86:	52.932440
23:	47.721010	55:	41.385040	87:	47.591550
24:	49.770250	56:	41.385040	88:	47.591550
25:	60.330750	57:	41.385040	89:	47.591550
26:	37.543610	58:	53.502260	90:	47.591550
27:	42.665270	59:	53.502260	91:	42.680730
28:	42.665270	60:	54.226320	92:	46.396660
29:	42.665270	61:	54.226320	93:	46.396660
30:	41.557470	62:	55.290780	94:	46.396660
31:	51.668480	63:	55.290780	95:	46.396660
32:	51.668480	64:	55.290780	96:	41.982080

Table 6. Degrees of centrality of individual hrebs.

Sum of Shortest Paths	Degree of Centrality	Hrebs
<35,40>	1	1,2,5,10,11,13,14,26,82,83,84
(40,45>	2	4,6,8,9,12,17,18,27,28,29,30,55,56,57,73,91,96
(45,50>	3	3,7,15,20,22,23,24,33,34,35,36,37,38,44,45,46,47,48, 49,50,51,69,74,75,76,77,78,87,88,89,90,92,93,94,95
(50,55>	4	16,19,21,31,32,39,40,41,42,43,52,53,58,59,60,61,85,86
> 55	5	35,54,62,63,64,65,66,67,68,70,7,72,79,80,81

If we replace the numbers in table 6 by the pertinent hrebs from table 1, we obtain the results in table 7.

Table 7. Degree of centrality of individual hrebs.

Degree	Hrebs
1	Kind, Vater, und, in, nicht, Arm, sehen, das, willig, brauch, Gewalt
2	der, ist, so, es, hat, reitet, Wind, spät, durch, Nacht, wohl, dürren, Blättern, säuselt, genau, ächzende, tot
3	Erlkönig, mit, schön, hält, ein, gehen, manch, birgst, bang, Gesicht, Kron, Schweif, Nebelstreif, bunte, Blumen, an, Strand, Mutter, gülden, Gewand, hörest, dort, scheinen, die, alten, Weiden, grau, Leids, getan, grausets, geschwind, erreicht, Hof, Mühe, Not
4	Töchter, fasst, was, sicher, warm, liebes, komm, gar, Spiele, spiel, leise, verspricht, willst, feiner, sollen, warten, jetzt, an
5	ruhig, bleibe, führen, nächtlichen, Reihn, wiegen, tanzen, singen, ein, am, düstern, Ort, liebe, reizt, gestalt

For the graphic presentations of the results of table 4 different methods can be applied.

1. First one orders all minimal paths according to magnitude and then joins the two hrebs with the smallest minimal path. One searches for other hrebs whose minimal paths belong to those already connected. One continues in this way until all hrebs are connected.
2. One uses the minimal path of each hreb and connects them. Since there can be several minimal paths for a hreb, we obtain a tree whose edges can contain more hrebs. They form cliques.

3.1 Method 1

First we order all minimal paths between hreb pairs in table 4 according to magnitude and obtain the results in table 8. Here, too, we can only show the beginning of the table since the result is very extensive.

Table 8. Hreb pairs ordered by minimal path length.

(10,13) 0.000810	(74, 77) 0.031250	(22,38) 0.062500
(13,26) 0.006050	(74, 78) 0.031250	(22,87) 0.062500
(10,26) 0.006860	(75, 76) 0.031250	(22,88) 0.062500
(5,11) 0.025300	(75, 77) 0.031250	(23,39) 0.062500
(7,23) 0.030240	(75, 78) 0.031250	(23,40) 0.062500
(25,54) 0.031250	(76, 77) 0.031250	(23,58) 0.062500
(27,28) 0.031250	(76, 78) 0.031250	(23,59) 0.062500
(27,29) 0.031250	(77, 78) 0.031250	(24,44) 0.062500
(28,29) 0.031250	(79, 80) 0.031250	(24,45) 0.062500
(31,32) 0.031250	(79, 81) 0.031250	(24,46) 0.062500
(33,34) 0.031250	(80, 81) 0.031250	(24,47) 0.062500
(33,35) 0.031250	(82, 83) 0.031250	(24,48) 0.062500
(34,35) 0.031250	(82, 84) 0.031250	(24,49) 0.062500
(36,37) 0.031250	(83, 84) 0.031250	(24,50) 0.062500
(39,40) 0.031250	(85, 86) 0.031250	(26,91) 0.062500
(41,42) 0.031250	(87, 88) 0.031250	(26,96) 0.062500
(41,43) 0.031250	(89, 90) 0.031250	(13,91) 0.068550
(42,43) 0.031250	(92, 93) 0.031250	(13,96) 0.068550
(44,45) 0.031250	(92, 94) 0.031250	(10,91) 0.069360
(44,46) 0.031250	(92, 95) 0.031250	(10,96) 0.069360
(44,47) 0.031250	(93, 94) 0.031250	(3,15) 0.073390
(45,46) 0.031250	(93, 95) 0.031250	(3,16) 0.073390
(45,47) 0.031250	(11, 14) 0.034680	(2,13) 0.091730
(46,47) 0.031250	(5, 14) 0.059980	(2,14) 0.091730
(48,49) 0.031250	(17, 27) 0.062500	(2,10) 0.092540
(48,50) 0.031250	(17, 28) 0.062500	(7,39) 0.092740
(49,50) 0.031250	(17, 29) 0.062500	(7,40) 0.092740
(52,53) 0.031250	(17, 89) 0.062500	(7,58) 0.092740
(55,56) 0.031250	(17, 90) 0.062500	(7,59) 0.092740
(55,57) 0.031250	(18, 27) 0.062500	(12,30) 0.093750
(56,57) 0.031250	(18, 28) 0.062500	(12,48) 0.093750
(58,59) 0.031250	(18, 29) 0.062500	(12,49) 0.093750
(60,61) 0.031250	(18, 55) 0.062500	(12,50) 0.093750

(62,63) 0.031250	(18, 56) 0.062500	(12,87) 0.093750
(62,64) 0.031250	(18, 57) 0.062500	(12,88) 0.093750
(63,64) 0.031250	(19, 31) 0.062500	(13,30) 0.093750
(65,66) 0.031250	(19, 32) 0.062500	(14,69) 0.093750
(65,67) 0.031250	(19, 85) 0.062500	(15,41) 0.093750
(65,68) 0.031250	(19, 86) 0.062500	(15,42) 0.093750
(66,67) 0.031250	(20, 31) 0.062500	(15,43) 0.093750
(66,68) 0.031250	(20, 32) 0.062500	(15,60) 0.093750
(67,68) 0.031250	(20, 91) 0.062500	(15,61) 0.093750
(70,71) 0.031250	(21, 33) 0.062500	(15,79) 0.093750
(70,72) 0.031250	(21, 34) 0.062500	(15,80) 0.093750
(71,72) 0.031250	(21, 35) 0.062500	(15,81) 0.093750
(74,75) 0.031250	(21, 52) 0.062500	(16,60) 0.093750
(74,76) 0.031250	(21, 53) 0.062500	(16,61) 0.093750

The pair (10,13) has the smallest distance, namely $d(10,13) = 0.000810$. We connect the two hrebs with two directed edges drawn as $\leftrightarrow$. Since this is the smallest distance in the whole graph, no edges can be formed from either 10 or 13. However, there can be incoming edges. One finds just below it $d(13,26) = 0.006050$, i.e. an edge goes from 26 to 13:

One seeks other hrebs whose minimal paths go to 10, 13 or 26. At another point, one finds $d(26,91) = 0.062500$ and $d(26,96) = 0.062500$. For 96 it is the shortest path, but for 91 there is another path of the same "length", namely to 20, and this is the smallest for 20. Thus in the next step we obtain:

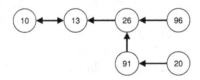

Hreb 20 has still two other paths of same length, namely to 31 and to 32, which in turn have the shortest path to one each other, i.e. $d(31,32) = 0.03125$. Thus in the next step we obtain:

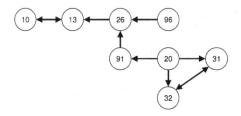

Continuing in this way one finally obtains the graph in figure 3 representing the *denotation-connotation flow* in the poem.

3.2 Method 2

This method refers to the output in table 3. For each hreb there is at least one coincidence (smallest direct distance), and frequently, several smallest coincidences are equal. One can begin with any hreb, but it is important that there is only one component. Those hrebs having mutually the same minimal coincidence and the same one to the others can be treated as cliques and considered as one unique vertex. The result is a tree in which cycles are considered as single edges. Figure 4 shows as a result the *manifest denotative structure of the text.*

4 Summary

The presented methods yield one way of capturing denotative and connotative structures in text. They are fruitful not only for text linguistics, as they allow us to discover textual mechanisms, but are also applicable in psycholinguistic research. From the linguistic point of view they allow us to study the regularities of denotation and its manifestation in consecutive sentences, e.g. to develop the topic-comment problems, to study the occupation of positions in sentence by hrebs in dependence on frequency, etc. The procedures shown above are merely a small extract from a broad research domain which is in development [18].

The presented graphs can be further evaluated, both qualitatively and quantitatively. It is very probable that they could become a basis for text classification, studies on style, the discovery of new properties of texts (e.g. an exactly defined term of coherence), psycholinguistic studies, the unveiling of the association world of the author or of the persons in drama etc. Since up to now only a very small number of texts has been analyzed, further developments can hardly be predicted. Perhaps the attained graphs represent the so-called "small worlds" whose properties are object of many scientific disciplines. Thus the presented graphs enable us to find contact to other, better developed sciences.

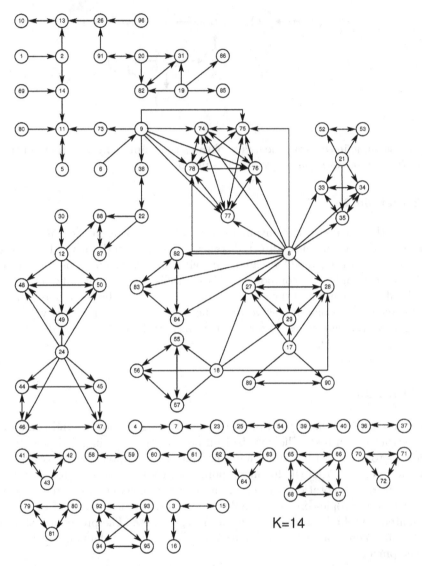

Fig. 3. The denotation-connotation flow in the poem.

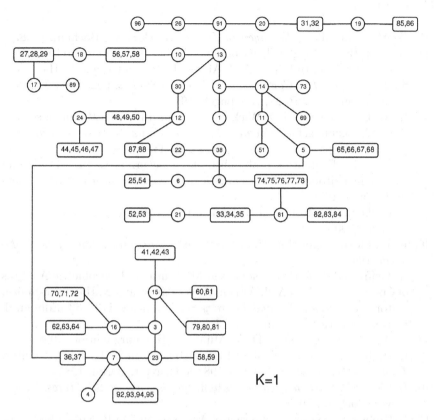

Fig. 4. The manifest denotative structure of the poem.

References

[1] G. Altmann. *Wiederholungen in Texten*. Brockmeyer, Bochum, 1988.

[2] G. L. M. Berry-Rogghe. The Computation of Collocations and their Relevance in Lexical Studies. In A. J. Aitken, R. W. Bailey, and N. Hamilton-Smith, editors, *The Computer and Literary Studies*, pages 103–112. Edinburgh University Press, Edinburgh, 1973.

[3] H.-M. Dannhauer and D. Wickmann. Quantitative Bestimmung semantischer Umgebungsfelder in einer Menge von Einzeltexten. *Zeitschrift für Literaturwissenschaft und Linguistik*, 2:29–43, 1972.

[4] C. Dolphin. Evaluation probabiliste des cooccurrences. In J. David and R. Martin, editors, *Études de statistique linguistique*, pages 21–34. Klincksieck, Paris, 1977.

[5] L. Egghe and R. Rousseau. *Introduction to Informetrics*. Elsevier, Amsterdam, 1990.

[6] R. W. Floyd. Algorithm 97: Shortest path. *Communications of the ACM*, 5:345, 1962.

[7] A. Geffroy, P. Lafon, G. Seidel, and M. Tournier. Lexicometric Analysis of Co-occurrences. In A. J. Aitken, R. W. Bailey, and N. Hamilton-Smith, editors, *The Computer and Literary Studies*, pages 113–133. Edinburgh University Press, Edinburgh, 1973.

[8] J. W. v. Goethe. *Werke*. DTV, München, Hamburg edition, 1998.

[9] L. Hřebíček. Text as a Unit and Co-references. In T. T. Ballmer, editor, *Linguistic Dynamics*, pages 190–198. de Gruyter, Berlin, 1985.

[10] L. Hřebíček. *Text in Communication: Supra-sentence Structures*. Brockmeyer, Bochum, 1992.

[11] L. Hřebíček. Text as a Construct of Aggregations. In R. Köhler and B. B. Rieger, editors, *Contributions to Quantitative Linguistics*, pages 33–39. Kluwer, Dordrecht, 1993.

[12] L. Hřebíček. *Text Levels. Language Constructs, Constituents and the Menzerath-Altmann Law*. WVT, Trier, 1995.

[13] L. Hřebíček. *Lectures on Text Theory*. Oriental Institute, Prague, 1997.

[14] L. Hřebíček. *Variation in Sequences. Contributions to General Text Theory*. Oriental Institute, Prague, 2000.

[15] A. Mehler. *Textbedeutung. Zur prozeduralen Analyse und Repräsentation struktureller Ähnlichkeiten von Texten*. Peter Lang, Frankfurt a. M., 2001.

[16] B. B. Rieger. Wort- und Motivkreise als Konstituenten lyrischer Umgebungsfelder. Eine quantitative Analyse semantisch bestimmter Textelemente. *Zeitschrift für Literaturwissenschaft und Linguistik*, 4:23–41, 1971.

[17] B. B. Rieger. Eine tolerante Lexikonstruktur. Zur Abbildung natürlich-sprachlicher Bedeutung auf unscharfe Mengen in Toleranzräumen. *Zeitschrift für Literaturwissenschaft und Linguistik*, 16:31–47, 1974.

[18] A. Ziegler and G. Altmann. *Denotative Textanalyse. Ein textlinguistisches Arbeitsbuch.* Edition Praesens, Wien, 2002.

Part IV

Corpus Linguistic and Text Technological
Modeling

Inferring Meaning: Text, Technology and Questions of Induction

Michael Stubbs

University of Trier
stubbs@uni-trier.de

1 Introduction

Corpus linguists, including lexicographers, use methods which are often called 'inductive'. That is, they study large corpora or large data sets (such as word-frequency lists) derived from these corpora, in order to identify patterns in the data. There is detailed discussion of a few statistical techniques (e.g. for identifying significant collocations), but little general discussion of the combination of automatic and intuitive methods which are used to make significant generalizations. It might be thought that, if linguists draw generalizations from large data sets, then they would generally agree about the resulting analyses, and certainly corpus work often reaches a remarkably large consensus across different studies. Findings from one corpus are regularly corroborated by studies of other independent corpora, and partly automated or computer-assisted analysis has led to major progress in the study of semantic and pragmatic data.

However, given the title of this book, I should say immediately that I will argue that there can be no entirely automatic semantic analysis. First, I discuss the historical and logical background to the concept of induction, from Francis Bacon in the 1600s, via David Hume in the 1700s, to Karl Popper in the 1980s. The broad consensus from this work is that induction does not exist, and that there are no automatic methods which can be used to infer reliable generalizations from repeated individual observations. Second, I discuss some problematic examples from corpus-based dictionaries in order to illustrate the uneasy balance in corpus lexicography between automatic and intuitive methods. As examples of the results of inductive inferences (in the rough sense), I discuss some definitions in modern corpus-based dictionaries, and the extent to which these definitions agree or disagree. It might be thought that they disagree surprisingly often in their definitions of individual words.

M. Stubbs: *Inferring Meaning: Text, Technology and Questions of Induction*, StudFuzz **209**, 233–253 (2007)
www.springerlink.com

2 Conventions and Terminology

I use the following conventions. "Double quotes" are used for the meanings of words and phrases. 'Single quotes' are used for quotes from other authors. Upper case is used for lemmas (lexemes). *Italics* is used for word-forms. For example, the word forms *crony* and *cronies* are two realisations of the lemma CRONY. I will also make a distinction which is very simple and could avoid much confusion, but which is only rarely made (though see [22]). The terms 'introspection' and 'intuition' are often used synonymously, but there is a clear distinction between using introspection as data (as neo-Chomskyan linguists have done since the 1950s) and using intuition to identify interesting problems and to analyse data. I assume that all corpus linguists reject introspection as the (only) source of data, but that none deny the essential role of intuition in formulating hypotheses and analysing data. One thing which becomes very clear in teaching students to interpret corpus data is that the ability to see patterns (e.g. in concordance lines) takes practice: recurrent patterns are not obvious, and recognizing them is more like a skill than knowledge. This already throws doubt on any mechanical view of induction. The same is presumably true in any observational science: chemistry students also have to be taught what it is important to observe.

3 An Introductory Example

Since the historical part of the paper may seem rather far from concrete linguistic data, I will start with a brief linguistic example of the general problem. Since the late 1980s, corpus studies have reinstated observation as central to linguistics on a scale previously unimaginable. Linguists can now work with data in the form of tens or even hundreds of millions of running words sampled from many different speakers and writers) or with large data-sets which are derived from these corpora with minimal human intervention. Now, one might think that, if linguists use such data and methods – in particular concordance software which allows large numbers of examples to be brought together and studied – then they would infer the same generalizations from the data, with a high degree of consensus across different studies. Findings from one study predict similar findings in other independent corpora, and it is certainly true that corpus studies are regularly corroborated in this way.

One very large set of parallel findings is available in the form of definitions of words in corpus-based dictionaries. However, if we compare the definitions in different dictionaries, which have been produced independently but with similar methods, we discover that the dictionaries sometimes differ in the meanings (perhaps especially the evaluative connotations) which they attribute to words. For example, here are the definitions of CRONY in four corpus-based dictionaries, all published in 1995.

- **crony**. Your cronies are the friends who you spend a lot of time with, an informal word. *Daily he returned, tired and maudlin, from lunchtime drinking sessions with his business cronies.* (Cobuild [7]).
- **crony**. (informal often derogatory) A close friend or companion. *He spends every evening drinking in the pub with his cronies.* (OALD [17]).
- **crony**. (usually plural) One of a group of people, who spend a lot of time with each other and will usually help each other, even if this involves dishonesty. *Nixon gave political power to many of his political cronies.* (LDOCE [12]).
- **crony**. (informal especially disapproving) A close friend or someone who works with a stated and usually dishonest person in authority. *The General and his cronies are now awaiting trial for drug-smuggling.* (CIDE [6]).

Although these definitions have much in common, they differ considerably in emphasis. Cobuild gives a neutral denotation as "friend", and notes that the usage is informal. The example of drinking companions perhaps implies something "disreputable" (but the word gets no "PRAGMATICS" label: see below). OALD uses a very similar example with drinking companions, but explicitly warns that the word is often (but not always?) derogatory. LDOCE goes further and gives "dishonesty" as part of the denotation(?) with a citation which refers to one of the most notoriously dishonest political figures of all time. CIDE also gives "dishonesty" as part of the denotation, and gives a citation involving a major crime. In summary, the definitions range from neutral "friend", with only an implicit hint of "disreputable activities", via an explicit warning that the word is "derogatory", to the statement that it implies "dishonesty", if not serious "criminality" and abuse of political power on a major scale.

Now, what is a foreign learner to make of this? Can the word be simply informal and casual, or is it insulting and therefore to be used only with great care? There is no clear dividing line between the connotations given by the dictionaries, but consider what would happen if these meanings were translated into German, and then back-translated into English. This is not an artificial example: I recently asked a lecture class of around seventy largely German-speaking university students of English if they knew the word CRONY. Only ten or so claimed to know it, so it is precisely the kind of word they might want to look up in a dictionary.

These four dictionary definitions are four little theories of meaning. They are generalizations from the corpus data (plus the intuitions of the lexicographers), which raise questions such as: Which theory is correct? Which is wrong? How can we test such semantic claims? What would be the evidence that one is wrong? Could we provide counter-examples? Can we at least defend a preference for one definition over another? Would this preference depend merely on the most frequent usage (implying that minority usages are wrong)? Are all four definitions correct, but for different speakers? Or are they all wrong, because each one is too narrow, and does not take into account variable usage across speakers? The last questions imply that even corpus-based dictionaries, which seem the ultimate example of linguistic description, are in fact **pre**scriptive, since they do not fully take into account variation in usage. And should foreign learners at least be warned of this variation, that many native speakers regard the word as insulting, and that it may be safest to avoid it altogether?

Chomsky [5, p. 51] early dismissed any attempts at automatic discovery procedures as far too demanding and quite unworkable for syntax, and therefore presumably all the more so for semantics. I will also argue here that any completely automatic procedures are impossible, (and that claims by corpus linguists in this direction are sometimes exaggerated). But it is worth bearing in mind that the scepticism of such procedures in the 1950s and 1960s was at least partly due to the very restricted amount of data which could be prepared in machine-readable form, and of the failure of early statistical models (which were often tarred with the brush of failed early attempts at machine translation).

So, corpus linguists – including lexicographers – look at large amounts of data, observe recurrent patterns, and use these observations as evidence of meanings. Such methods, which use large data-sets to infer patterns, are often referred to as inductive, and they would seem to provide a much firmer empirical basis for linguistic description than the small amounts of introspective data used in neo-Chomskyan work. However, they raise several questions which are largely unresolved (and indeed hardly discussed).

4 Some Traditional Distinctions

I assume that no-one these days believes in automatic methods which can reliably lead, purely objectively, from (repeated) empirical observations to significant generalizations. Intuition and inspired guesswork are always involved in selecting the initial data (e.g. designing a corpus), deciding what problems to investigate, and identifying the interesting patterns. Since corpora are typically designed according to a sociolinguistic theory of language variation, theory is involved from the beginning. Thus, even though corpus linguists often talk of 'raw data', and even though they have methods of avoiding some of the assumptions of pre-corpus grammar, I assume also that no-one these days believes in the possibility of a neutral observation language. (But see http://www.linguistlist.org/, July 2002, for a debate between Mukherjee and Pullum: disputes between these approaches are by no means settled.) These questions have a history of at least 2,000 years, and I can here make only a few traditional distinctions as a preface to concrete corpus examples. Deduction and induction are often distinguished as follows.

A deductive argument starts from premises, and draws conclusions which must be true if the premises are true. Given two premises, 'all students like beer' and 'Bertha is a student', then it must follow that 'Bertha likes beer'. Deduction concerns the validity of the conclusion, given the truth of the premises (that all students really do like beer, and that Bertha really is a student). It can say nothing about how the premises are established, or whether they are well defined (e.g. does alcohol-free beer count as beer, or if Bertha is a mature part-time student, does she count as a valid case?) Deductive logic concerns only the validity of the argument which relates premises and conclusion.

An inductive argument starts from a number of specific observations (hopefully a large and representative number) and proposes a generalization which is true of similar cases. Thus if all the students we have seen like beer, then we have reason to believe that other students like beer. However, the observation does not lead logically to this conclusion, and indeed it is not clear how far we can extend the generalization: to all students, to most, or only to many? Other problems include whether our sample was a good one: perhaps the students we observed were not typical.

Deductive reasoning takes place within a closed system, in the sense that all the information is already contained in the premises: implications are merely made explicit by argument. It studies ways in which sentences follow logically from other sentences, and thereby relates propositions (premises) to other propositions (conclusions). The conclusions would be true in all possible worlds, because they depend on the meaning of the words. Deduction can lead to new knowledge only in the sense of a new perspective on old knowledge which is already contained (implicitly) in the premises. However, one of its weaknesses is that it can tell us nothing about the truth of the premises.

Inductive reasoning claims to go beyond the particular starting point to a generalization about cases which we have not observed. It relates individual observations to general statements. This is its strength: we have confidence in the starting point since we have observed these cases to be true, and it tells us something new by going from the particular to the general. However, this is also its weakness, since we cannot be certain about what we might see in the future. In addition, a generalization is not an explanation: it says we will observe more of the same, but does not explain why. (We might predict that the next student we meet also likes beer, but we do not know whether this is due to peer pressure, students having more money and spare time than is good for them, their needing alcohol due to depression brought on by too much work, or to depression brought on by too little work and subsequent fear of exams.)

Deduction and induction are often assumed to be symmetrically related. Deduction starts from premises, and goes from the general to the specific. Induction starts from observations, and goes from the specific to the general. However, this opposition is only apparent. Both deduction and induction assume reliable starting points: either self-evident premises or observed facts. Deductive logic simply assumes the truth of the premises, and treats only their consequences in a possible world. But inductive logic assumes the reliability of the initial observations in the real world, and since all observations are already interpretations and open to all kinds of potential errors, they can never be certain. If the initial evidence is unreliable, then the conclusions cannot be reliable [19, pp. 221-223].

When we talk of drawing generalizations from a finite sample of observations, we ought to distinguish between three rather different situations. First, if we have simply observed all the members of a group, then it is quite possible to summarize the observations quantitatively (e.g. 90 per cent of this class of 100 students like beer), and this is not open to any problems of generalizing to a larger population. Second, if we repeatedly observe some phenomenon (e.g. students drinking beer), we might think it likely that we will see further similar cases in future, without making any claims about numbers or proportions. That is, we will have precedents for such similar cases. (This is sometimes called 'eduction'.) Third – and this is the difficult case which is usually meant by induction – we can observe only a (small) finite sample of a (very large) population, but we wish to make predictions about this large open-ended population.

5 Some History

Modern ideas about inductive reasoning are often traced back to the early 1600s, when Francis Bacon argued that scientific progress must be based on systematic data collection and observation (though he himself admits that some of his main points had been made by Plato). Bacon rejected dogma and authority as sources of knowledge, and criticized deductive reasoning as being similar to spiders making webs of knowledge out of their own substance [21, p. 26, p. 55]. In its stead, he proposed methodically recording observations, and then proceeding gradually and cumulatively towards general principles. He also clearly understood the difference between positive and negative observations, commenting that 'major est vis instantiae negativae' (the force of the negative instance is greater):

"The human understanding [...] forces everything to add fresh support and confirmation; and although more cogent and abundant instances may exist to the **contrary**, yet either does not observe or despises them [...]. It was well answered by him who was shown in a temple the votive tablets by such as had escaped the peril of shipwreck, and was pressed as to whether he would then recognise the power of the gods, by an enquiry; 'But where are the portraits of those who have perished in spite of their vows?' [...] It is the peculiar and perpetual error of the human understanding to be more moved and excited by **affirmatives** than by **negatives**, whereas it ought duly and regularly to be impartial; nay, in establishing any true axiom, the **negative instance** is the **more powerful**." [1, aphorism 46. Emphasis added.].

This is the point about the asymmetry of confirming and falsifying data that Popper much later built into a demarcation criterion for science. A confirming instance is just one more instance which neither proves, nor even makes more probable, a conclusion, since one single counter-example may falsify a hypothesis. One cannot prove that a statement is true (unless in the closed fashion of a simple deduction), but one can prove that a statement is false. (Though see Popper on why he is not a 'naive falsificationist': e.g. [19, p. xxxiii].)

Not everyone has shared the view that Bacon's ideas were particularly original. In a very long, often very funny, review of an edition of Bacon's works (which is often a vitriolic attack on Bacon's character, rather than a review of the book), Lord Macaulay [13] questions the originality of the concept of induction. Here is one of his more ironic statements:

"The vulgar notion about Bacon we take to be this, that he invented a new method of arriving at truth, which method is called Induction. [...] The inductive method has been practised ever since the beginning of the world by every human being. It is constantly practised by the most ignorant clown, by the most thoughtless schoolboy [...]. That method leads the clown to the conclusion that if he sows barley he shall not reap wheat. By that method the schoolboy learns that a cloudy day is the best for catching trout. [...] Not only is it not true that Bacon invented the inductive method; but it is not true that he was the first person who correctly analysed that method." [13, pp. 406-408.].

Macaulay continues in this vein for another page or two. (Quinton [21] provides a more balanced short account of Bacon's ideas.) Bacon is also standardly criticized for his naive faith in the possibility of reliable unbiased observation (he thought that we could start from our intuitions (NB!) that some observations and sense-perceptions were self-evident), and for his insistence on the careful and rather timid plodding accumulation of data, as opposed to the leaps of imagination and guesswork which lead to real progress in science [19, pp. 222-223].

In 1758, David Hume had already made the point that there is nothing new about the idea that we learn from experience: 'none but a fool or madman will ever pretend to dispute the authority of experience' and 'it is certain that the most ignorant and stupid peasants – nay infants, nay even brute beasts – improve by experience'. Hume admitted that we cannot avoid jumping to inductive conclusions. It is an unavoidable mental habit, and a perfectly reasonable thing to do, indeed often the only thing to do. But this is a matter of necessary everyday custom and habit, not of logic. He makes these comments in his famous discussion of different types of inference, where he distinguishes clearly between the psychological certainty which induction seems to bring, and the impossibility of inductive generalizations providing logical certainty, since any predictions about the future are open to potential counter-example. Any observations we have made were made in the past, but there is no logical reason to assume that the future will resemble the past, since there can always be new cases and new observations:

"As to **past** Experience, it can be allowed to give direct and certain information of those precise objects only, and that precise period of time, which fell under its cognizance: but why this experience should be extended to **future** times, and to other objects, which for aught we know, may be only in appearance similar; this is the main question on which I would insist. [. . .] The consequence seems nowise necessary. [. . .] If we be, therefore, engaged in arguments to put trust in **past** experience, and make it the standard of our **future** judgement, these arguments must be **probable** only [. . .]. All our experimental conclusions proceed upon the supposition that the **future** will be conformable to the **past**." [9, Section 4. Emphasis added].

Hume continues in this vein for another page or two, making the same point several times in slightly different words, that all inferences from experience assume 'that the future will resemble the past', but that there is no proof that this will be so, since 'the course of nature may change'.

Hume has been read in two rather different ways. He is traditionally interpreted as simply rejecting induction as a rational procedure. Alternatively, he is interpreted as merely arguing that induction must be rejected as a rational procedure, only if reason is interpreted in very narrow deductive way. These different readings of Hume himself are not of direct concern in this article, but Noonan [16, pp. 116-131] summarizes the various positions and argues for this second view.

In the 1960s, these points from Bacon and Hume were developed by Popper [19], who refers [19, p. 62] rather ironically to Hume's 'problem of tomorrow', which he regards as a simple philosophical muddle. First, it is rather likely that the future will not resemble the past. Second, Hume thought there was a paradox in saying both that the laws of nature may change, and also that the laws of nature are just these things which we think can never change. Popper points out that there is no paradox: it simply means that we formulated the laws wrongly in the first place, and shows again that our theories are always open to correction and counter-example. Popper (e.g. [19, pp. 11-158]) agrees with Hume [9, pp. 31-32] that there are countless regularities in nature on which we rely in practice, but that we cannot logically reason from singular observations to general laws of nature. This landed Hume in what he saw as another clash between the invalidity of induction and the principle of empiricism. Since Hume was unwilling to abandon empiricism, he concluded pessimistically that we have to rely on habit, but cannot rely on reason, and this drove him into an irrationalist position. Popper [19, pp. 32-33] accepts both the argument against induction and the principle of empiricism (that theories are accepted or rejected on the basis of observational evidence), but changes the role which observation plays. Observation is essential, but it cannot prove that a theory is true, only that it is false. Popper argues at length that there is no such thing as induction, since hypotheses are always provisional conjectures (a) which are influenced by prior knowledge and expectations, and (b) which may turn out to be false due to refutation by counter-example.

So there are in fact three stages which should be distinguished. The problem is how we get (1) from exploratory data analysis (2) to hypotheses (3) which we can test. First, we explore a mass of messy data (e.g. a large corpus plus associated concordances, word lists, statistics on frequent collocations, etc.) and these facts somehow suggest a theory. But how we arrive at the theory (argues Popper) is irrelevant to its possible truth. In our search for patterns we certainly get ideas from observations, but we never draw true inductive inferences, since we always start from expectations. Second, we formulate generalizations and hypotheses (e.g. dictionary entries about word meaning). These hypotheses do not emerge from pure logic, since they also depend on the categories which we use to classify and interpret the world (e.g. a distinction between denotation and connotation) and on our assumptions (e.g. that words have relatively stable meanings in a speech community). Third comes a process of formulating and testing consequences from these hypotheses. With reference to dictionaries, I am not sure if such testing is ever carried out systematically, or indeed whether it could be carried out in practice across large comprehensive dictionaries.

In summary, I think there is nowadays general acceptance of Popper's view that there is no such thing as pure induction. McGuire [15, p. 399] provides a useful summary: knowledge is always an underrepresentation (since there is always selective attention to data), a misrepresentation (since it is influenced by the knower), and an overrepresentation (since it is based on inferences which go beyond the given data).

6 Some Lexicographic Examples

These questions of research method are unresolved despite some 400 years of intense discussion. Phrased rather negatively, it would be valuable if corpus linguists were at least more aware of these questions. Words such as *deduction* and *induction* do not appear in several widely-used introductions to corpus methods (including my own, I must admit [26, 27]). Phrased positively, corpus linguists could use their unique combination of very large data sets, computer-assisted quantitative methods and human intuition to make some conceptual progress on the problem. (Relevant methods, especially cyclic procedures of data analysis, have been described by Sinclair [24], Sinclair, Mason et al. [25], and Barnbrook [2].) So, from a rather abstract historical discussion, I return now to the concrete questions I raised at the beginning.

Below I will be using the codings for pragmatic connotations used by the Cobuild [7] dictionary. In an excellent article, Channell [4], the linguist who developed the pragmatic coding framework for the dictionary, discusses the methods used to discover these connotations, and provides clear examples of how the lexicographers worked in practice. The essential method involved using concordance lines to display recurrent patterns in the use of a given word. Channell then illustrates how statements about evaluative connotations can be 'based in systematic observation', which makes it possible to 'produc[e] a sound description' [4, p. 39]. Concordance data provide facts which are not accessible to introspection, and are 'not visible from the study of single examples' [4, p. 40]. The main part of the article makes no simplistic claims about automatic methods (and does not use the term 'induction' at all). However, the summary section at the end makes claims which cannot be taken literally: *'Without recourse to intuitions*, quantitative data show clear evidence of where there is an evaluative polarity to an item' [4, p. 54, emphasis added]. This implies more automatism than is possible, and a more guarded statement would be more accurate: that the concordance software can be instructed to find the appropriate data (possibly with further help from the kind of software described by Sinclair et al. [25]), but that these data still require the lexicographers' intuition to extract the significant patterns.

Given the broad historical discussion so far, the following section may seem disappointingly modest, but it does ask a specific question: How far do dictionaries agree in their definitions of words? Lexicographers have large corpora and associated data-sets, plus the hermeneutic procedures described by Channell [4]. Do these procedures lead to consistent results? Inter-subjective agreement would not of course prove that the analysis is correct: the analysts may all have been misled in the same way, Hume would point out that 'the course of nature may change', and Popper would point out that corroboration does not even increase the probability of a generalization being true. But disagreement would point to a potential problem.

We have no independent statements of what the meaning of a word is, but we can compare definitions of the same words in the four highly comparable dictionaries used above for comparing the definitions of CRONY. They are all corpus-based, all intended for advanced foreign learners, and all published in 1995:

- **CIDE**: Cambridge International Dictionary, based on the Cambridge Language Survey corpus of 100 million words.
- **Cobuild**: Collins Cobuild English Dictionary, 2nd ed., based on the Bank of English corpus of over 200 million words.
- **LDOCE**: Longman Dictionary of Contemporary English, 3rd ed., based on the Longman Corpus Network and the British National Corpus of 140 million words.
- **OALD**: Oxford Advanced Learner's Dictionary, 5th ed., based on the British National Corpus of 100 million words, and an American English Corpus of 40 million words.

The remainder of the paper discusses cases where these dictionaries disagree in their analyses of individual words, and therefore discusses the uneasy balance in corpus studies between automatic and intuitive methods.

7 The Frequency of Disagreements

I will start with the simplifying assumption that dictionaries may tend to
agree most easily over the denotations of words, but less easily over their
connotations, (though I will also question below whether this distinction can
be maintained). All(?) dictionaries use labels of one kind or another (such as
"informal" and "derogatory") for evaluative connotations, but these labels are
notoriously unstandardized. Cobuild [7] attempts to systematize such description by placing the label PRAGMATICS next to words which have features
of usage which need to be specially signalled. This label is used for several
rather distinct purposes, including conversational markers (e.g. *anyway*), and
words where the core semantic meaning already denotes something good or
bad (e.g. *dreadful*). However, the dictionary also uses a range of syntactic
forms (discussed in detail by Barnbrook [2]), in order to explicitly mark evaluative connotations and speaker attitudes with phrases such as 'if you say
x, you want to emphasize it', or 'if you say x, you approve/disapprove of it',
or 'if you say x, it is because you are irritated'. If we take words labelled
in this way, and compare their definitions in the four dictionaries, we would
have an initial rough sample for discussion. So, from these four corpus-based
dictionaries, a sample of words and phrases was selected as follows:

1. I started on Cobuild page 5, and took the first word (i.e. word sense) with
   the PRAGMATICS label, where the entry has a further explicit evaluative
   descriptor as defined above.
2. I selected words in this way every hundred pages (pages 5, 105, 205, etc.).
   If there was no such word on the page, I went to the next page.
3. I compared the definitions in the four dictionaries. (It was of course possible that the given sense was not listed at all in a dictionary.)

Given my starting point, all entries selected from Cobuild have, by definition, an explicit evaluative label. However, I do not distinguish further
between the dictionaries below. (And the main comparisons are amongst only
the 1995 printed versions of the four dictionaries. Different printed and CD-
ROM editions of the dictionaries, between 1995 and 2000, often have distinct
differences of emphasis in their definitions and give different citation examples.) Here are some brief comments on each of the words in this small sample.

1. page 6: **absolute** (sense 2)
   All four dictionaries explicitly label this word as "emphatic" or "expressing a strong opinion". One dictionary adds that it "emphasizes your opinion [. . .] especially when you think [that something is] very bad, stupid",
   etc. (*absolute disgrace*). But the other three dictionaries give mixture of
   negative, neutral and positive citations (*absolute nonsense, absolute minimum, absolute trust*).
2. page 106: (do something) **behind someone's back** (*back* sense 10)

All four dictionaries give the denotation of "doing something without a person's knowledge or agreement".

But two of the dictionaries add that it is "disapproving" (e.g. *saying nasty things behind his back*).

3. page 205: **brood** (sense 2)

All four dictionaries give the denotation of "a family of young children". One dictionary adds: "when you want to emphasize that there are a lot of them". The other three dictionaries label the usage "jocular" or "humorous".

(For what it is worth, my intuitive judgement is that it could be rather risky and potentially insulting to use the word "humorously".)

4. page 305: **cohort** (sense 1)

All four dictionaries give the denotation of a person's companions or supporters, and explicitly label it as "disapproving" or "derogatory" (i.e. rather similar to *cronies*). Three of the dictionaries label this usage as especially American, or give a citation which implies an American usage.

5. page 405: **cut something out** (sense 4)

All four dictionaries agree that if you tell someone to *cut it out*, it is because their behaviour is "annoying" and/or you are "irritated".

6. page 505: **would not dream of** (*dream* sense 9)

All four dictionaries either label this explicitly as "emphasizing" that the speaker would never do something because they think it is (morally) wrong or unpleasant, or imply this in the citation.

It is not strictly within my comparisons, however another dictionary gives no such restriction: *I wouldn't dream of going without you*. This seems more accurate than the four dictionaries under comparison.

7. page 605: **far be it from me** (*far* sense 17)

Two dictionaries give a neutral gloss: "I certainly would not want to do this".

The other two agree that the speaker is about to criticize someone, but one thinks that the speaker wants to appear hostile, whereas the second thinks that s/he wants to pretend to agree.

8. page 705: **gerrymandering**

All four dictionaries agree that the word denotes "altering political boundaries to give advantage to a political party". All use the word "unfair", which implies speaker attitude, and two dictionaries explicitly label it "disapproving" or "derogatory".

9. page 805: **hole** (sense 5)

For the sense of referring to a place (e.g. usually where someone lives) as a *hole*, all four dictionaries label the usage as "unpleasant" and "informal".

10. page 905: **not a jot** (*jot* sense 2)

Two dictionaries label the phrase "old-fashioned". One labels it "informal" (which seems logically inconsistent with "old fashioned"), and one has neither of these labels.

11. page 1007: **on the make** (*make* 3 sense 9)

Three dictionaries label the phrase "disapproving".

One of these goes further and labels it as possibly implying "illegal and immoral" methods. But one dictionary labels the phrase merely "informal".

12. page 1106: **nepotism**

All four dictionaries agree on the denotation: using power to gain advantage for friends or relatives.

Two dictionaries label this practice "unfair" and as signalling the speaker's "disapproval". The other two do not use any such evaluative descriptors: everything is left to implication (see below).

13. page 1205: **party politics** (sense 2)

Here the differences between the dictionaries are not entirely distinguishable from the issue of whether a word has distinct senses: a literal denotation (= "relating to political parties") and an extended usage which is an "accusation" and which "criticizes" people for doing or saying something which they do not believe.

Only one dictionary gives these attitudinal labels. Two dictionaries give only the first literal sense. One gives only the second sense. One gives both.

14. page 1306: **principled**

One dictionary labels the word "approving"; the other three dictionaries imply this with phrases such as "honest and moral", or "esp good".

Only one dictionary labels it "formal".

15. page 1405: **repetitious**

All four dictionaries label the word "disapproving" or give citations with the word *boring*.

16. page 1505: **self-important**

(This item occurred by chance in my small sample: it is one of the examples discussed in detail by Channell [4].)

Three dictionaries label the word "disapproving". The fourth implies this in the citation (*a self-important, pompous little man*).

17. page 1605: **spendthrift**

All dictionaries give denotations such as "spending money wastefully or extravagantly". Two dictionaries label it "disapproving". The other two use words such as "careless" and "wastes money", and seem to assume that the evaluative connotations of these words are clear.

18. page 1705: **talk down to**

All four dictionaries agree on the denotation: talking to someone as if they are not very intelligent. One dictionary explicitly labels the phrase "disapproving". The other three dictionaries imply this with phrases such as "too simple".

19. page 1805: **two-dimensional** (sense 2)
    One dictionary labels the phrase "critical"; a second labels it "disapproving"; a third implies this with "not very interesting"; the fourth implies it even less directly: "does not seem real".

20. page 1905: **whatever you say** (*whatever* sense 7)
    Two dictionaries agree that this means "you do not believe or accept what someone has said" or "do not really agree" with someone.
    The other two dictionaries have no entry for the phrase.

8 Interpretation

Dictionaries can differ widely in how they represent word-meanings, including how they divide word-meanings into different senses, whether they present senses as separate or as specialized cases of a more general meaning, and so on. (See Kilgarriff [10, 11], who concludes 'I don't believe in word senses'.)

My topic here is more specific: the extent of agreement over attitudinal meanings. First, dictionaries differ in the labels they use, simply because there are no standard terms for presenting pragmatic information, but only a rather small and crude set (e.g. "emphatic" and "formal"), which are not always clearly distinguishable (e.g. "disapproving" and "derogatory"). Second, dictionary entries differ in whether connotations are explicitly labelled, or only implicitly encoded in value-loaded words. In examples above, definitions use words such as *unfair*, which certainly implies disapproval. Third, there are cases which Channell [3] argues are more serious than this type of implication. Consider the definition in one dictionary (Channell's point, but my example) of *nepotism* as "the practice of giving the best jobs to members of your family when you are in a position of power". The lexicographer presumably takes it for granted that this practice is a bad thing, and has (covertly) encoded something that s/he disapproves of. However, there is no evaluative label, the definition contains positive words (*best, power*), and some readers might regard it as behaviour which is completely rational and only to be expected. In another case, as a citation for *principled*, one dictionary gives *a principled stand against federalism*, which seems to imply that federalism is a bad thing (and that the lexicographer may be sharing particularly British prejudices against this form of government)!

This set of twenty definitions is a very small, and clearly not a random, sample. Given my starting point, all the Cobuild definitions obviously contain explicit descriptors of speaker attitude. Nevertheless, this small sample shows a surprising number of differences between the dictionaries. Depending on how strictly one interprets the conventions used by the four dictionaries, there is good agreement across all four of them in perhaps half of the cases. There is one case where two dictionaries did not have the word or phrase at all, and in the other ten or so cases, there is either distinctly different information (e.g. the same phrase is labelled "old-fashioned" and "informal"), or at least distinctly different emphases in the connotations given (e.g. a speaker wants to appear "hostile" or wants to "pretend to agree"). How might one explain these differences?

1. Perhaps the lexicographers looked at different corpora, which were not entirely comparable samples, and drew legitimate generalizations from different data. In the case of relatively infrequent words and their even less frequent combinations in longer phrases (e.g. *brood?*), perhaps there were not enough examples even in a large corpus to allow valid generalizations. This would throw doubt on claims that the dictionaries are based on representative samples of language use. This explanation is, however, not very convincing, since all four dictionaries are based on large mixed corpora.

2. A second explanation might be that native speakers do not always agree in their use of words, (which would explain why corpus samples differ), and that the search for meanings shared across a discourse community is misconceived. This could imply in turn that language use is more variable than is admitted in dictionaries of general English. This explanation sounds superficially similar to the frequently heard excuse in neo-Chomskyan linguistics of why native speaker intuitions often fail to agree: 'it's grammatical in my dialect'. However, the reasons for this variability in word meanings may be more interesting: I return to this below.

3. This could lead to a third, methodological, problem. If analyses are based on highly variable data, they should logically be formulated as probabilistic statements. First, dictionaries only seldom adopt this strategy (cf. [7] on *foreigner*: "some people believe this word is slightly offensive"). Second, it is difficult to see how such definitions can be refutable in any clear way, since the concept of individual counter-examples does not apply. And, third, if variable data are given a categorial description, this implies that the dictionary is prescribing one usage as correct. If there is a difference between this meaning and my usage, does this imply that I am using the word wrongly?

Barnbrook [2, pp. 36-39] summarizes some of the main issues. Is meaning in the mind of the lexicographer or in the usage of the speech community? Do lexicographers decide, on the basis of their native speaker competence, what the meaning of a word is, and then search for corpus examples which illustrate this meaning? (Is the lexicographer the source of the meaning?) Or do they discover the meaning in the corpus data (induce it from the data)? In what sense can semantic information be '**derived**' from '**reliable** sources' and '**based directly** on representative corpus data' [2, p. 46, emphasis added]? In addition, does a word have a 'correct' meaning, which can be illustrated by such data and then recorded in dictionaries?

9 A Final Example: Size of Context

There is a fourth possible explanation of disagreements between dictionaries. Here is a final illustration of the hermeneutic method, but also of a problem, which is simple in this case, but unresolved in general: how much context (co-text) is relevant to deciding meaning, especially connotation? Here are some examples (from the 100-million word British National Corpus) of the word *horde*, followed by *of* and a noun phrase. At first sight, there seem to be examples of both neutral and negatively evaluated uses, respectively:

- horde of children; horde of courtiers; horde of souls; horde of tiny crablets; horde of young girls; horde of young men; horde of volunteers
- horde of disturbed bats; horde of goblins; horde of hooligans; horde of the damned from hell; horde of troublesome workmen

However, collocates in a larger context reveal the apparently neutral and positive examples as often very negatively evaluated indeed:

- **horde of children**: 'He left it [his carriage] [...] where it attracted the interest of a **horde of children** [...] Garbage was piled high in corners, and Maggie watched Sarah stepping carefully so as not to tread in the filth.'
  (A second example of *horde of children* collocates with *awfully crowded*. An example of the plural *hordes of children* collocates with *a slum*.)
- **horde of courtiers**: '[T]he scene was one of frenetic confusion, servants scurrying around, shouting and gesticulating [...] the situation was not improved by a **horde of courtiers** standing around also issuing their instructions to a vast army of retainers [...].'
- **horde of souls**: '[A] vast **horde of souls** were rumbling towards heaven. There were whole companies of white-trash, clean for the first time in their whole lives, and bands of black niggers in white robes, and battalions of freaks and lunatics shouting and clapping and leaping like frogs'.

(These three examples are all from fiction: the first two from novels by Pamely Pope and Doherty Crown, the third from a famous short story by Flannery O'Connor.) Similarly, the *horde of tiny crablets* is described as having an extravagant and wasteful breeding strategy; the *horde of young girls* is mobbing a limousine carrying a pop group; the *horde of young men* is a critical reference to the speaker's rivals for the attentions of a young lady; and the *horde of volunteers* refers to volunteer soldiers in an incident in Scottish history.

In this case, the four dictionaries agree on the unpleasant connotations:

- **horde**: A large, noisy and excited crowd. *A horde of students on bikes made crossing the road difficult.* (CIDE [6]).

- **horde**: (sometimes derogatory) A very large group, especially of people; a huge crowd. *Fans descended on the concert hall in their hordes.* (OALD [17]).
- **horde**: A large crowd moving in a noisy uncontrolled way. *Hordes of people milling around the station.* (LDOCE [12]).
- **horde**: If you describe a crowd of people as a horde, you mean that the crowd is very large and excited and, often, rather frightening. ... *a horde of people was screaming for tickets.* (Cobuild [7]).

The four definitions, plus the citations, differ in emphasis, ranging from a crowd which is "excited" (not necessarily a bad thing), "noisy" (sounds like a nuisance), *milling around* (implies aimless, useless activity), "uncontrolled" (might be dangerous), to causing "difficulties" and "frightening". That is, they differ in how explicit the connotations are made and how strong they are claimed to be. In terms of methodology, we do not know exactly how much context the lexicographers have used in phrasing their definitions. More fundamentally, we do not know when connotations should be made part of the denotation, or when they should be generated by inference from common-sense knowledge (that excited, aimless and uncontrolled crowds can become dangerous).

10 A Final Attempt at Explanation

Given practical constraints (including restrictions of space, the conservativeness of their users, and ultimately commercial pressures) dictionaries are forced to present their definitions as though words have definitive meanings and/or distinct senses. This is, however, a misleading model of semantics. Sampson [23, pp. 180-207] presents a detailed account of why this is so, and why linguistic and encyclopedic knowledge cannot be neatly separated. Due to cultural changes and the need for new concepts, the meanings of words change, many words encode concepts which have been recently institutionalized in our culture, and due to their different experiences and cultural beliefs, speakers simply reach different conclusions about word meanings. Meanings are therefore unpredictable and creative and, in this sense, Hume is right: all our observations were made in the past, but there is no logical reason to assume that the future will resemble the past.

Perhaps the problem of recording attitudinal meanings is not then, after all, separable from the question of how many distinct senses of a word a dictionary should list. Pustejovsky [20] also argues that linguistic and encyclopedic knowledge cannot be neatly separated, and formalizes a position which seems close to Sampson's. He argues that words do not have a fixed number of distinct senses, but that these senses can be generated by a fixed number of rules which operate in context. For example, we know that *Susan has finished her book* can be interpreted in two ways ("finished reading" or "finished writing"). This does not mean that *finish* has two senses, but that we know things about books (and about Susan, for example, whether she is an avid reader or a budding author).

Take again the case of CRONY. Speakers draw inferences about the meanings of words, and these inferences are based partly on what they know about people's behaviour (or possibly men's behaviour: do women have cronies?). If people (men?) are friends, then they spend time together, often go drinking together, form alliances, acquire commitments to each other, and may then be led into supporting dubious, and sometimes criminal, activities. That is, the word makes sense only within larger cultural frames, which generate implications. The same points hold for words and phrases such as *brood* (how large is a large family?), *nepotism* (when does helping one's relatives become illegal favouritism?), or *party politics* (when does working for one's own party turn into ignoring the general good?). Since the meanings depend on assumed common sense knowledge of cultural schemas, and on shared evaluations of what is mildly disapproved of versus regarded as immoral or illegal, they cannot have fixed meanings.

These observations start to explain why it is particularly difficult to handle pragmatic meaning in dictionary entries, but they do not solve the problem. Indeed, they produce two different problems. (1) When do such inferences become conventionalized? When do connotations become part of the denotation, and therefore when do words such as CRONY become insulting independent of context? (2) If words make sense only within larger cultural frames, then this implies that their meanings should be represented within the kind of frame semantics which has long been recommended by Fillmore (e.g. [8]).

11 Conclusions

Despite his overall argument for 'empirical linguistics', Sampson's conclusion about semantics is pessimistic and uncompromising:

"[O]ne might expect to find [...] areas which cannot be treated scientifically at all. The outstanding example [...] is word meaning. [...] [A]nalysis of word meaning cannot be part of empirical science. [...] Word meanings are not among the phenomena which can be covered by empirical, predictive scientific theories." [23, p. 181, p. 206].

It would be more helpful to distinguish between different stages of research. It may be misguided to try and 'establish a rigorous scientific analysis of word meanings' [23, p. 197]. Nevertheless, there are ways of collecting empirical observational data on meaning, even if there can be no automatic analysis of the data. There are automatic procedures which can select data and put them into a convenient form for the human analyst, but the interpretation of these data can be only partly automated or computer-assisted, since it requires the intuition and experience of the lexicographers.

In a fascinating book, Macfarlane and Martin [14] argue that a specific technological discovery decisively influenced what we often call the Scientific Revolution, that is, the development of modern scientific methods and findings from the 1600s. This was the discovery of how glass could be used to make scientific instruments. Their most telling points are (1) that, without glass, we would have no test-tubes, retorts, thermometers or barometers, and that many forms of experiment and measurement would be impossible, and (2) that, without glass, we would have no lenses, and therefore no magnifying glasses, microscopes, telescopes or cameras, and no instruments for observing small things, distant things and fast-moving things. As a consequence, micro-biology, astronomy and many other disciplines could simply not have developed. In other words, there was a close relation between the development of a material technology, a scientific method and a whole mode of thought.

Glass instruments made it possible to collect many new observations, and the authors point out that the telescope and the microscope were invented in the late 1500s, only a few years before Francis Bacon was making his points about empirical methods: his *New Atlantis* was written around 1610. They carefully point out that glass is an enabling, and probably necessary, technology for the development of science: as they say, it is the only substance which directly influences how humans see the world, and reveals things which were previously 'invisible to the naked eye' [14, p. 81]. However, they emphasize that it was not a sufficient cause and there were many other factors involved in the development of a sceptical scientific method.

How does this relate to corpus linguistics? Computer-readable corpora and access software also allow linguists to see things that they have never seen before. They are no longer restricted to observing their own individual introspections or short individual texts ('the extent of language that can comfortably be accommodated on the average blackboard': [18, p. 8]), but can now observe large-scale patterns of language behaviour across large text-collections, which are evidence of the mental lexicon of thousands of speakers across a speech community. Technologies alter what can be observed, suggest problems, make scholars satisfied with particular answers, and therefore alter descriptions and theories. For linguistics, this was true of the invention of written language and of tape-recorders. Computers are particularly good at repetitive tasks, and it is types of repetition which are particularly significant in corpus semantics. Since corpora plus software are now one of the 'technologies of the mind' [14, p. 31] of contemporary linguistics, linguists need to be clear about the relations between these technologies and their observations, generalizations and theories.

Acknowledgements

For help with data collection I am grateful to Bettina Starcke. For discussion of some of the data and of issues relating to practical dictionary-making, I am

grateful to Joanna Channell (one of the Assistant Editors responsible for the
Pragmatics coding in Cobuild [7]) and Gwynneth Fox (Editorial Director of
Cobuild [7]).

References

[1] F. Bacon. *Novum Organum (Translation by B. Montague, 1854)*. Parry
    & MacMillan, Philadelphia, 1620.

[2] G. Barnbrook. *Defining Language*. Benjamins, Amsterdam, 2002.

[3] J. Channell. Cultural Assumptions and Shared Knowledge in English
    Dictionaries. Unpublished lecture, University of Trier, 1995.

[4] J. Channell. Corpus-based Analysis of Evaluative Lexis. In S. Hunston
    and G. Thompson, editors, *Evaluation in Text*, pages 38–55. Oxford Uni-
    versity Press, Oxford, 2000.

[5] N. Chomsky. *Syntactic Structures*. Mouton, The Hague, 1957.

[6] CIDE. *Cambridge International Dictionary of English*. Cambridge Uni-
    versity Press, Cambridge, 1995.

[7] Cobuild. *Collins COBUILD English Dictionary*. Harper Collins, London,
    1995.

[8] C. J. Fillmore and B. T. S. Atkins. Starting where the Dictionaries
    Stop: The Challenge of Corpus Lexicography. In B. T. S. Atkins and
    A. Zampolli, editors, *Computational Approaches to the Lexicon*, pages
    349–93. Oxford University Press, Oxford, 1994.

[9] D. Hume. *An Enquiry Concerning Human Understanding*. Harvard Clas-
    sics. Collier & Son, Edinburgh, 1910.

[10] A. Kilgarriff. Dictionary Word Sense Distinctions. *Computers and the
    Humanities*, 26:365–87, 1993.

[11] A. Kilgarriff. "I don't believe in word senses". *Computers and the Hu-
    manities*, 31:91–113, 1997.

[12] LDOCE. *Longman Dictionary of Contemporary English*. Longman, Lon-
    don, 1995.

[13] L. T. B. Macaulay. Lord Bacon. In *Critical and Historical Essays*, pages
    280–429. Longmans, Green & Co. 1878, 1837.

[14] A. Macfarlane and G. Martin. *The Glass Bathyscaphe*. Profile, London,
    2002.

[15] W. J. McGuire. *Constructing Social Psychology*. Cambridge University
    Press, Cambridge, 1999.

[16] H. W. Noonan. *Hume on Knowledge*. Routledge, London, 1999.

[17] OALD. *Oxford Advanced Learner's Dictionary*. Oxford University Press,
    Oxford, 1995.

[18] M. Phillips. *Lexical Structure of Text*. Discourse Analysis Monograph 12.
    University of Birmingham Printing Section, English Language Research,
    University of Birmingham, 1989.

[19] K. R. Popper. *Realism and the Aim of Science*. Routledge, London, 1983.

[20] J. Pustejovsky. The Generative Lexicon. *Computational Linguistics*, 17(4):409–41, 1991.

[21] A. Quinton. *Francis Bacon*. Oxford University Press, Oxford, 1980.

[22] M. Rundell. Corpora, Evidence and Intuition. Corpora List, 1 November 2001. http://helmer.hit.uib.no/2001-4/0080.htm.

[23] G. Sampson. *Empirical Linguistics*. Continuum, London, 2001.

[24] J. Sinclair. *Corpus, Concordance, Collocation*. Oxford University Press, Oxford, 1991.

[25] J. Sinclair, O. Mason, J. Ball, and G. Barnbrook. Language Independent Statistical Software for Corpus Exploration. *Computers and the Humanities*, 31:229–55, 1998.

[26] M. Stubbs. *Text and Corpus Analysis*. Blackwell, Oxford, 1996.

[27] M. Stubbs. *Words and Phrases*. Blackwell, Oxford, 2001.

Linguistic Information Modeling: From Kilivila Verb Morphology to RelaxNG

Dieter Metzing and Jens Pönninghaus

Bielefeld University
{Dieter.Metzing,Jens.Poenninghaus}@uni-bielefeld.de

Summary. We will explore the role of an advanced type of document grammar, RelaxNG, in the context of different approaches to the formalization of linguistic regularities based on corpora and XML annotations. Our domain of exploration will be Kilivila verb morphology. The following topics will be focused on: Which kind of regularities in the domain can be expressed given formal limitations of document grammars, i.e. tree grammars? Which linguistic analyses may be taken as a basis for document grammar development? In which way can a document grammar be sensitive to properties of annotations and raw data (document validation and data validation)? Which kinds of formalization may be helpful in the (semi-automatic) development of a document grammar in the case explored? In the first part we will consider aspects of Kilivila verb morphology from the point of view of linguistic analyses. In the second part different strategies for the development of a RelaxNG based document grammar will be examined.

1 From XML and Corpora to Linguistic Formalizations

The introduction of the Extensible Markup Language (XML) as a standard for the representation and/or the interchange of electronic documents has led to far reaching changes in disciplines concerned with the analysis, representation, generation or archiving of natural language data and textual information; for these disciplines XML-based language resources and advanced processing tools are becoming more and more important. Since a crucial property of XML documents is their structural and/or semantic markup added to original data and since this information augments their processing and application value, it is also important to control the validity of markup, i.e. document quality. And since the process of adding markup may be resource-intensive, this control is even more important. However, probably due to limited capabilities of Document Type Definitions (DTDs) [3], XML document grammars are widely perceived as being almost exclusively suitable for 'storage format' prescriptions, and their role in the validation of XML annotated content is often ignored. In this article we want to go a step further and use a new type of document

D. Metzing and J. Pönninghaus: *Linguistic Information Modeling: from Kilivila Verb Morphology to RelaxNG*, StudFuzz **209**, 255–276 (2007)
www.springerlink.com

grammar, RelaxNG, for an extended concept of validation such that not only single annotation tags are considered but also their co-occurrence constraints and their appropriateness to primary data. We exemplify this concept with respect to XML annotated verb structures and morphological co-occurrence constraints.

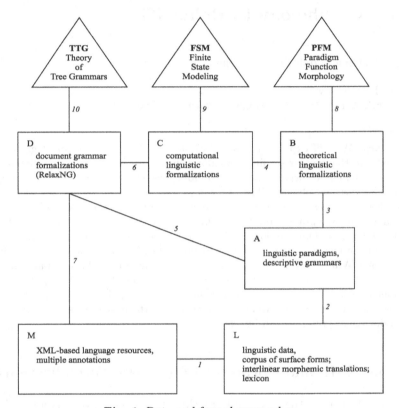

Fig. 1. Data and formal approaches.

In this article we want to place the development of a document grammar in a context relevant for different linguistic disciplines such as corpus linguistics, descriptive linguistics, theoretical linguistics, and computational linguistics. If we assume that XML-based language resources are becoming more important, then this will probably have also consequences for these linguistic disciplines, their relationships and interdisciplinary cooperation.

The context into which we place the development of a document grammar is shown in Fig. 1. On the one hand there is data, i.e. linguistic data (L) and XML-based language resources (M). On the other hand there are formalizations of different types (A,B,C,D) related to different theoretical backgrounds (Paradigm Function Morphology (PFM), Finite State Morphology (FSM),

Theory of Tree Grammars (TTG)). The different representations considered are connected through different relationships (1-10). Note that we chose only some relationships, not all possible or desirable ones. Note also that document grammars typically validate markup, not primary data. Link 7 in Fig. 1 refers to formalizations that combine data and its markup. The following kinds of relationship between representations may be distinguished: a relationship of conversion (1), a relationship of systematization and grammar development (2), relationships of (computational) linguistic formalization (3,4), the relationship of constraint exportation (5,6), the relationship of validation (7) and of theoretical foundation (8-10).

2 Corpora of Language Data

Kilivila, an Austronesian language, has been extensively explored by Gunter Senft (Senft, 1986, 1993 [16, 17]) and different aspects of the language have been described on the basis of large corpora of audio and video data, annotated according to widely accepted conventions, e.g. the 'interlinear morphemic translation' (Lehmann 1983 [11]). The corpus data have the following structure shown by an example:

Example 1. **Mtona tau eyosali yata kanunuva, ekepapi.**
  m-to-na   tau e-yosali ya-ta(la)   kanunuva e-kepapi
  Dem-CP.male-Dem man 3.-stretch- CP.flexible-one sheet   3.-hold.tight
  DEM   N   V   NUM   N   V
  This man is tightly holding one sheet, he holds it tightly.
  (DEM: demonstrative; CP: classifier particle)

Kilivila corpora are currently accessible as a browsable corpus of the Max Planck Institute for Psycholinguistics.[1] The conversion of a Kilivila corpus into an annotated XML corpus has been tested as part of a text-technological project at the University of Bielefeld and an XML version of the corpus has also been developed at the MPI for Psycholinguistics (cf. Broeder et al. 2005 [4]).

  This simple example shows that language documentation is multilevel documentation from the beginning and even more so when more than morphological structures are considered (as for example in multimodal communication). Moreover the different levels have to be related in a systematic way. This requirement continues to play an important role in XML-based language documentation. Different approaches have been developed, e.g. Annotation Graphs (cf. Bird and Liberman [2]), NITE (cf. Carletta et al. [5]) and the Multiple Annotation Approach (cf. Witt [21, 22]).

---

[1] http://corpus1.mpi.nl/BC/IMDI-corpora/

3 Systematized Regularities and Formalizations

A subpart of Kilivila morphology seems to be complex enough for our purpose: Kilivila verb morphology and requirements for document grammar development. From the point of view of linguistic information modeling the following question comes into focus: May a document grammar be an adequate framework for the expression of the observed linguistic properties, regularities and constraints? In order to answer this question we analyzed at first a systematization of Kilivila verb morphology worked out by Claudia Wegener [20] (cf. Fig. 2). According to Penton et al. [15] this may be called a "linguistic paradigm, a tabulation of linguistic forms to illustrate systematic patterns and variations". In their paper they show how XML technologies can be used to store and render linguistic paradigms. Our goal, however, is to go beyond these technologies and to address questions of information modeling and document grammar development.

Wegener in her linguistic paradigm of Kilivila verb morphology distinguishes 8 position classes: Emphasis, Tense-Aspect-Mood, Person, Root, Object, Plural, Object, Emphasis. At each position special affixes or a placeholder for the root are listed: emphasis marking affixes, tense-aspect-mood marker in order to distinguish different types of action, marker for person and number and affixes for object marking. There are, obviously, three types of constraints:

a. affix position constraints,
b. affix correlation constraints,
c. affix root constraints.

That means affixes may only occupy certain positions relative to the root; affixes may constrain the choice of subsequent affixes, and certain verbs (roots) allow for object marking. These patterns of Kilivila verb morphology are not an exception, similar or even more complex patterns may be found in other languages as well (cf. Inkelas [8]).

With respect to a document grammar, why should information about the constraints distinguished be relevant for a document grammar? There are several answers to this question. First, this information may be used to validate morphological corpus annotations; annotation and linguistic data may not always correspond in the expected way and either this is due to an annotation error or to forms of language usage not yet considered. Second, this information may be integrated into annotation tools. In both these applications, the constraints observed will require a type of document grammar powerful enough for complex constraints.

4 Position Class Morphology and its Formalization

The linguistic paradigm of the Kilivila verb morphology may lead to variants of position class morphology. "Position classes as a whole are typically used in languages with rich inflectional domains in the word" (McDonough [13]). However, without a foundation in terms of a morphological theory "position classes are exclusively formal or virtual entities: they are artifacts of a particular kind of analysis, one that assumes prior, even diachronic, knowledge of morphophonemic structure" (McDonough [13]). It will be shown that such a foundation may be found in Stump's Paradigm Function Morphology (Stump [18]).

The linguistic paradigm in Fig. 2 is a concise representation of structural paradigmatic properties. It could be replaced by a tabulation of all possible syntagmatic realizations. How could this be done? Imagine Fig. 2 as a matrix of cells without content; content can be placed into these empty cells only by paradigm functions. "A paradigm function is a function which, when applied to the root of a lexeme L paired with a set of morphosyntactic properties appropriate to L, determines the word form occupying the corresponding cell in L's paradigm" (Stump [18, p. 32]).

The approach of Paradigm Function Morphology will be illustrated by a very simple example taken from the Kilivila data introduced above. A third person verb form, **eyosali**, will be produced given the root, features, a paradigm function and realization rules.

Step 1: Take a root and the complete set of morphosyntactic features needed.

$$\text{yosali}, \{\ldots, \text{PERS:3}, \text{NUM:SG}, \ldots\}$$

Step 2: Apply the paradigm function of the language. A paradigm function is a function in the set of form/property-set pairings, the format being:

$PF < X, \sigma >=< X', \sigma >$, where X is the root (or stem) of a lexeme L and σ is a complete set of morphosyntactic properties for L. The function applies to $< X, \sigma >$ to yield $< X', \sigma >$, where X' is the surface-form of L. Paradigm functions are conceived of as realization rules, these rules have the following general form:

$$\text{RR}_{\eta,\tau,C}(< X, \sigma >) =_{def} < X', \sigma >, \text{ e.g. A4 RR}_{A,\{\text{PERS:3}\},V}(< X, \sigma >) =_{def} < eX, \sigma >$$

The application of the paradigm function results in a sequence (i.e. functional composition) of realization rule applications:

$$\begin{aligned}
&\text{PF}(< \text{yosali}, \{\ldots, \text{PERS:3}, \text{NUM:SG}, \ldots\} >) \\
&= \text{RR}_{\mu}(\ldots(\text{RR}_{A4}(< \text{yosali}, \{\ldots, \text{PERS:3}, \text{NUM:SG}, \ldots\} >))\ldots) \\
&= \text{RR}_{\mu}(\ldots(< eyosali, \{\ldots, \text{PERS:3}, \text{NUM:SG}, \ldots\} >)\ldots) \\
&= < eyosali, \{\ldots, \text{PERS:3}, \text{NUM:SG}, \ldots\} >
\end{aligned}$$

Step 2.1: Realization rules are organized into rule blocks. Each 'slot' in a word's sequence of inflectional affixes corresponds to a distinct block of realization rules. A rule block is a set of competing rules, the most specific one will be chosen. If no matching rule is defined, an identity function is applied instead. We assume that block A is specialized for the person-subject position of a verb and rule A4 is a function realizing the third person affix.

The result of RR_{A4} is an affix-root concatenation (called stem), and other rule blocks may be applicable to this result obtained so far. Note for example that the number feature < NOM: SG > has not yet been considered.

Step 2.2: The result may be modified by morphophonological rules, rules associated to realization rules. In our case, there is affix variability (e vs. i). If a rule were known, it would be applied in step 2.2.

We can visualize the process of affix concatenation and of the application of realization rules as a process of "filling-in content into cells" (cf. Table 1 and Table 2). When a surface form has been realized by the paradigm function, the form and the features are filled into a cell of a paradigm. The application of the paradigm function to the set of all feature combinations yields the paradigm of a lexeme L.

Table 1. Initially:root and property-set.

	3	4	
	PERS	ROOT	
		<yosali,σ >	

Table 2. Contribution of RR_{A4}.

	3	4	
	PERS	ROOT	
	<eyosali,σ >	<yosali,σ >	

The linguistic paradigm of Kilivila verb morphology can be reconstructed in a formal way in the framework of the Paradigm Function Morphology; its functional "machinery" is powerful enough for the constraints observed in Kilivila verb morphology, and it has been developed on a very broad basis of morphological regularities and irregularities found in many typologically different languages.

5 Paradigm Function Morphology and its Formalization

As has been pointed out by Karttunen "formal precision and unambiguous notation are clearly important for Stump but there is no discussion in the book about what the formal power of Realizational Morphology might be." (Karttunen [10]).

1 (Emph)	2 (TAM)	3 Person	4 root	5 (Object)	6 (Pl)	7 (Object)	8 (Emph)
*	{ { b (IA,I) / l (CA) / m (HA, O, I) } }	a 1.	X	{ { m 2.Sg / mi 2.Pl } }	*	(simia all.of.them)	({ la / ga } Emph)
(o Emph)	{ { bu (IA,I) / lu (CA) / mu (HA, O, I) } }	ku 2.	X	{ { gu 1.Sg / da Dual.Incl / ma Dual.Excl / dasi 1.Pl.Incl / masi 1.Pl.Excl } }	(si Pl.)	(simia all.of.them)	({ la / ga } Emph)
*	{ { b (IA,I) / l (CA) / m (HA, O, I) } }	{ i / e } 3.	X	{ { gu 1.Sg / m 2.Sg / da Dual.Incl / ma Dual.Excl / dasi 1.Pl.Incl / masi 1.Pl.Excl / mi 2.Pl } }	(si Pl.)	(simia all.of.them)	({ la / ga } Emph)
*	{ { bi (IA,I) / la (CA) / ma (HA, O, I) } }	ta Dual.Incl — ta 1.Pl.Incl	X	{ { m 2.Sg / mi 2.Pl } }	* — si Pl	(simia all.of.them)	({ la / ga } Emph)
*	{ { bi (IA,I) / la (CA) / ma (HA, O, I) } }	ka Dual.Excl — ka 1.Pl.Excl	X	{ { m 2.Sg / mi 2.Pl } }	* — si Pl	(simia all.of.them)	({ la / ga } Emph)

complementary

IA	Incompleted action	X	arbitrary root
I	Irrealis	*	not allowed
CA	Completed Action	()	optional
HA	Habitual Action	{}	choice
O	optative		

Fig. 2. Kilivila verb morphology.

Karttunen's view is a computational linguistic view. He analyzes representations of Paradigm Function Morphology (PFM) in the context of formal language theory, finite state modeling and automata: PFM's lexical representations constitute a regular language and realization rules as well as morphophonological rules represent regular relations. As a consequence regular expressions corresponding to PFM's representations can be specified and compiled into a finite state network. This approach has two advantages. First, theoretically guided morphological analyses can be verified with respect to their formal properties and their empirical basis, verified in a bi-directional way (from an abstract representation to a concrete corpus instance or in the other direction). Second, there are application aspects: the computational linguistic approach may facilitate XML-based language documentation or the development and application of language technology tools.

When we compare the linguistic and the computational linguistic approach, we observe the following differences. In the PFM analysis morphological surface forms are reconstructed formally but still manually, on the basis of types of rules and feature sets. Rules are grouped into blocks, and when several rules may be applicable the most specific one is chosen (Panini's Principle), an aspect to be considered when rule sets are built. Metaphorically speaking: behind each morphological surface form there are groups of rules, applicable to roots and feature sets. In the Finite State Approach (FSA) a design (script) for how rules combine with roots, stems and feature sets is set up once manually and compiled automatically into a network of states and arcs. This network connects strings of an "upper language" (root + feature set) to strings of a "lower language" (surface form + , optionally, feature set). Although Panini's Principle has no immediate implementation in the finite state approach, Malouf [12] describes an algorithm that faithfully simulates it. Rules are translated in such a way, that at most one recognizer for a single finite state implementation in a competing rule block accepts a certain feature string. The algorithm modifies the feature recognizers for all less specific rules to exclude those feature set descriptions (strings representing σ) that are accepted by more specific rules. The recognizers of the most specific rules remain unchanged. However, from the modeling point of view, the network resulting from compiling such a script is only "machine legible".

We observe that there are levels of (connected) representations stressing different properties. The linguistic paradigm systematized the paradigmatic options; the PFM analysis emphasized the role of specific functions and features realizing morphological forms; the FSM implementation is a way to reconstruct the PFM representations as elements of a clearly defined formal language. Additionally, it has automatic verification and computational usability to offer. The FSM approach will be exemplified with respect to an example of Kilivila morphology.

6 From Linguistic Paradigms to XML Corpus Validation

In order to confirm the appropriateness of annotations with respect to sound-ness and completeness, the linguistic description of Kilivila Verb Morphology has to be tested. It is desirable to have automatic procedures to test the description and to facilitate incremental development or refinement. Since linguistic paradigms seem to be a widely accepted construct (cf. Penton et al. [15]) and are quite suitable for our purpose, we took them as starting point in the development of automatic procedures.

Our primary goal is to validate data and markup and possibly to facili-tate data acquisition by guiding the process of annotation of 'raw' data. XML document grammars provide for both. The main disadvantage of XML docu-ment grammars lies in the traditional view of structural validation as opposed to data validation. XML validation is frequently seen as validation of struc-tural conformance, neglecting the relation between markup and data. Recent document schema languages, as XML Schema [7] or RelaxNG [9], provide features for tightly coupling data content and structural markup. As XML schema validation is properly integrated into the XML framework and increas-ing numbers of editors are available that support schema based validation as well as schema guided annotation, we explore means to derive highly restric-tive document grammar instances from semi-formal descriptions. The derived formal grammars incorporate language specific morphological knowledge and effectively act as morphological analyzers.

In the following we assume XML markup to conform to conventions ex-emplified below:

Example 2.

```
<verb> <m feature='value'>surface form</m><m> ... </m></verb>
```

or in order to handle complex values:

```
<verb><m><fs><f name="per"><v>2</v></f></fs><g>ku</g></m>...</verb>
```

Each m-element can be seen as the root of a tiny tree, with attributes and textual data attached. Each verb annotation corresponds to a sequence of such tiny m-trees, a construction which typically maps to productions like w→m* in a Document Type Definition.[2] This very general production does not enforce language specific regularities, for example feature-value combinations or affix order.

Our main goal now is to restrict the sequence of realizations of m-elements, such that only instances are accepted that describe and match possible verb forms of the Kilivila language (cf. Section 10). These restrictions derive either from the tabular description itself and are not explicitly mentioned or they are explicitly given as informal statements. Figure 2 features several constraints:

---

[2]See for example the P4-DTD of the Text Encoding Initiative [19].

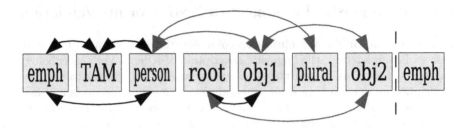

Fig. 3. Affix positions and co-occurrence constraints.

1) Affixes not mentioned in any cell do not participate in any verb form. 2) In order to generate a verb candidate only realization descriptions from within a single row are acceptable. 3) Each position is realized at most once for each verb form. 4) Chosen subdivisions in positions 3 and 6 must match.

Not all co-occurrence constraints that are known to apply to the domain of Kilivila verbs are expressed in the table. For example, the table does not reflect the impact of verb valency on affix realization and subsequent restrictions.

As shown in Fig. 3, the constraints between position classes form cross-serial dependencies, so that it is impossible to have a simple mapping to nested XML structures. Note that the 'person'-position (cf. column 3 in Fig. 2) participates in many constraints, which explains its selection as a principal component for the table itself.

The problem of grammar derivation can be reformulated as creation of a mapping from the cross serial dependencies between affix positions to a document grammar, modeling dependencies within and between annotation subtrees. The apparent mismatch between mild context sensitivity and context freeness can be solved as there is only a finite number of possible realizations for each affix position and no recursion occurs, so there exists a fixed upper bound to the total number of sequences allowed. Regarding the grammar, there remains the problem of modeling different instantiations of m-elements depending on some left and right context.

The solution to this difficulty is sketched in Fig. 4. In the upper section there is a visualization of the tree structure of a verb annotation in which the realization of an m is dependent on the realization of the descendants (i.e. attributes and text) of the neighboring elements. The trees in the middle part of the figure differ in so far as the m-nodes are subclassified (indices a, b, c, d,...) and their realization is now only dependent on the indices of their siblings. The critical information encoded in subtrees in the two upper trees is now attributed to the node itself. The actual counterpart in the grammar is found in the distinction and separation between element names and element types ('compact representation' in the lower left hand part of the figure). While Document Type Definitions (DTDs, [3]) do not allow multiple content models for an element name x, less restricted document grammars such as

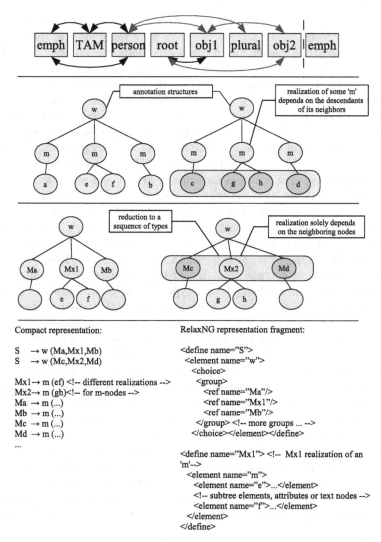

Fig. 4. Constraints on annotations and mapping to grammars.

XML Schema or RelaxNG distinguish element name and element types. This allows to constrain the content model for some element x depending on some 'context'. That is, multiple definitions for the same element name are possible and specific definitions can be referenced from within other content models, thus providing means to simulate context dependent elements.

The lower left hand side of Fig. 4 lists a compact representation of a grammar instance[3], which accepts the trees from the middle of the figure. In comparison to string based context free grammars, we observe a bracketed structure on the right hand side of each rule, which represents the actual content model. The preceding item defines the observable element name. The left hand side non-terminal is some arbitrary identifier for the type of the element. The non-terminal 'S' maps to two different realizations of an element w, and their realizations are determined by regular expressions over some other types. The lower right hand side of the figure shows a fragment of a RelaxNG grammar instance which is equally expressive. This formal device allows us to define different types of subtrees for some XML element m and assign unique identifiers. These identifiers can then be used in different expressions.

Another important difference between XML-DTDs and modern document grammars is the more sophisticated typing of actual data content, which in our case of morphological annotations is necessary to limit some annotation to only apply to proper surface forms. XML Schema and RelaxNG enable e.g. limiting textual content to some strings as given by an enumeration, regular expression or predefined data type. RelaxNG is capable of handling co-occurrence constraints between textual content and attribute-values, which allows us to restrict the types of m-elements to contain only the combination of features, values and surface forms as they are given by the tabular description. So each element of a set in a cell of our Kilivila table can be assigned a type in the grammar, which expands to a subtree carrying appropriate attribute names, values and textual content, e.g. binding the annotation for second person to the Kilivila affix **ku** instead of arbitrary characters (#PCDATA in DTDs), which allows to restrict the assignment of the markup for second person to only that surface form (cf. Fig. 9).

We now have an appropriate target device and the property of 'finiteness' of the set of all allowed sequences, which have to be covered by the grammar. Since the underlying description may change (frequently), e.g. due to descriptive improvements triggered by additional language material, we will refrain from manual development of grammars but will explore methods to derive them.

In following sections we will discuss different approaches to the development of document grammars representing regularities and constraints of the description in Fig. 2, including mentioned additional constraints.[4]

---

[3]We only sketch the general 'idea' and refer e.g. to [14] for a more formal introduction to regular tree grammars and differences between XML Schema and RelaxNG.

[4]We thank Uwe Mönnich who suggested an approach involving Monadic Second Order Logic formulas and their translation into tree automata. To the best of our knowledge this is a mathematically elegant but 'painful' way, as adaption to the problem domain and to colloquial XML as the back-end domain has to yet to be realized.

7 Derivation by Finite State Compilation Techniques

There are at least two different strategies involving finite state compilation techniques. First, regular tree grammars and their automata have mathematical properties similar to regular string grammars. It seems worthwhile to explore the feasibility of their direct application. This approach is currently not investigated as, to the best of our knowledge, there has been considerable research on theoretical aspects of unranked tree automata but there are no implementations available which provide the necessary operations to actually construct and apply them. This approach will only be feasible if high-level operators, similar to those developed for regular string automata, and optimized implementations are available.

Another, but less powerful finite state approach is based on well known finite state string transducers and their compilers. In a first step the morphological description of verbs in Kilivila is translated into a finite state representation in the xfst (Xerox finite state tools) rule format (cf. Beesley & Karttunen [1]) mapping the tuple representation into a relation between strings. The transducer resulting from the compilation of the rules will translate a string representation of a base form and some feature-value pairs into corresponding surface forms – and vice versa.

Example 3 (Input-Output of a Kilivila transducer).
    upper band: <yosali#Emph:- PER:3 TAM:- OBJ:- Num:Sg ALL:- EmphS:->
    lower band: eyosali
                iyosali

The transducer underlying example 3 generates and accepts two surface forms since there is no selection criterion described, which could form a processable rule for the generation of the transducer.

Naturally, this kind of morphological analyzer/generator is bound to the string representation of the features. This automaton is, in principle, appropriate for validation of the XML data if the gap between tree structured representation and string based, order sensitive description is bridged, e.g. by some adaptor (e.g. a script) that reads the XML-data, aggregates distributed feature-value information and surface forms into a proper string representation and pipes this to the automaton. However, we opt for a tight integration that does not require runtime adaptors, for such a kind of adaptor would require information about 1) the grammar of the XML documents to be processed, 2) the structure of the string description (usually the upper band of a two-band string transducer) and 3) the current position within the document in order to locate the candidate annotation. In the case of semi-automatic annotation, a similar adaptor has to be built which maps in the opposite direction producing XML fragments. If the underlying morphological analysis changes three distinct components, namely the transducer and both adaptors have to be modified in a consistent but independent manner. Since the document grammar is expressive enough to substitute the transducer we favor to eliminate runtime components like scripts and transducer implementations which require runtime environment as well as work-flow outside common XML editing.

Although we abandon the string transducer as a back-end computational device, it is highly valuable as a source to document grammar construction. The directed graph underlying the transducer encodes linguistic knowledge, with cross-serial dependencies disentangled into different continuations of some common prefix path. Each path through the graph describes a relation between the description of and a surface form itself. Two obstacles remain: First, the relationship between the description and the surface form is holistic. There is no detectable cause-consequence or feature-surface relation between the substrings because those are encoded in the rules themselves, but not in the resulting network. Therefore it is necessary to create some device (within the rules) which allows us to attach the appropriate feature-values to the corresponding substrings of the surface form in order to analyze network paths and to assign feature-value pairs to matching surface forms. The second problem is to map the obtained string descriptions to XML element declarations.

8 Derivation by Restricting Domains

An alternative group of approaches conceives Kilivila verb morphology and constraints between the position classes as a constraint satisfaction problem over finite domains. They use characterizations based on sets of possible values instead of relations between parts of strings. Each tuple of feature-values and a surface form from the tabular description can be seen as an element from the domain of some position class variable. Interpositional restrictions between position classes map to constraints over the binding or assignment of variable values. Informally, the domain of a variable corresponds to a column in the table and constraints limit the bindings of those variables.

The final goal is to derive exactly all solutions to the constraint system, such that each solution encodes a sequence of affixes, which in turn can be straightforwardly mapped to some definitions for a document grammar. However, considering the requirement of complete coverage of the solution space as well as the effort to be made in order to map items in the problem space to a description acceptable for constraint solvers, we did not pursue this approach any further. But we implemented solution strategies which assume that in each description there is only an arbitrary but finite number of position variables with finite domains and a finite set of binary constraints encoding the restrictions required.

The first solution strategy requires a dominating 'control' variable. Each possible value of the control variable enables a set of domain restrictions on position class variables, which determine subsets of the initial domains. So each value of the control variable results in a sequence of sets of possible realization-description pairs. A valid verb and its description can then be constructed by freely choosing from each of these sets. The main disadvantage of this approach is the requirement to express constraints in terms of values for the control variable and a cascade of domain restrictions. We already identified the 'person'-attribute as a principal component. Its values may serve as a starting point to be augmented to address the additional constraints, e.g. by extending the domain (e.g. by adding a modifier to existing values) to handle constraints which arise from verb valency. An advantage of this approach is its value for the generation of a document grammar, since each value of the control variable encodes a sequence of sets, resulting in a very compact grammar representation with ample use of choice groups. And contrary to the finite state automata approach, we can immediately take advantage of subsequence and subset sharing.

In order to handle rapid change in language description, we developed a device that automatically propagates consistent 'paths' (sequences of variable bindings) through the space of all possible variable bindings. Thus it is no longer necessary to determine manually a control variable and the relevant domain restrictions.

9 Derivation by Path Propagation

In this approach we apply search within the space of all possible variable bindings.A path constitutes a valid verb realization where each component is 1) chosen from the domain of the respective position class variable and 2) all constraints between path components (i.e. position classes) are satisfied. We propose to include domain knowledge directly into the search procedure to significantly reduce search costs and mainly to preserve means to handle unrestricted value domains. In order to generate all allowed paths, any element from the domain of the first variable can be chosen without violating any constraints. Each binary constraint in which this realization participates can be enforced for the current path by restricting the domain of the other participant. If, for example, the emphatic prefix o is chosen for the first position class, then this implies that the person affix needs to match ku indicating second person. The search space for the continuation of the current path is reduced to exactly that affix tuple for the third position variable. If a different search order is chosen, for example to take advantage of highly restricting position classes early in the search process, then restriction rules have to be reformulated to reflect the order of determination of variables, as constraints are treated as implications in the search process since for example the realization of ku does not require the realization of o. Figure 5 gives an overview of the process[5]. After selection of the first and second path component, the

---

[5] Please note that for reasons of conciseness, the figure does not match the Kilivila position classes.

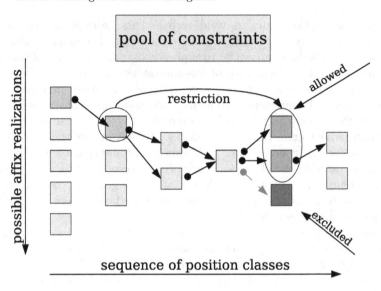

Fig. 5. Visualization of path propagation.

constraint from the selected realization of position 2 to position 5 restricts the acceptable realization candidates of the position class variable for position 5. This results in the exclusion of a single path from position 4 to 5.

The result of the process of path propagation is a set of tuples of variable bindings, which describes all acceptable realizations. These bindings can be either directly used to generate a document grammar or post-processing can be applied, for example to identify common subsequences. A description obtainable from such post-processing seems more desirable to us, as it is more compact than a prefix-tree-like one and reflects the enforced constraints within the observable grammar structure. Another strategy which is actually implemented is to group realization description based on the constraints which were enforced during their generation. Any kind of post-processing is however completely optional and does not alter the language accepted by the grammar.

The resulting grammar is a perfect acceptor for morphological annotations that match the reformulations of the descriptions that were given in Table 2 and the prose constraints. Rejection of linguistically well-formed annotations is a good indication for either a deficiency in the transfer to set constraints or a deficiency in the linguistic description itself.

10 Conclusion

In the first part of this article we focused on linguistic information modeling with respect to structural or formal approaches to constraints of Kilivila verb morphology. In the second part we concentrated on linguistic information modeling in RelaxNG. This advanced type of document grammar is powerful enough to represent in a precise way complex linguistic constraints and to connect markup to raw data: RelaxNG grammars are able to validate Kilivila verb annotations and to guide the annotation process given suitable XML-editors.

Different approaches to a semi-automatic derivation of a RelaxNG document grammar have been tested and described in this article. We want to emphasize that modern XML document standards are not only suited to definition of serialization or storage formats, document grammars may also control seemingly complex constraints among markup as well as the association of markup with raw data, a control exercised in the domain of morphology generally by finite state analyzers for sequential representations.

We think it is important to bridge the gap between standards of XML-based language documentation on the one hand and structural or formal linguistic analyses on the other, for empirical linguistic research is more and more based on corpora and on XML-structured annotations, annotations controlled by advanced grammatical or semantic frameworks: document grammars or formal ontologies. We suppose that both domains of analysis are also becoming more important for current approaches to metadata for descriptive grammars (cf. Good [6]; an integration of a RelaxNG component into an augmented version of his approach to descriptive grammars seems desirable).

References

[1] K. R. Beesley and L. Karttunen. *Finite State Morphology*. CSLI Publications, 2003.

[2] S. Bird and M. Liberman. A Formal Framework for Linguistic Annotation. *Speech Communication*, 33(1-2):33–66, 2001.

[3] T. Bray, J. Paoli, and C. Sperberg-McQueen. Extensible Markup Language Recommendation (XML) 1.0, 2004.

[4] D. Broeder, H. Brugman, and G. Senft. Documentation of Languages and Archiving of Language Data at the Max Planck Institute for Psycholinguistics in Nijmegen. *Linguistische Berichte*, 201:89–103, 2005.

[5] J. Carletta, S. Evert, U. Heid, J. Kilgour, J. Robertson, and H. Voormann. The NITE XML Toolkit: Flexible Annotation for Multimodal Language Data. *Behavior Research Methods Instruments & Computers*, 35(3):353–63, 2003.

[6] J. Good. The Descriptive Grammar as a (Meta)Database. E-MELD Workshop on Linguistic Databases and Best Practice, 2004. http://www.emeld.org/workshop/2004/jcgood-paper.html.

[7] XML Schema Part 1: Structures. Second Edition. W3C Recommendation, October 2004.

[8] S. Inkelas. Nimboran Position Class Morphology. *Natural Language and Linguistic Theory*, 11:559–624, 1993.

[9] ISO. Information Technology – Document Schema Definition Languages (DSDL) – Part 2: Grammar-based validation – RELAX NG. published as ISO/IEC 19757-2:2003, 2003.

[10] L. Karttunen. Computing with Realizational Morphology. In *Computational Linguistics and Intelligent Text Processing: 4th International Conference, CICLing 2003 Mexico City*, volume 2588 of *Lecture Notes in Computer Science*, pages 203–214. Springer, 2003.

[11] C. Lehmann. Directions for Interlinear Morphemic Translations. *Folia Linguistica*, 16:193–224, 1982.

[12] R. Malouf. Disjunctive Rule Ordering in Finite State Morphology. Paper presented at the 41st Meeting of the Chicago Linguistics Society, April 2005. http://bulba.sdsu.edu/~malouf/papers/cls05-abs.pdf.

[13] J. McDonough. Athabaskan redux: Against the position class as a morphological category. In W. Dressler, O. E. Pfeiffer, M. Pöchtrager, and J. R. Rennison, editors, *Morphological Analysis in Comparison*, volume 201 of *Current Issues in Linguistic Theory*, pages 155–178. John Benjamins Publishing, 2000.

[14] M. Murata, D. Lee, and M. Mani. Taxonomy of XML Schema Languages using Formal Language Theory. In *Extreme Markup Languages*, Montreal, Canada, 2001.

[15] D. Penton, C. Bow, S. Bird, and B. Hughes. Towards a General Model for Linguistic Paradigms, 2004.

[16] G. Senft. *Kilivila: The language of the Trobriand Islanders*. Mouton de Gruyter, 1986.

[17] G. Senft. These 'Procrustean' Feelings – Some of my Problems in Describing Kilivila. In G. Reesink, editor, *Topics in Descriptive Austronesian Linguistics*, pages 86–105. Vakgroep Talen en Culturen van Zuidoost-Azië en Oceanië, Rijksuniversiteit te Leiden, Leiden, 1993.

[18] G. T. Stump. *Inflectional Morphology – A Theory of Paradigm Structure*. Cambridge University Press, 2001.

[19] Text Encoding Initiative Consortium, Oxford, Providence, Charlottesville, Bergen. *TEI P4: Guidelines for Electronic Text Encoding and Interchange*, 2002.

[20] C. Wegener. Verbflektion im Kilivila. Technical report, Universität Bielefeld, 2002.

[21] A. Witt. Meaning and Interpretation of Concurrent Markup. In *Joint Conference of the ALLC and ACH (ALLCACH2002)*, Tübingen, 2002.

[22] A. Witt. Multiple Hierarchies: New Aspects of an Old Solution. In S. Dipper and M. Götze, editors, *Heterogeneity in Focus: Creating and Using Linguistic Databases*, volume 2 of *Interdisciplinary Studies on Information Structure (ISIS), Working Papers of the SFB 632*. Universitätsverlag, Potsdam, Germany, 2005.

Appendix

Benefit from Kilivila Document Grammar

The figures 6 to 10 show screenshots of editing markup for a Kilivila verb in an emacs editor with RelaxNG capabilities as provided by nxml-mode[6]. The document element **corpus** contains arbitrary markup (e.g. **ann**-element) and **verb**-elements which are subject to conform to the derived grammar fragment. Violations are underlined to highlight. Figure 6 shows a morphologically well-formed Kilivila verb which is regarded invalid since its markup is incomplete. In the next screenshot the prefix annotation is accepted while the verb annotation remains invalid. In Figure 8 all affixes and the root **lukwe** are fully annotated and all known constraints are met. After modifying the surface form from **ku** to **ka** in Fig. 9, the verb markup is recognized to be invalid which illustrates the use of co-occurrence constraints between textual data and attributes within an **m**-element. The annotator changed person from second to first in Fig. 10, which introduced violations against constraints between position 3 and 5 (no first plural inclusive object with first person) and position 3 and 6 (no plural with first person). This tiny example reflects potential benefits in further data acquisition and data validation.

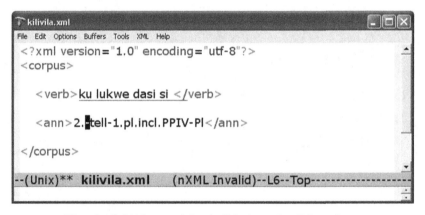

Fig. 6. Guided annotation/validation – invalid markup.

---

[6]http://www.thaiopensource.com/nxml-mode/

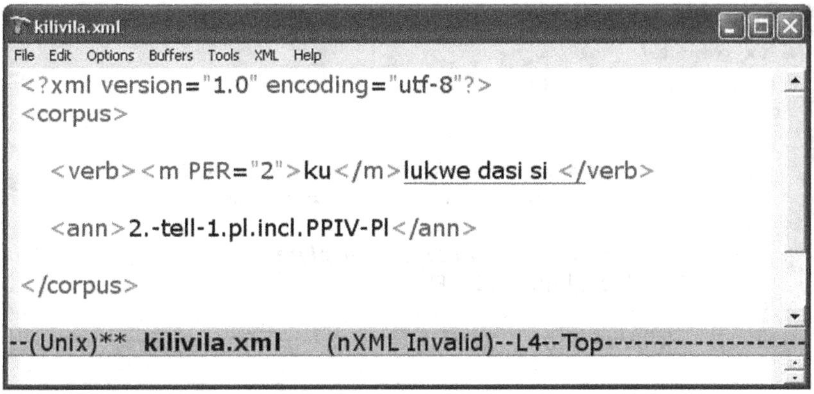

Fig. 7. Partially annotated data.

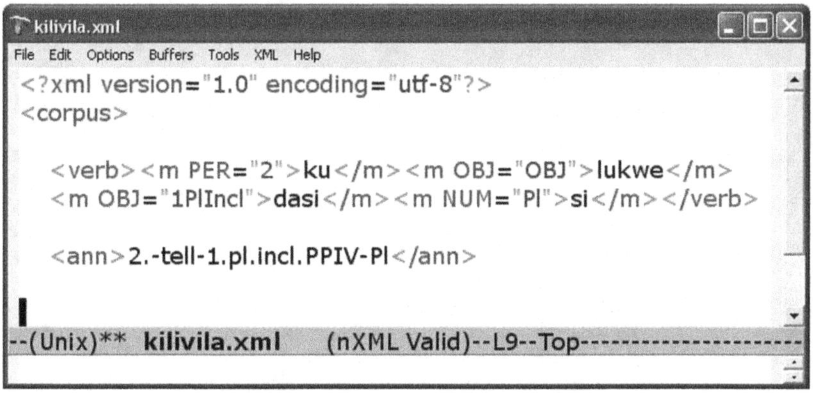

Fig. 8. Valid annotation.

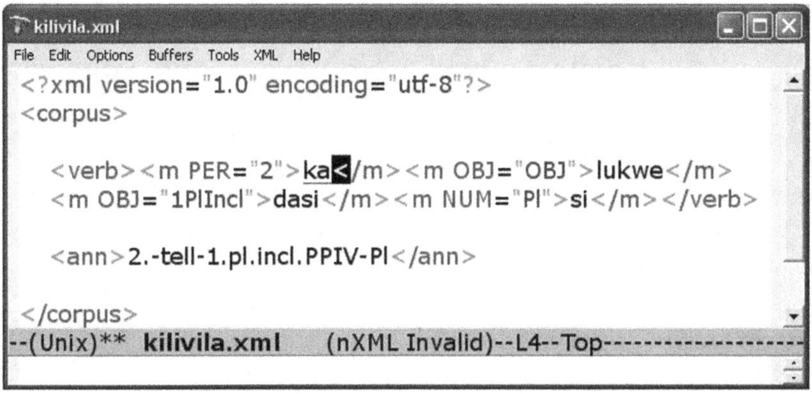

Fig. 9. Invalid due to surface form modification 'ku' → 'ka'.

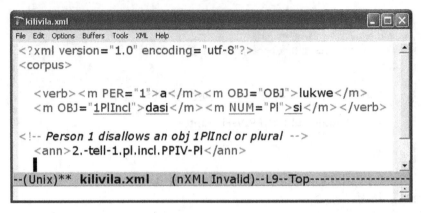

Fig. 10. Violation of constraints between person and object/plural affix.

Affix Discovery by Means of Corpora: Experiments for Spanish, Czech, Ralámuli and Chuj

Alfonso Medina-Urrea

Universidad Nacional Autónoma de México
amedinau@iingen.unam.mx

1 Introduction

Although the focus on morpheme discovering techniques originated within those linguistic schools which inherited from Franz Boas the concern for the unknown languages of the New World, automatic, unsupervised morphological segmentation remains a field of interest for the computational processing and engineering[1] of natural languages, as well as for the plain exercise of getting to know them intimately.[2]

Given the diversity of morphological strategies across languages to create new words or to inflect them, it should be obvious that morpheme discovery

---

[1]Applications requiring some processing of human language may have to deal directly or indirectly with morphological phenomena. For example, stemmers, or so called lemmatizers, morphological analyzers, etc., can be used for many purposes, among them for information extraction and retrieval, text mining, electronic dictionary processing, etc. If resources like these are to be developed for the lesser known natural languages (which happen to be the majority of human languages, many of which exhibit unimagined morphological structures), unsupervised morpheme discovery is bound to be a required development stage.

[2]There are several prominent approaches to word segmentation. The earliest one is due to Zellig Harris, who first examined corpus evidence for the automatic discovery of morpheme boundaries for various languages (Harris [9]). His approach was based on counting phonemes preceding and following a possible morphological boundary: the more variety of phonemes, the more likely a true morphological border occurs within a word. Later, Nikolaj Andreev designed a method based on character string frequencies which applied to various languages. His work was oriented towards the discovery of whole inflectional paradigms and applied to Russian and several other languages (see Cromm [1]); and that of Josse de Kock and Walter Bossaert [2, 3] in the seventies for French and Spanish. Other prominent approaches deal with bigram statistics (see for instance Kageura [11]) and minimal distance methods (Goldsmith [6]).

A. Medina-Urrea: *Affix Discovery by Means of Corpora: Experiments for Spanish, Czech, Ralámuli and Chuj*, StudFuzz **209**, 277–299 (2007)
www.springerlink.com

must go beyond the application of techniques to segment words.[3] Nevertheless, the method explored here focuses on these techniques, since it deals with one of the most widely disseminated morphological strategies: affixation. But even within this reduced scope, when dealing with corpora, problems arise about traditional linguistic categorial concepts[4] such as *affix*, which is seen here basically as a morph – *i.e.* the version of a morpheme determined by its context of occurrence. Therefore, rather than making strong pronouncements about some string of characters' status as an affix, the method essentially grades strings according to their likelihood of representing a true affix or valid sequence of affixes for that language.

This approach differs somewhat from minimal distance methods, which seek to find *the best* morphological model. As many would agree, the ultimate objective of segmenting words is the discovering of not only affixes, but also of affix paradigms. This is, however, a task not to be underestimated, mainly because there are many kinds of paradigms. In fact, the better organized ones coexist with the lesser organized paradigms. And, even though these may be traced by automatic means, minimal distance methods (Goldsmith [6]) should prefer, by definition, the more compact paradigm types. For the sake of not excluding any valuable item, the method sketched bellow gathers every possible candidate in a format that can later be used for human evaluation.

Although interesting comparison of methods exists for various languages (Hafer & Weiss [8], Kageura [11], Medina-Urrea [14], among others), new comparison experiments must be conducted, that take into account more languages and the very diverse objectives that morphological segmentation may have, including those requiring least productive morphemes not to be excluded.

In this paper, some results of experiments to discover affix subsystems of four languages, two Indo-European languages, from the Romance and Slavonic branches, and two unrelated American languages, from the Uto-Aztecan and the Mayan families: a non European variant of Spanish, standard Czech, a variant of Ralámuli or Tarahumara and a variant of Chuj. Essentially, results of experiments conducted previously are gathered and briefly discussed (Medina-Urrea [14, 15]; Medina-Urrea & Hlaváčová [18]; Medina-Urrea & Alvarado García [16]; and Medina-Urrea & Buenrostro Díaz [17]). More importantly, an attempt is made to compare such dissimilar experiments in order to evaluate, by means of precision and recall measurements, how well are affixes discovered.

In the next section the method will be presented, as described in Medina-Urrea [14], which portrays some features of what we call an affix a priori. Then, the notion of an affix catalog will be formalized as a simple measurement tool to capture quantitative structural features of (graphical) words – rather than

---

[3]Numerous works deal with morphological complexity of natural languages. Take for instance the classical work by Sapir [22]. In the handbook by Spencer & Zwicky [24] many morphological phenomena are exemplified.

[4]This should not be surprising; see for instance Rieger [20, p. 156].

a kind of morphological rule-based model. Later, catalogs of automatically generated affix candidates for one of each of the four languages' interesting affix subsystems are briefly examined. Finally, the method evaluation section and some concluding remarks are presented.

2 The Method

The idea of measuring affixality of word fragments by means of the three indexes that will be next presented (number of squares, economy of signs and entropy) arose from the comparison of various word segmentation measurements (also in Medina-Urrea [14]). As opposed to the very well known bigram statistics (*e.g.* mutual information, log likelihood, etc.), the three indexes chosen seem to capture some aspect or another of what is traditionally understood as an affix: they are fewer, more frequent, contain less information than other kinds of signs, etc.

Before presenting the indexes, some basic concepts will be introduced. Let v_i be the word-types in text sample Ψ, where $0 \leq i \leq \Omega$ and Ω is the number of types. Each possible segmentation of each v_i can be represented by means of a double colon '::' (so $a{::}b$ represents a type where the segmentation between word fragments a and b is put into focus). To remind us that the fragments belong to word-types, let them carry the subscript i so that $a_i{::}b_i = v_i$. Furthermore, let j be another subscript, which corresponds to each of the possible segmentations in a word-form v_i. So, if v_i is m_i characters long, then it contains $m_i - 1$ possible segmentations, $a_{i,j}{::}b_{i,j}$, $1 \leq j \leq m_i - 1$. Thus, the segmentations of $v_i = $ '*ejemplo*' can be represented:

$$a_{x,1}{::}b_{x,1} \ \text{(e::jemplo)}$$
$$a_{x,2}{::}b_{x,2} \ \text{(ej::emplo)}$$
$$a_{x,3}{::}b_{x,3} \ \text{(eje::mplo)}$$
$$\vdots$$
$$a_{x,m_x-1}{::}b_{x,m_x-1} \ \text{(ejempl::o)}$$

Essentially, each segmentation j of each word-type i is examined automatically in order to calculate the following three measures.

2.1 Number of Squares

Although the notion of what a *square* is can be traced back to de Saussure, let us turn to Greenberg's words to illustrate it.[5] He characterized a square as the set of word-forms which:[6]

---

[5]This is indeed a very old notion, which researchers keep rediscovering or just renaming to reflect their approach – Goldsmith [6] calls them *signatures*; from a graph-theoretical perspective, Johnson & Martin [10] characterize them as *hubs*.

[6]See Greenberg [7, p. 20].

exists when there are four expressions in a language which take the form AC, BC, AD, BD. An example is English *eating:walking::eats:walks*, where A is *eat-*, B is *walk-*, C is *-ing*, and D is *-s*. One of the four members may be zero, as in *king:kingdom::duke:dukedom*, where C is zero.

Thus, a square is a set of four word fragments, two left ones (a_1 and a_2) and two right ones (b_1 and b_2), such that combining either of the left segments with either of the right segments, a word-type results ($a_1::b_1$, $a_1::b_2$, $a_2::b_1$, $a_2::b_2$). One of the segments may be a null string, ø, in order to allow structures like {*ver::stehen, ver::teilen, ø::stehen, ø::teilen*}.[7] Thus, each possible segmentation of each type v_i can be examined to determine the number of squares[8] in which each of its sides appears, by searching for matches[8] within all v_i of the corpus. Let us call this number $c_{i,j}$ (*i.e.* the number of squares found in segment j of word-type i).

2.2 Economy Principle

Another important measure of word structure deals with the principle of economy of signs. Some experiments based on this principle – either maximum or minimum approaches – are described in de Kock & Bossaert [2, 3], Medina-Urrea [14], Goldsmith [6], Gelbukh, Alexandrov & Han [5], etc. In essence, the principle deals with the fact that affixes can combine with bases to produce a number (virtually infinite) of lexical signs. It is clear that affixes do not combine with every base. Certain ones combine with many bases, others with only a few. It makes sense to expect more economy where more combinatory possibilities exist.

The attachment of affixes to bases refers to the syntagmatic dimension. But the paradigmatic dimension should also be considered:[9] in their attachment to bases, affixes alternate with other affixes (occur in complementary distribution).[10] If there is a relatively small set of alternating signs (paradigms) which attach to a large set of unfrequent signs (to constitute syntagma) the

---

[7]The requirement of squares can be varied in different ways, ranging from an incomplete square to what can be called a *hexagon*, *i.e.* six segments contained in three words (Kock & Bossaert [3, p. 19-20]).

[8]There are many possible ways of doing this, like finding cycles of length four in a matrix or building graphs with in-degree and out-degree greater than one (Johnson & Martin [10]).

[9]It should not be surprising that both these dimensions are relevant also within semiotic approaches, see Rieger [20].

[10]Let us define *alternation* as a process by which a segment belonging to one of the two ends of the word is *fixed* in order to search for all the possible segments belonging to the other side and which happen, when combined with the fixed one, to form words-types of the corpus. If one lets one side alternate, one obtains a set of segments to that side (*alternants*), if one lets the other one alternate, another set is obtained. Saying that one of the sides alternates, will necessarily mean that the other one remains fixed (even if that is not explicitly said).

relations between the former and the latter must be considered even more economical. This is naturally pertinent for both derivation and inflection; *i.e.* this is as much true for lemma affixes, as it is for affixes of textual and spoken discourse.

The earliest attempt to capture this experimentally by means of computers is due to J. de Kock & W. Bossaert [2, 3]. Based on traditional linguistics, de Kock observed that the number of signs at all levels of language should be lower than the number of things named, so that "the code is organized in such a way that a sign can serve in more than one instance without creating any ambiguity."[11] In short, a small inventory of signs – benefiting speakers of a language – of one level results in a large inventory – benefiting recipients or hearers – of signs of the next level. Thus, word-types (syntactic signs) are inflected or derived from other word-types by means of signs of the morphological level. The latter being necessarily fewer but more frequent than the former.

Formally, if a word-type v_i is divided into two segments, $a_i::b_i$, and one of the segments occurs in many other types, while the other occurs in only a few other, and if the first one belongs to a small set of frequent segments, while the other to a potentially infinite set of low occurring segments, then a morphological cut can be proposed between these segments. Moreover, the former one would be an affix while the latter a base.

For example, take the segments in Figure 1. Suppose that each of the left word fragments combine with every right one to form word-types found in a Spanish corpus (*compra, comprada, comprado, comprando, ... compró; ... canta, cantada, ... cantó; ... controló; ...*).

The fact that the right word fragments constitute a small set and occur very frequently, whereas the left ones have a much lower frequency and belong to a very large set (potentially infinite in an open-ended corpus) is a very reasonable clue that the right hand set is an affix set.

Let $A_{i,j}$ be the set of word fragments which occur in complementary distribution attached to right word fragment $b_{i,j}$; in other words, the set of segments found when we let $a_{i,j}$ alternate (thus $a_{i,j} \in A_{i,j}$); and $B_{i,j}$ the set of segments which alternate to the right of $a_{i,j}$ (thus $b_{i,j} \in B_{i,j}$). Let $|A_{i,j}|$ be the number of members belonging to $A_{i,j}$, and $|B_{i,j}|$ be the number of members in $B_{i,j}$.[12]

By comparing the sizes of these sets we can argue for the morphological character of the segmentation in question. Some restrictions can be applied to that sharpen these numbers. In essence, we will attempt to eliminate from both sets anything likely to be a base by not counting those instances in which the string is more frequent than its accompanying segment (*i.e.* bases should be a lot less frequent than their accompanying affixes). Let $A^p_{i,j}$ be the set of left segments which are likely to function as prefixes and $B^s_{i,j}$ the one containing the right segments likely to behave as suffixes. Thus, the set $A^p_{i,j}$ is

---

A	B
compr	a
cant	ada
alivi	ado
rest	ando
ray	ar
sum	aron
seleccion	aste

$$\vdots \; \vdots$$

arrest	es
elabor	é
nad	o
anhel	ó
contrat	
apel	
mand	
colabor	
control	

$$\vdots$$

$$\infty$$

Fig. 1. Left and right word segments.

a subset of $A_{i,j}$ consisting of only those members likely to behave as prefixes. Similarly, the members of $B^s_{i,j}$ are those members of $B_{i,j}$ likely to behave as suffixes. Furthermore, let $|A^p_{i,j}|$ be the number of elements in $A^p_{i,j}$ and $|B^s_{i,j}|$ the number of elements in $B^s_{i,j}$.

But before eliminating anything likely to be a base, let us consider the fact that $B_{i,j}$ (the right-hand alternants) may contain segments which begin the same way, while $A_{i,j}$ (the left-hand segments) may contain several ones which end the same way. This means that all phonemes common to the variable segments which are adjacent to the invariant segment may in fact belong to the latter one. In other words, if within the set of supposed affixes, there is one whose accompanying bases share affix-adjacent characters, one can suspect that those characters are actually part of the affix and not of the bases with which it appears. Similarly, if within the set of supposed bases, there is one whose accompanying affixes share base-adjacent characters, those characters may be suspected to be part of the base.[13] They should therefore be eliminated. That means that in the example above, the forms ending with the segments $\sim a$, $\sim ada$, $\sim ado$, $\sim ando$, $\sim ar$, $\sim aron$, $\sim aste$ would be counted as only one form.[14]

---

[13]Cf. de Kock & Bossaert [3, p. 21].

[14]The sizes of the updated sets correspond to M_g and M_d in the work of de Kock & Bossaert [3, p. 22].

Also, it can be required that both sets contain members with more than one occurrence (each segment must occur in at least two different word-types). Actually, when this is not required, no complete squares can be found. In this experiment, the presence of at least one square is required, since a lack of squares implies no morphological cut.

By comparing the sizes of these sets one can get an idea of how economical a segmentation is: the greater the difference in number of the word fragments considered bases versus those hypothesized as affixes, the more economical it is. If more base-like segments alternate to the left $(A_{i,j} - A^p_{i,j})$ than do affix-like ones to the right $(B^s_{i,j})$, one may consider the right segment $b_{i,j}$ to be a suffix. Conversely, if more base-like segments alternate to the right $(B_{i,j} - B^s_{i,j})$ than do the affix-like ones to the left $(A^p_{i,j})$, one may be justified to accept the left segment $a_{i,j}$ as a prefix. In this way, we get two measures of how economical the segmentation is, depending on the hypothesized type of affix:

$$k^p_{i,j} = \frac{|B_{i,j}| - |B^s_{i,j}|}{|A^p_{i,j}|} \tag{1}$$

represents how economical a prefix is. It will be much greater than one when we are dealing with such type of affix and a fraction of one when we are dealing with a suffix. Also,

$$k^s_{i,j} = \frac{|A_{i,j}| - |A^p_{i,j}|}{|B^s_{i,j}|} \tag{2}$$

represents how economical a suffix is. Similarly, it will be much more than one when dealing with a suffix and less than one when dealing with a prefix. Although their notation is quite different, these quotients are basically what de Kock explored in the seventies, so I will also refer to them as the de Kock-Bossaert indexes.

2.3 Shannon's Entropy

The last index that will be examined deals with the information content of the bases accompanying affixes. High entropy measurements have been repeatedly reported as more or less successful indicators of boundaries between bases and affixes (Hafer & Weiss [8], Frakes [4], Oakes [19], Medina-Urrea [14, 15], Medina & Buenrostro Díaz [17], Medina-Urrea & Hlaváčová [18], etc.). These measurements are relevant because, as it was pointed out by Greenberg,[15] "both in the technical sense of information theory and in the non-technical meaning of information, the utterance of a member of a root class of morphemes gives more information."[16] Thus, shifts of amounts of information

---

[15] As mentioned before, Harris relied on phoneme counts before and after a given word segmentation, a matter undoubtedly related to entropy measurement. But, unlike Greenberg, he did not specifically refer to information theory.

[16] See Greenberg [7, p. 91].

can be expected to correspond to the amounts of information that a reader or hearer obtains from text or spoken discourse. Frequent word fragments contain less information than those occurring seldom. Hence, affixes attach to bases containing highest amounts of information.

This is the case for a wide range of affixes, including those whose structural evidence – like that behind the amount of squares or the economy principle – is not fully provided by a corpus, either because the corpus is too small or not representative enough of the language or because the affixes in question are old and unproductive [15].

Thus, if an affix is supposed to contain mostly grammatical information – as opposed to a base which may contain much more (the occurrence of a particular base is bound to surprise us more than that of an affix) –, a local peak of entropy in the middle of a word would signal the beginning of a base, whereas a local minimum that of a affix.

Given the word-type $a_{i,j}::b_{i,j}$, the set $B_{i,j}$ can be seen as a reservoir of word fragments likely to be picked out in order to form a word by attaching it to $a_{i,j}$. The probability of picking a member of $B_{i,j}$ once $a_{i,j}$ has been selected would be:

$$p(b_{k,j} \mid a_{i,j}) = \frac{f(b_{k,j})}{f(a_{i,j})} \tag{3}$$

where $k = 1, 2, 3, \ldots |B_{i,j}|$ and each $b_{k,j} \in B_{i,j}$. Shannon's entropy [23] can be easily calculated for every set $B_{i,j}$ (all possible segments of all word-types of the corpus):

$$H(a_{i,j} :: B_{i,j}) = - \sum_{k=1}^{|B_{i,j}|} p(b_{k,j}) * log_2(p(b_{k,j})) \tag{4}$$

Thus, for each possible segmentation of each word-type v_i, an entropy value is calculated corresponding to the content of information of all word fragments adjacent to each word fragment considered an affix. Let us call this number $h_{i,j}$ (*i.e.* the entropy associated to one of the fragments of word-type i at segmentation j). Notice that, as with the economy index, there are two entropy values for each segmentation of every word-type, depending on which side of the v_i is taken to be the affix candidate (and which side corresponds to the set of bases that occur with such affix).

3 Affix Catalogs

Upon examination of a segmentation, it may be concluded that it is morphological and, therefore, constitutes the boundary between a base and an affix. However, the word fragment considered to behave as an affix will very likely appear again somewhere else with higher or lower values. These different values correspond to different measurements of the abstract affix unit and the simplest way to combine them is to average them. Moreover, each affix type has a set of averages, one for each kind of index. The important thing is that the information collected from each presumed morphological segmentation must be stored somewhere. Thus, the need of a structure, which will be called catalog.

Formally, a catalog Υ can be seen as containing γ affix candidates, each one represented by the ordered set $\langle s_x, \Omega_x, \bar{c}_x, \bar{k}_x, \bar{h}_x, AF_x \rangle$, where s_x is a member of the set $\{s_1, s_2, s_3, \ldots, s_\gamma\}$, the character strings which represent the affixes of a corpus; Ω_x is the frequency of candidate k as the best affix[17] within a word-type, $\Omega_x \in \{\Omega_1, \Omega_2, \Omega_3, \ldots, \Omega_\gamma\}$; $\bar{c}_x$ is the average of squares associated to the string, $\bar{c}_x \in \{\bar{c}_1, \bar{c}_2, \bar{c}_3, \ldots, \bar{c}_\gamma\}$; $\bar{k}_x$ the average of economy values for the same, $\bar{k}_x \in \{\bar{k}_1, \bar{k}_2, \bar{k}_3, \ldots, \bar{k}_\gamma\}$; $\bar{h}_x$ its averaged entropy value, $\bar{h}_x \in \{\bar{h}_1, \bar{h}_2, \bar{h}_3, \ldots, \bar{h}_\gamma\}$; and AF_x the value which expresses the estimated segment's affixality, $AF_x \in \{AF_1, AF_2, AF_3, \ldots, AF_\gamma\}$. In this manner, Υ can be also described as a set of γ ordered relations:

$$
\begin{aligned}
\Upsilon = \{ &\langle s_1, \Omega_1, \bar{c}_1, \bar{k}_1, \bar{h}_1, AF_1 \rangle, \\
&\langle s_2, \Omega_2, \bar{c}_2, \bar{k}_2, \bar{h}_2, AF_2 \rangle, \\
&\langle s_3, \Omega_3, \bar{c}_3, \bar{k}_3, \bar{h}_3, AF_3 \rangle, \\
&\ldots \langle s_\gamma, \Omega_\gamma, \bar{c}_\gamma, \bar{k}_\gamma, \bar{h}_\gamma, AF_\gamma \rangle \}
\end{aligned} \tag{5}
$$

This simple scheme can be used to store either prefixes, Υ^p, or suffixes, Υ^s. Furthermore, the affix entries of Υ can be ordered according to their affixality values, AF_x $(1 \leq x \leq \gamma)$, in such a way that higher ranks correspond to lower affixality values. This concentrates true affixes and valid affix sequences at the top of the catalog.

Regarding the estimation of each AF_x, it can simply be defined by:

$$
AF(s_x) = k_x c_x h_x \tag{6}
$$

that is, the quality of s_x of being an affix is directly proportional to some measure of economy (k) and to a number of squares (c), both calculated from the segment's boundary with its adjacent strings, as well as to a measure (h) of how surprising these adjacent strings (supposed bases) are.

This generalization holds for the measures calculated for each separate word fragment as a token, but the same relationship holds among the averages of all the values calculated for all occurrences of this affix among the word-types, resulting in an affixality index for the affix item in the catalog (affix-type):

$$
AF(s_x) = \bar{k}_x \bar{c}_x \bar{h}_x \tag{7}
$$

---

[17]The number of word-types in which the affix candidate obtained the highest value within the word-type (*i.e.* with respect to all other segmentations j of the type in question.)

Finally, each index is normalized separately and the affixality value is estimated by their arithmetic mean (rather than multiplication in order to avoid extremely small values):

$$AF^n(s_x) = \frac{\frac{c_x}{\max c_i} + \frac{h_x}{\max h_i} + \frac{k_x}{\max k_i}}{3} \tag{8}$$

This calculation is performed for every segmentation j of all the word-types i of the corpus.

The procedure basically takes each word-type of the corpus and determines the best segmentation for each one, using formula (8). Each best segmentation represents a hypothesis postulating a base and an affix. Thus, the presumed affix (and the values associated with it) are stored into the catalog. The more frequent a presumed affix is detected as the best candidate, the more likely it is really an affix.[18]

4 Affix Discovering for Various Languages

In this section I will report the experiments mentioned above. Namely, catalogs for each language will be presented in order to look into the Spanish verbal inflection subsystem, the Czech derivational prefixes, the Ralámuli derivational suffixes and the Chuj verbal inflection subsystem.

4.1 Spanish Suffixes

The first language examined with this method was Spanish, namely Mexican Spanish (Medina-Urrea [14, 17]), for which it exists a carefully compiled, representative and well balanced corpus for the twentieth century[19] (Lara [12, 13]).

Spanish suffixes are more interesting than its prefixes because the former constitute a compact, organized system of items carrying inflection and syntactic information (*e.g.* verbal and noun concordance). Prefixes are few and mainly inherited from Latin and Greek (a much more interesting set of prefixes will be presented in the next section for the Czech language), so I will focus on the suffixes.

Table 1 exhibits the top thirty suffix candidates of Spanish. These candidates are presented in the second column. The third column exhibits the number of word-types where the candidate came out as the best possible suffix

---

[18]Other possibilities, like selecting several best segmentations per word or including some threshold criteria to filter forms with low values, are discussed in Medina-Urrea [14].

[19]The Corpus del Español Mexicano Contemporáneo (CEMC) was the first statistical base in the Spanish speaking world for a long term lexicographical project.

Table 1. Catalog of Spanish Suffixes.

rank	suffix	freq.	squares	economy	entropy	affixality
1.	~ó	1,428	0.73710	0.91920	0.87200	0.84280
2.	~o	6,314	0.68600	0.97880	0.80170	0.82220
3.	~s	12,013	1.00000	0.99680	0.46090	0.81920
4.	~a	7,687	0.57530	0.98180	0.88880	0.81530
5.	~os	4,554	0.47750	0.97540	0.82350	0.75880
6.	~as	4,324	0.42160	0.97790	0.86450	0.75470
7.	~en	945	0.41070	0.89910	0.90600	0.73860
8.	~ar	1,633	0.21780	0.96210	0.91490	0.69820
9.	~ado	1,429	0.20610	0.96190	0.90700	0.69170
10.	~ando	976	0.18360	0.95440	0.91620	0.68470
11.	~e	2,363	0.42000	0.94820	0.68170	0.68330
12.	~é	639	0.41040	0.81980	0.81530	0.68180
13.	~aba	828	0.18210	0.95650	0.90240	0.68030
14.	~aron	736	0.17790	0.96040	0.89350	0.67730
15.	~ada	1,135	0.16540	0.94910	0.91590	0.67680
16.	~arse	665	0.14620	0.95410	0.90720	0.66920
17.	~ados	941	0.14770	0.95490	0.90080	0.66780
18.	~aban	551	0.14340	0.93950	0.90020	0.66100
19.	~adas	813	0.13160	0.94490	0.90410	0.66020
20.	~an	1,775	0.19500	0.94340	0.83540	0.65790
21.	~ara	370	0.10980	0.91510	0.91510	0.64670
22.	~ará	387	0.12100	0.92950	0.87390	0.64150
23.	~arlo	316	0.09269	0.92910	0.88490	0.63560
24.	~arla	270	0.07950	0.91850	0.90710	0.63500
25.	~arme	244	0.08683	0.91340	0.89160	0.63060
26.	~andose	260	0.07949	0.91360	0.88550	0.62620
27.	~arán	256	0.08995	0.91120	0.87590	0.62570
28.	~ido	445	0.10380	0.85670	0.91400	0.62480
29.	~ita	453	0.09631	0.89650	0.87290	0.62190
30.	~aría	231	0.08063	0.88690	0.89200	0.61980

of that word-type. The fourth, fifth and sixth columns contain the normalized measurements of word squares, economy and entropy. The last column exhibits the affixality index, which – as mentioned above – was calculated as the arithmetic average of the entropy and economy values of the prior columns. Finally, the first column shows the rank of the candidates according to this index: the lower the rank, the greater the affixality index.

Most items are inflectional suffixes or chains of inflectional suffixes (depending on their context, some may be derivational). Since pronoun enclitics are graphically suffixed to infinitives, gerundives and imperatives, some candidates are chains of an inflectional item and enclitic (*e.g.* ~*ar.se*, ~*ar.lo*, ~*ándo.se*). Furthermore, notice that the shorter entries are the most polysemous and represent items of both nominal and verbal inflectional subsystems.

Longer items and chains are readily identifiable as one sort or another of affix or affix sequence.

Most of the complex verbal inflection subsystem occurred within the first 749 catalog entries, gathered towards the top and mixed with derivational items and valid sequences of affixes. As the rank increases, the lesser productive items (such as subjunctive markers) appear along older and rarer derivational entries and residues that could be considered mistakes.

The size of the verbal inflection system varies mainly according to how allomorphs, homophones (homographs) or polysemous items are counted, and whether or not all affix chains with thematic vowels are taken into account: should ~a.r.ía.mos, ~e.r.ía.mos and ~i.r.ía.mos be counted separately or is ~r.ía.mos enough? In this experiment, all these combinations were counted. Thus, 15 of 163 verbal suffixes and suffix combinations[20] were missing within the first 749 catalog entries (a recall of 91%). In the evaluation section, a lower catalog size of 500 will be used to calculate recall for the sake of comparison with the other languages.

4.2 Czech Prefixes

In a recent experiment (Medina-Urrea & Hlaváčová [18]), this method was applied to a list of 166,733 lemmas from the Czech National Corpus. Although this language is highly inflecting, our experiment focused only on the prefix system. Partial results of the experiment are shown in Table 2, which shows the top 30 entries of the catalog.

It was motivated by the need to improve a morphological analyzer which recognizes words by eliminating prefixes. In order to do this, however, the analyzer must know a priori the typical prefixes of Czech. As noted by Hlaváčová in [18], these happen to be very many and oddly productive. They may be, for instance, numerals modifying nouns, e.g. sedmi~ (entry 24 in Table 2). It must be noted that many items that appear prefixed are really nouns. Thus, it may be argued that we are dealing more with composition than with affixation. However, according to the structural relationship they keep with the lexical system to which they belong, these items are in fact simply behaving like prefixes.

Since there are no false prefixes at least within the first hundred candidates, there is a precision measure of 100% within these, and of 54% within the first five hundred ones. Regarding recall, Hlaváčová compiled a set of the most productive and traditional Czech prefixes (45 items) in order to determine how many of the indispensable ones were missed by the method. Approximately 49% of them occurred within the first hundred catalog entries and 89% appeared within the first five hundred.

---

[20]Enclitics were ignored, although the perspective of analyzing them as verbal inflection marks in concordance with direct and indirect objects is not so delicate (see Rini [21]).

Table 2. Catalog of Czech Prefixes.

rank	prefix	freq.	squares	economy	entropy	affixality
1.	severo~	75	0.02227	0.97350	0.93030	0.95190
2.	proti~	457	0.10710	0.92840	0.96750	0.94790
3.	jiho~	76	0.02194	0.94590	0.92150	0.93370
4.	mezi~	199	0.04668	0.92270	0.92150	0.92210
5.	super~	263	0.05129	0.85720	0.96530	0.91130
6.	dvoj~	233	0.05450	0.86290	0.94750	0.90520
7.	mimo~	154	0.03380	0.87910	0.93020	0.90470
8.	troj~	136	0.04416	0.85820	0.94390	0.90100
9.	mnoho~	103	0.04946	0.91330	0.88820	0.90080
10.	osmi~	97	0.07297	0.92890	0.87210	0.90050
11.	spolu~	267	0.05657	0.89610	0.90220	0.89920
12.	video~	138	0.03361	0.93000	0.86800	0.89900
13.	východo~	47	0.01397	0.92600	0.87140	0.89870
14.	devíti~	59	0.05196	0.96070	0.83340	0.89710
15.	při~	1,361	0.72720	0.91030	0.88220	0.89630
16.	více~	151	0.07567	0.88610	0.89850	0.89230
17.	radio~	102	0.02021	0.86240	0.92040	0.89140
18.	šesti~	113	0.08147	0.92990	0.84430	0.88710
19.	nad~	437	0.05547	0.77360	1.00000	0.88680
20.	celo~	123	0.02970	0.87090	0.90150	0.88620
21.	šéf~	45	0.01141	0.93750	0.83320	0.88530
22.	pěti~	168	0.08477	0.88590	0.88450	0.88520
23.	západo~	44	0.01283	0.88780	0.88070	0.88430
24.	sedmi~	82	0.06226	0.94320	0.81720	0.88020
25.	několika~	67	0.10170	0.95710	0.80290	0.88000
26.	pseudo~	149	0.03044	0.81950	0.93850	0.87900
27.	třiceti~	39	0.03653	0.94370	0.81130	0.87750
28.	velko~	172	0.03735	0.91650	0.83540	0.87600
29.	elektro~	168	0.02725	0.80200	0.94720	0.87460
30.	od~	2,393	0.61670	0.81390	0.93510	0.87450

4.3 Ralámuli Suffixes

Ralámuli or Rarámuri, better known as Tarahumara, is a Uto-Aztecan language spoken in northern Mexico. It is more an agglutinative language than a fusional one. Word formation is mainly accomplished by means of suffixation. As could be expected, stems are followed by derivational suffixes, and these by inflectional ones. Since stems can be the result of other morphological processes, there might be morphemes to be discovered towards the beginning of words, but they are not necessarily affixal (Medina-Urrea & Alvarado García [16]). Furthermore, given that Ralámuli has very little inflection, we applied the method to examine suffixes of a derivational nature.

Table 3. Catalog of Ralámuli Suffixes.

rank	suffix	freq.	squares	economy	entropy	affixality
1.	~ma	35	1.00000	1.00000	0.88030	0.98050
2.	~re	77	0.79960	0.81100	0.86060	0.82370
3.	~sa	33	0.63640	0.93060	0.75590	0.77430
4.	~ra	62	0.66130	0.64610	0.85080	0.71940
5.	~si	28	0.75000	0.52570	0.83450	0.70340
6.	~na	25	0.41140	0.72240	0.79840	0.64410
7.	~go	4	0.21430	0.90650	0.64930	0.59000
8.	~é	49	0.16620	0.43580	1.00000	0.53400
9.	~ame	51	0.25210	0.30640	0.85910	0.47250
10.	~gá	18	0.40480	0.37810	0.61360	0.46550
11.	~ka	19	0.25560	0.28060	0.84130	0.45920
12.	~á	67	0.13860	0.31330	0.91950	0.45710
13.	~ré	11	0.16880	0.41020	0.73430	0.43780
14.	~ga	50	0.18290	0.28340	0.80650	0.42430
15.	~a	281	0.10520	0.18960	0.97250	0.42250
16.	~ba	8	0.21430	0.30220	0.74000	0.41880
17.	~ayá	8	0.21430	0.44320	0.57570	0.41110
18.	~í	42	0.10200	0.26480	0.80540	0.39070
19.	~či	39	0.10260	0.27510	0.74000	0.37260
20.	~e	164	0.15240	0.29100	0.64290	0.36210
21.	~mi	4	0.07143	0.30220	0.69910	0.35760
22.	~áame	12	0.15480	0.00000	0.90570	0.35350
23.	~yá	20	0.34290	0.11080	0.57420	0.34260
24.	~i	139	0.03905	0.10320	0.84220	0.32820
25.	~ira	11	0.06494	0.10990	0.79570	0.32350
26.	~o	41	0.00000	0.00000	0.96810	0.32270
27.	~ne	3	0.14290	0.40290	0.41950	0.32170
28.	~wa	9	0.04762	0.13430	0.74000	0.30730
29.	~agá	6	0.16670	0.20140	0.53640	0.30150
30.	~sí	4	0.35710	0.00000	0.53740	0.29820

The text sample[21] represents the dialectal variant from San Luis Maji-machi, Bocoyna, Chihuahua. For today's corpora standards, this sample is an extremely small one, consisting of 3,584 word-tokens and 934 word-types (everything fits in a 21Kb plain text file). Although we can hardly assume this sample's representativeness of the language, we still proceeded to apply the method in order to see how appropriate it is for small corpora. Table 3 shows partial results of procedure.

Even though Ralámuli has relatively few inflectional forms, the larger cata-log exhibits more items containing inflectional material than were expected.[22]

---

[21]Mainly memoirs and short stories collected by Patricio Parra.

[22]This is certainly due to the fact that input texts are constituted by linguistic acts in the pragmatic act of narrating a story. Words appear therefore inflected.

In fact, if inflectional suffixes are to be considered somehow more affixal than derivational ones, it should not be surprising to find the four most prominent Tarahumara inflection affixes appear at the top of the table: $\sim ma$, $\sim re$, $\sim sa$, and $\sim si$, which mark tense, aspect and mode.

Using her own field work experience and taking into account the work of other experts, Alvarado García determined the 35 most prominent nominal and verbal derivational suffixes for this language. 25 of these occurred within the first 100 catalog entries (a recall measure of 71% within this limit). The other entries are chains of suffixes (including sequences of derivational and inflectional items) and residual forms.[23] The 10 derivational suffixes which did not appear in the catalog are essentially verbal derivational forms, or modifiers of transitivity or some semantic characteristic of verbal forms. This might mean that the small sample used is more representative of nominal structures, rather than of verbal ones. Finally, it is worth stressing that a significant part of the known Ralámuli derivational system – essentially the nominal subsystem – was retrieved from a ridiculously small set of texts, which hardly constitutes a corpus of this language.

4.4 Chuj Verbal Inflection System

Chuj belongs to the Mayan family of languages and it is spoken in both sides of the border between Mexico and Guatemala. The experiment conducted is fully described in Medina-Urrea & Buenrostro Díaz [17]. This language is particularly interesting for the present paper because its verbal inflection system is constituted by both prefixes and suffixes. To illustrate this, Tables 4 and 6 show these affixes along with their affixality values and ranks from the catalogs partially shown in Tables 5 and 7.

The text sample used is also very small – somehow bigger than the one for Ralámuli – and contains 15,485 word-tokens, about 2,300 types (in a 86Kb plain text file). It was compiled by Buenrostro Díaz during various of her field work visits to the region. Given its reduced size and the fact that it is composed of only five narrations, it also cannot properly be considered a balanced and representative corpus of the language. Results are nevertheless interesting because, given her grammatical interests, Buenrostro Díaz put special emphasis in compiling a collection of texts representative of verbal structures.

Obviously, using dictionary entries without inflection (lemma sets), rather than text in context, would be a much better way to obtain derivational items.

[23]The examination of residual items was especially difficult. Questions about lexicalized affixes (possibly fossilized items) and about the relationship between syllable structure and affix status emerged. These matters remain to be revised by Ralámuli experts. Meanwhile, for evaluation purposes (see evaluation section below), entries with unexpected syllabic structure were not counted as acceptable suffixes nor valid chains of them.

Table 4. Paradigm of Chuj Verbal Inflection Prefixes.

tense		grammatical person		
		absolutive	ergative	
7. tz~ 0.74	1	1. in~ 0.91	1. in~ 0.91	13. w~ 0.63
2. ix~ 0.86	2	27. ač~ 0.50	8. a~ 0.71	— ø~ —
24. x~ 0.51	3	— ø~ —	3. s~ 0.83	17. y~ 0.56
12. ol~ 0.65	1	74. onh~ 0.33	5. ko~ 0.77 / 43. ku~ 0.42	58. k~ 0.36
— ø~ —	2	251. ex~ 0.20	22. e~ 0.54	183. ey~ 0.24
	3	— ø~ eb' —	3. s~ eb' 0.83	17. y~ eb' 0.56

Chuj Prefixes

Buenrostro Díaz's proposal of the Chuj's prefixed verbal inflection paradigm appears in Table 4. Every item of the table is listed with its rank to the left and with its affixality index to the right (they were obtained from the catalog partially shown in Table 5).

Tense markers appear first, one of which is the null prefix ø. Then some marker indicating the grammatical person follows, which might appear either in its absolutive or ergative form (ergatives to the right attach to vowel initial stems and those to the left attach to consonant initial ones). The first and fourth lines show respectively singular and plural forms for first person ($ko\sim$ and $ku\sim$ are allomorphs), the second and seventh show second person and the third and fifth third person.

Regarding Table 5, all tense markers occur within the first 24 entries. Also, within the first 30 candidates, a couple of the absolutive and most ergative personal markers appear. Furthermore, there are chains of tense and person marker prefixes: $\sim tz.in$, $\sim ol.in$, $\sim ix.in$, $\sim ix.s$, $\sim tz.s$, $\sim tz.onh$, and $\sim ol.ač$. In fact, personal markers are word initial only because one of the tense markers is the null affix ø-.

In Table 5, there appear ten residual entries (either verb stems, non-readily recognizable or fragmented prefixes). Hence, precision for this table would be the proportion of right guesses, 66.6%.

With respect to recall, every prefix listed in Table 4 appeared in the prefix catalog. But, since some of them obtained a rank greater than the size of Table 5, they cannot be found there. This implies a precision measure of 100% for the first 251 catalog items, and of 90% for the first 74 (there are two of 20 items of Table 4 with rank above 74).

Chuj Suffixes

Buenrostro Díaz's proposal for Chuj's inflectional suffix system appears in Table 6. Essentially, these suffixes mark voice, mode and end of utterance (the

Table 5. Catalog of Chuj Prefixes.

rank	prefix	freq.	squares	economy	entropy	affixality
1.	in~	93	0.56200	0.83990	0.98280	0.91130
2.	ix~	181	1.00000	0.80210	0.90880	0.85540
3.	s~	187	0.43190	0.66620	0.98740	0.82680
4.	kak'~	1	0.18460	1.00000	0.59290	0.79650
5.	ko~	71	0.32980	0.66030	0.87830	0.76930
6.	xsči'~	1	0.04615	1.00000	0.51070	0.75540
7.	tz~	349	0.69420	0.59450	0.88610	0.74030
8.	a~	164	0.18550	0.41110	1.00000	0.70550
9.	tzin~	48	0.17210	0.44820	0.93380	0.69100
10.	olin~	26	0.17990	0.47320	0.88180	0.67750
11.	xal~	2	0.07692	0.67500	0.66010	0.66750
12.	ol~	185	0.47400	0.52790	0.78070	0.65430
13.	w~	70	0.44750	0.73820	0.52560	0.63190
14.	olač~	26	0.18880	0.47210	0.76630	0.61920
15.	tzs~	49	0.08069	0.42520	0.81010	0.61770
16.	ixin~	29	0.13050	0.36740	0.81830	0.59290
17.	y~	127	0.33540	0.54740	0.57100	0.55920
18.	k'a~	11	0.01678	0.36360	0.74520	0.55440
19.	ma~	31	0.01241	0.22020	0.87220	0.54620
20.	al~	15	0.03692	0.21230	0.87400	0.54320
21.	na~	9	0.01880	0.30950	0.77610	0.54280
22.	e~	63	0.03272	0.16630	0.91730	0.54180
23.	ak'~	12	0.04231	0.23990	0.78810	0.51400
24.	x~	43	0.04401	0.25860	0.75680	0.50770
25.	tzonh~	16	0.06731	0.22490	0.78660	0.50570
26.	b'ati~	1	0.03077	0.75000	0.25540	0.50270
27.	ač~	9	0.03248	0.19440	0.80340	0.49890
28.	ixs~	24	0.02949	0.16880	0.81540	0.49210
29.	ay~	23	0.07692	0.43870	0.53450	0.48660
30.	k'e~	3	0.05128	0.25000	0.70220	0.47610

Table 6. Paradigm of Chuj Verbal Inflection Suffixes.

	voice		modal/ temporal	thematic vowel
	63. ~aj	0.4129	6. ~ok 0.7479	11. ~i 0.6703
	68. ~chaj	0.4018	18. ~nak 0.5977	12. ~a 0.6549
passive	872. ~b'il	0.1212		
	1,016. ~nax	0.0949		
	— ~*ji*	—		
antipassive	19. ~an	0.5958		
	28. ~wi	0.5531		
	161. ~waj	0.2629		

thematic vowels distinguish transitive verbs from intransitive ones and signal end of phrase). Again, items appear surrounded by the rank and affixality value they obtained in the suffix catalog which is partially shown in Table 7.

Among the interesting entries of the latter table, a system of directionals can be partially seen. These are movement verbs which function as verbal classifiers: ~kan, ~b'at, ~el, and ~k'och. The paradigm is not complete. However, since these classifiers are more derivational than inflectional, they must be very productive in order to compete with verbal inflection items for slots among the top entries of the catalog.

Table 7. Catalog of Chuj Suffixes.

rank	suffix	freq.	squares	economy	entropy	affixality
1.	~kan	68	1.00000	1.00000	0.90290	0.95150
2.	~nej	24	0.41190	0.98110	0.76980	0.87540
3.	~ta'	70	0.64010	0.75260	0.82030	0.78650
4.	~b'at	63	0.56900	0.67170	0.86590	0.76880
5.	~al	82	0.28940	0.53560	1.00000	0.76780
6.	~ok	68	0.46950	0.55850	0.93740	0.74790
7.	~ab'	49	0.29590	0.50780	0.90590	0.70690
8.	~il	62	0.24570	0.46340	0.93470	0.69900
9.	~ač	16	0.34620	0.71490	0.67950	0.69720
10.	~xi	37	0.54160	0.70030	0.68290	0.69160
11.	~i	205	0.58760	0.46370	0.87690	0.67030
12.	~a	142	0.18360	0.37910	0.93060	0.65490
13.	~kot	48	0.41030	0.55640	0.74040	0.64840
14.	~el	68	0.18890	0.43240	0.86430	0.64830
15.	~tak	19	0.16800	0.37850	0.90620	0.64240
16.	~in	46	0.16470	0.34980	0.89170	0.62070
17.	~kani	8	0.10100	0.48810	0.71660	0.60240
18.	~nak	18	0.34830	0.41430	0.78120	0.59770
19.	~an	233	0.33710	0.36460	0.82710	0.59580
20.	~alan	13	0.36980	0.36500	0.82250	0.59380
21.	~ab'i	9	0.26070	0.54200	0.64330	0.59260
22.	~ni'	7	0.17580	0.61660	0.56680	0.59170
23.	~k'oč	28	0.23350	0.29220	0.86700	0.57960
24.	~ak'	43	0.18520	0.34370	0.79220	0.56800
25.	~ak'tej	6	0.09615	0.52520	0.58790	0.55660
26.	~koti	18	0.06410	0.42930	0.68360	0.55640
27.	~ila	9	0.34620	0.54070	0.57080	0.55580
28.	~wi	14	0.04945	0.42080	0.68530	0.55310
29.	~ik'	12	0.53210	0.48130	0.60150	0.54140
30.	~o	123	0.03096	0.11440	0.96340	0.53890

Regarding the verbal paradigm of Table 6, almost all its members do occur in the suffix catalog, except passive voice marker ~ji (which is shown in boldface). Therefore, we can measure recall within the first 1,016 catalog entries to be 92% (eleven of twelve) and 75% for the first 500 entries (nine of twelve). Taking prefixes and suffixes together, recall would be of 96.55% (28 items of 29) for the first 1,016 items and 91% within the first 500 catalog entries (29 of 32).

5 Evaluation

Evaluating discovery results for such different affix subsystems of such a set of unlike languages, represented by such dissimilar corpora, is certainly a challenge. Nevertheless, granted that plain errors should be expected, given the nature of the method and the reduced size of two of the text samples, some comparison scheme must be considered in order to evaluate results.

For the sake of simplicity and since the experiments described above dealt with some specific affix subsystems such as inflectional or derivational, verbal or nominal, prefixal or suffixal, I will base the following considerations on the size of those subsystems. Recall, that these were the research matter of separate experiments, so they were gathered from linguistic tradition (Spanish verbal inflection and Czech's set of traditional prefixes) or from published work or specialist's field experience (Ralámuli verbal and nominal derivational prefixes and the Chuj verbal inflection paradigms).

Hence, it makes sense to pick a window of the size of each relevant subsystem to look into the top of the relevant affix catalog and determine the proportion of errors within that window in order to calculate precision. Notice that items belonging to other subsystems of the same language are bound to appear there and that they could not, of course, be considered errors. Thus, what is measured is precision for the window, not precision for the subsystem in question.

A recall measure, on the other hand, would deal specifically with how much of the subsystem sought is not retrieved. In this case, the evaluation window must be larger, mainly because these languages have other subsystems competing to appear towards the beginning of the catalog, but also because some parts of the subsystem may cohabit with other complex subsystems or simply be less productive (Spanish unambiguous subjunctive forms, e.g. ~ásemos, ~iéremos). Upon examination of results, a window of five hundred was selected (the Ralámuli catalog has fewer items than that).

Obviously, a smaller window means greater precision and lower recall, whereas a greater window means lower precision and greater recall. Therefore, the window sizes selected will maximize both measurements – the smaller window (subsystem size) for precision, the bigger one (of 500) for recall. Nevertheless, it should be clear that precision will decrease considerably as the window grows because rarer items are mixed with plain mistakes and unrecognizable, residual forms (which were counted as errors). Conversely, recall

Table 8. Evaluation measurements.

	Spanish	Czech	Ralámuli	Chuj	
subsystem	verbal inflection	prefixes	derivational suffixes	prefix verbal inflection	suffix verbal inflection
sample tokens	2,000,000	—	3,584	15,485	
sample types	79,000	166,733	934	2,300	
subsystem size$^a(n)$	163	45	35	20	12
right guesses^b	156	45	28	15	12
presumed errors^b	7	0	7	5	0
unretrieved items^c	39	5	10	0	3
precision	0.96	1.00	0.80	0.75	1.00
recall	0.76	0.89	0.71	1.00	0.75

^{a} Allomorphs, homographs and polysemous items count separately; null affixes are excluded.

^{b} In relevant catalog within subsystem size.

^{c} Members of subsystem not found in relevant catalog within first 500 catalog candidates.

will decrease as the window is diminished because rare and lesser productive members of the subsystem examined will fall outside the smaller window (whereas other subsystem's more productive members also compete for the upper catalog slots).

With all of this in mind, let us look into Table 8, which shows numerical data, as well as precision and recall measurements, for the languages examined.

The first row shows the name of the specific affix subsystem focused. The second and third rows characterize the corpora used: their size in number of word-tokens and number of word-types (the Czech sample is a set of lemmas, so there is no number of tokens). The fourth row shows the size of the subsystem sought in those corpora.[24] Then, based on a window of the size of the subsystem in focus for each corpus, the correct guesses and presumed errors were counted within that window (rows five and six). The seventh row contains the number of subsystem members not found within the much larger window of five hundred catalog items.

The last two rows exhibit the precision and recall measures. As mentioned above, precision is the proportion of correct guesses within the first n entries of the relevant catalog, such that n is the size of the subsystem sought. Ad-

---

[24]It is worth stressing that determining subsystem size is indeed a problem not to be underestimated. For each of the experiments reported here, the specialists had to study the subsystem in order to know its size, which surely varies from perspective to perspective. Here, null morphemes were excluded and allomorphs, homographs, and polysemous items were counted separately. In the case of Spanish, antiquated inflections (for informal 2nd person plural, very productive in the Castilian dialect), and those followed by enclitics, which are graphical affixes, were not counted.

ditionally, recall is the number of members of the subsystem actually found within the first 500 hundred catalog items.

The evaluation measures look rather good because, as mentioned above, window sizes were selected to maximize them. The important remark to make, for the four languages, is that there is a sense in which affixal items tend to be concentrated towards the top of their catalogs and that most of the subsystems in which the items play a part can be retrieved within the first five hundred catalog entries. Undoubtedly, other languages and more corpora should be examined with the idea of making sounder result comparisons. Meanwhile, the method seems good enough to approach some unknown language, structurally similar to those presented here, and start examining its morphological structure, although the language sample may not be as representative as it may otherwise be desirable.

6 Final Remarks

In this paper, an affix discovering method was examined and results of its application to four very distinct languages, represented by very unequal corpora, were presented and evaluated. There are, of course, certain problems with the approach that leave a lot of room for improvement. Namely, it would be much improved if affix and word contexts were to be accounted for by the method. This would at the very least make residue evaluation easier. Also, the mechanical and quantitative examination of affix sequences is necessary in order to study the affitactics of these languages from an unsupervised perspective.

The catalogs examined in this paper should not be considered *the* morphological models of the languages examined. They should perhaps be seen more as windows to complex phenomena which can be described in different ways, according to the preferred language theoretical perspective. They are tools for the discovery of the unknown, more related to text mining than to rule-based formalism design.

Whether or not the method proposed here is the best way to discover affixes or whether or not morphological models are to be constructed in later stages of language research, it can be observed from this kind of experiment that, like Edward Sapir [22, p. 29] once put it eloquently, there appears to exist a potential energy that glues morphemes into the bricks of language. This glutinous energy may seem elusive, but somehow, from what we have seen, it can be measured by means of magnitudes portraying economical sign structure and carried information.

Acknowledgements

Thanks to Jaroslava Hlaváčová, Cristina Buenrostro Díaz and Maribel Alvarado García – with whom I have processed the Czech, Chuj and Tarahumara materials (results of those experiments appeared in [15, 16, 18]) – for their invaluable collaboration. This work was supported by grant PAPIIT UNAM IN 400905.

References

[1] O. Cromm. Affixerkennung in deutschen Wortformen. Eine Untersuchung zum nicht-lexikalischen Segmentierungsverfahren von N. D. Andreev. Abschluß des Ergänzungsstudiums Linguistische Datenverarbeitung, Frankfurt am Main, 1996.

[2] J. de Kock and W. Bossaert. *Introducción a la lingüística automática en las lenguas románicas*, volume 202 of *Estudios y Ensayos*. Gredos, Madrid, 1974.

[3] J. de Kock and W. Bossaert. *The Morpheme. An Experiment in Quantitative and Computational Linguistics*. Van Gorcum, Amsterdam, Madrid, 1978.

[4] W. B. Frakes. Stemming Algorithms. In W. B. Frakes and R. Baeza, editors, *Information Retrieval, Data Structures and Algorithms*, pages 131–160. Prentice Hall, New Jersey, 1992.

[5] A. Gelbukh, M. Alexandrov, and S. Y. Han. Detecting Inflection Patterns in Natural Language by Minimization of Morphological Model. In *Congreso Iberoamericano de Reconocimiento de Patrones, CIARP-2004*, LNCS, 2004.

[6] J. Goldsmith. Unsupervised Learning of the Morphology of a Natural Language. *Computational Linguistics*, 27(2):153–198, 2001.

[7] J. H. Greenberg. *Essays in Linguistics*. The University of Chicago Press, Chicago, 1967.

[8] M. A. Hafer and S. F. Weiss. Word Segmentation by Letter Successor Varieties. *Information Storage and Retrieval*, 10:371–385, 1974.

[9] Z. S. Harris. From Phoneme to Morpheme. *Language*, 31(2):190–222, 1955.

[10] H. Johnson and J. Martin. Unsupervised Learning of Morphology for English and Inuktitut. In *Proceedings of the 2003 Human Language Technology Conference of the North American Chapter of the Association for Computational Linguistics*, 2003.

[11] K. Kageura. Bigram Statistics Revisited: A Comparative Examination of Some Statistical Measures in Morphological Analysis of Japanese Kanji Sequences. *Journal of Quantitative Linguistics*, 6:149–166, 1999.

[12] L. F. Lara and R. Ham Chande. *Investigaciones lingüísticas en lexicografía*, chapter Base estadística del Diccionario del Español de México, pages 5–39. Volume 89 of *Jornadas* [13], 1st edition, 1974.

[13] L. F. Lara, R. Ham Chande, and M. I. García Hidalgo. *Investigaciones lingüísticas en lexicografía*, volume 89 of *Jornadas*. El Colegio de México, A. C., Mexico, 1st edition, 1979.

[14] A. Medina-Urrea. Automatic Discovery of Affixes by Means of a Corpus: A Catalog of Spanish Affixes. *Journal of Quantitative Linguistics*, 7(2):97–114, 2000.

[15] A. Medina-Urrea. *Investigación cuantitativa de afijos y clíticos del español de México. Glutinometría en el Corpus del Español Mexicano Contemporáneo.* PhD thesis, El Colegio de México, Mexico, April 2003.

[16] A. Medina-Urrea and M. Alvarado García. Análisis cuantitativo y cualitativo de la derivación léxica en ralámuli. In *Primer Coloquio Leonardo Manrique*, Mexico, Conaculta-INAH, September 2004.

[17] A. Medina-Urrea and E. C. Buenrostro Díaz. Características cuantitativas de la flexión verbal del chuj. *Estudios de Lingüística Aplicada*, 38:15–31, 2003.

[18] A. Medina-Urrea and J. Hlaváčová. Automatic Recognition of Czech Derivational Prefixes. In *Proceedings of CICLing 2005*, volume 3406 of *Lecture Notes in Computer Science*, pages 189–197. Springer, Berlin/Heidelberg/New York, 2005.

[19] M. P. Oakes. *Statistics for Corpus Linguistics.* Edinburgh University Press, Edinburgh, 1998.

[20] B. B. Rieger. Computing Granular Word Meanings. A Fuzzy Linguistic Approach in Computational Semiotics. In P. Wang, editor, *Computing with Words*, pages 147–208. John Wiley & Sons, New York, 2001.

[21] J. Rini. *Motives for Linguistic Change in the Formation of the Spanish Object Pronouns.* Juan de la Cuesta, Newark, Delaware, 1992.

[22] E. Sapir. *Language: An Introduction to the Study of Speech.* Harcourt, Brace & Company, New York, 1921.

[23] C. E. Shannon and W. Weaver. *The Mathematical Theory of Communication.* University of Illinois Press, Urbana, 1949.

[24] A. Spencer and A. M. Zwicky. *The Handbook of Morphology.* Blackwell, Oxford, 1998.

Licensing Strategies in Natural Language Processing

Jürgen Rolshoven

rols@spinfo.uni-koeln.de

Summary. This article discusses strategies for licensing within the framework of generative grammar as well as their application in the LPS linguistic processing system. LPS is a Linguistic programming language developed at the Institute for Linguistic Data Processing at the University of Cologne. It also is a computer system which employs this language for natural language processing, in particular for machine translation. In the introduction we give a brief sketch of formal grammar development and derive the idea of licensing from this development. We also describe the generation of structures in linguistic processing systems by means of the object-oriented linguistic programming language LPS. The third part discusses optimization strategies through the competing of variant structures evaluated by means of licensing. The concluding fourth part discusses licensing, specifically as a topic of computational linguistics with the aim of distinguishing its placement within the domain of performance or competence.[1]

1 Introduction

1.1 Formal Grammars and Natural Languages

Natural Language Processing (NLP) and the development of Computer Science are related in two ways: practically, they are related via data processing machines, and theoretically via the creation of formal languages developed for programming these machines. The formalisms for defining such languages were transferred to the description of natural languages. In the mid 1950s Chomsky achieved renown both in the fields of theoretical computer science and linguistics. Setting out from a typology of grammar-types he went on to distinguish the different types of formal languages. Formal languages however are not only defined by grammar but also by recognizing automata. In his work, Chomsky also demonstrated the correspondence of specific types of automata to grammar and language types. The question of language-analysing

---

[1] I would like to thank to Gustav Vella for discussion and helpful suggestions with this paper.

J. Rolshoven: *Licensing Strategies in Natural Language Processing*, StudFuzz **209**, 301–320 (2007)
www.springerlink.com

automata is insofar of relevance in computational linguistics in that language analysis is one of its tasks.

Chomsky distinguishes among 4 types of languages: Type 0, Type 1, Type 2, and Type 3 languages. Type 3 languages are subsets of Type 2 languages, Type 2 languages are subsets of Type 1 languages, and Type 1 languages subsets of Type 0 languages. This is the so called Chomskian hierarchy of formal languages. The rules in these grammars are rewrite rules. They use the arrow operator: to the left we have the symbol or symbols to be substituted, to the right we define the symbol through which another symbol or chain of symbols can be substituted.

The formal language hierarchy is evoked by constraints and substitution rules. It follows that the subject matter here is the syntax of rules. The rules of the grammars which produce Type 0 languages are not constrained. It is of particular importance that the left part of a rule may be longer than the right part of the rule. Apart from this an empty chain in the left part of the rule is not permissible.

Type 1 languages are characterised by rules, in which the right part has to be at least as long as the left part. Consequently empty strings are not possible. Type 2 languages are distinguished by the fact that only one symbol is permissible to the left of the arrow. These type of languages are context-free. Type 3 languages are produced by rules in which the left part of the rule either is itself a terminal symbol or else precedes or follows a non-terminal symbol. If the former is the case, then the type 3 language is said to be right-regular, in the latter case it is said to be left-regular.

Formal languages are typified by the constraints of their production rules. The question as to the attribution of a word to a particular language can be seen from two points of view. On the one hand, one can show that a grammar is capable of producing such a word, and on the other hand, it can be proved that an automaton associated with a language would either accept or reject the word.

The construction of such automata is of great relevance for formal and natural languages. Formal languages, programming languages in particular, are built to be efficiently processed by simple automata (cf. Wirth [25]). Looking back at the development of formal languages since the 1950s, one notices that grammatical formalisms have become increasingly constrained. In the 1960s [5] and 1970s, Type 0 grammars [1, 2, 3, 4] changed to Type 1 grammars – their productive strength was also tamed by constraints on context-sensitive rules tending towards Type 2 grammars [6]. From the point of view of computational linguistics this development was very welcome, since it simplifies the modeling of the automata that recognize natural language.

The development of X-Bar syntax [14] put further constraints on production rules. Ever since, the grammar model within the generativist framework has been based essentially on two components: the production component, characterized by X-Bar schemas; and a component designed to handle context sensitivity.

1.2 Licensing

The concept of licensing is related in particular to the Principles and Parameters model in Generative Grammar. One assumes that there is an X-bar schema component for the production of syntactic structures and a further component used mainly for checking context dependencies composed of submodules such as the theta module, the case module, the binding module, and the control module.

These subcomponents determine the quality of structures. They either accept them (license them) or reject them: a so-called PRO – eg. as the subject of the embedded sentence in *he tries PRO to climb the mountain* – has to be licensed. It may only appear in a syntactic configuration in which no case can be assigned. Only if this condition is fulfilled will a PRO be licensed, the configuration will otherwise be rejected as ungrammatical (cf. Haegeman [13, p. 441]).

In terms of LPS, licensing comprises traditional generative concepts, while extending them in two ways:

1. Licensing is a *generate and test* process in which each configuration hypothesis is evaluated (cf. Section 2 on page 305).
2. Models of competence abstract strongly from the actual conditions of communication. Although normal communication usually contains many errors, it does not collapse. The simulation of this fault tolerance and robustness is a weak point in NLP systems. This stems from the relationship of these systems to recognizing automata whose task it is to decide if a string belongs to a language or not. In our concept of licensing a model is put forward which records partial as well as complete structures. We are thus dealing with persistent syntactic structures. In terms of competence these structures are licensed. They form a stock of canonical structures to be matched with, in a linguistic sense, incomplete sentences. Through a tolerant matching process which allows discrepancy, input sentences can be completed by means of the pattern structures available (cf. Section 4 on page 317).

The interaction of both components is similar to generate and test procedures. The X-Bar schema controls the generator, while test processes evaluate the results.

For linguistic modelling and work in computational linguistics such a generate and test approach has several advantages. The test-, or licensing conditions are defined statically. They apply to structure and not to the construction processes. They are declarative and not procedural. Even from a cognitive point of view, declarative statements are much easier to control. Declarative statements characterize the linguistic system, the competence or – in Saussurean terms – the *langue*, not individual use, i.e. performance, or the *parole*.

If we thus abstract to a large extent from the *parole*, another advantage of declarative and system-oriented workflow comes to light: computational modelling can abstract from the parsing and generating modes of language usage. The generative knowledge relevant to the system can be leveraged for both modes – parsing and generating – and thus does not need to be modelled separately.

Yet it is necessary to safeguard the communication between the declarative system knowledge and the corresponding syntactic-semantic structures (of the *parole*) to be built. For this purpose, the following solution turns out to fulfil more than just the requirements in modern software technology: Since the end of the 70s, generative-oriented linguistics has developed the concept of modules of linguistic knowledge [7, 11]. This concept was taken from the computer science of that period – compare for instance programming languages such as MODULA [24]. However this concept was not pursued to the full in its linguistic adaption.

Here, we would not only like to pursue the concept in greater depth but also attempt to extend it within the concept of *object orientation*. Linguistic knowledge is expressed in terms of object classes. A class is a blueprint for an object, describing which components it contains. In object-oriented programming, algorithms are called *methods*. A class has no real existence as such; it is the (abstract) typing of an object. Objects, however, only exist in a computer's memory during runtime. *Objects* are produced according to a description provided by the class. Object-oriented terminology calls this *instantiation*. Instantiation can also be seen as the birth of an object according to the class blueprint. Thus an object is called the *instance* of a class. Those components of an object which are typified by data structures are also known as declarative variables, the methods are equally known as procedural variables. When methods are called we can also speak of "message passing".

The classes are instructions for creating nodes in tree-structures. In other words, nodes are instances of classes. The process described here is realized as an object-oriented linguistic programming language, which is interpreted by a custom interpreter specifically designed for linguistic data processing.

The properties of a class (its components) can be passed on to other classes. This is called *inheritance*. Classes that pass on their properties allow further properties (components) to be added to the inheriting classes. By this means they become specialized. If, in turn, these too pass on properties, the whole chain produces a class hierarchy. From these classes we can derive objects, with highly specialized classes resulting in highly specialized objects. Classes are grouped by their functional similarity and are organized in modules or packages. This grouping constrains inheritance beyond module boundaries by requiring an explicit request for such an action. Isolating classes like this helps avoid unwelcome side-effects.

Linguistics and the cognitive sciences have adopted the notion of modularisation from Computer Science. In the following we will argue that introducing the concept of object orientation to linguistics can also be very helpful. One has to state however that this concept is not that unfamiliar to linguistics. By the time the concept of object orientation had been implemented in Computer Science, it was recalled that, to a certain extent, one was adhering to a much older tradition: that of the Aristotelian theory of categories. Structuralist taxonomies in Linguistics are a reflection of this philosophical tradition.

The implementation of the concept of object orientation with a special focus on licensing conditions is the topic of the following section.

2 Implementation

2.1 Modularization and Object Orientation in Computational Linguistics

Introducing the concept of object orientation into the organization of linguistic knowledge makes linguistic description compact and transparent. Linguistic object orientation adopts the fundamental terms of object orientation and we shall thus also use the notions of classes, inheritance and instantiation.[2] By means of inheritance, linguistic knowledge can be hierarchically structured. Sibling classes inherit from parent classes. Sibling classes express specialization, parent classes express generalization. We also make use of Multiple inheritance. The concept of inheritance promotes a typological approach. Typological similarities are expressed through more abstract parent classes, language-specific features are expressed through specialized sibling classes. As a consequence the subject matter is also naturally presented to the linguist in a clearly structured manner.

As in classical object orientation, linguistic object orientation also includes components, data structures, and methods. Through instantiation objects are created from classes. The objects thus created are the nodes of a phrase structure tree. These nodes are thus realizations or instances of linguistic classes. In the linguistic object orientation described here, instantiation is tightly related to the generation of tree structures. Linguistic classes have an elementary data structure, comprising of a cluster or list of attribute-value pairs. As an example we here can consider an attribute labelled *category*. This attribute takes on specific values previously declared in a value set, for instance N for noun, V for Verb, D for Determiner. Further Attributes are *bar* for the bar index of a category in compliance with the X-bar schema (taking the values 0, $'$ and $''$, as well as *case*, *gender*, or *number*.

For the German language, a determiner class can be defined in the following manner

---

[2]The concept of *object orientation* in linguistics and computational linguistics is not that well established. For more details cf. Daelemans et al. [10] and [9]. For more on the concept within the LPS framework cf. Lalande [17], Möller [19] and Rolshoven [21, 22].

```
CLASS D⁰ [neutr,sing,nom];
END;
```

Such a class provides the blueprint for a T^0 or D^0 node, for instance in a tree structure of the following form:

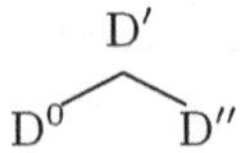

Fig. 1. A linguistic tree structure.

This blueprint begins with the keyword *CLASS*; the keyword indicates that a class is being defined. The keyword is followed by a list of attribute-value pairs. The list is closed by a semicolon. The class in turn is closed with the keyword *END* followed by a semicolon.

```
CLASS D⁰ [Gen,Num,Kasus];
END;
```

For further projections of D, the following classes apply

```
CLASS D' [Gen,Num,Kasus];
END;
CLASS D" [Gen,Num,Kasus];
END;
```

2.2 Inheritance

The classes D^0, D' and D'' differ from each other with respect to the value of an attribute – the bar attribute. They otherwise are identical. It thus stands to reason, that this specific difference should be expressed as a model of inheritance. We therefore set out from the assumption of a shared underlying class:

```
CLASS D[Gen,Num,Kasus];
END;
```

This class does not contain a *bar* attribute and consequently has no specifications with respect to the bar index; these will be included in the following classes:

```
CLASS D⁰ < D;
END;
CLASS D' < D;
END;
CLASS D" < D;
END;
```

In these classes the "$<$" symbol has to be interpreted in the following way (for the first class): The D^0 inherits all properties (as the case may be) all attributes of the underlying D class described further up. The same applies for the classes D' und D''. Through inheritance the following D^0 class emerges:

D^0 [Gen,Num,Kasus]

The class which inherits, stands before the "$<$" symbol, the class passing on follows the symbol. The class inheriting is an extension of the class passing on attributes. More than one class can potentially pass on attributes. The hierarchy described so far can be graphically depicted as follows:

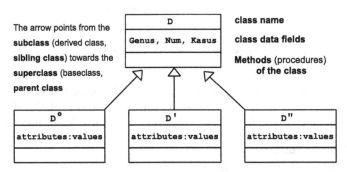

Fig. 2. LPS class hierarchy in UML.

Class hierarchies start with very basic classes and end with very specialized ones, serving as blueprints for nodes in phrase structure trees. Phrase structure trees are built as a result of language analysis and synthesis, they are attributed to the domain of *parole*. Classes express properties of the langue. In a framework of hierarchical classes, *langue* does not only refer to the system of a specific language but also to the language capacity itself: this is particularly the case with the abstract classes. The concept of inheritance is a very helpful means for organizing linguistic knowledge i.e. for describing the language system. Within the domain of *parole*, classes serve the purpose of providing blueprints by means of which the nodes of tree structures may subsequently be derived. We thus have a tool at our disposal – an artificial tool – for processing natural languages.

In computer science nodes which are built according to the blueprint of a class are called objects. We adopt this terminology and thus call the following approach *Linguistic Object Orientation*.

2.3 Methods

The classes presented so far are comprised of attribute-value pairs, hence only of static components. Together with static components, classes can also contain dynamic components. Dynamic components are called methods. Methods trigger actions, in our case linguistic actions. These actions could be the definition of grammatical agreement and case assignment. We shall explain methods, taking case assignment as an example. Let us consider the following phrase-structure tree[3]:

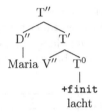

Fig. 3. A phrase-structure tree.

In such a configuration, the specifier is assigned nominative case, when T^0 carries the feature 'finite'. Case is assigned in a spec-head relationship. This is formally defined as the following Prolog (cf. Clocksin & Mellish [8]) clause[4]:

```
case(TNull,Specifier) :-
    HasValue(TNull,'finit'),
    SetValue(Specifier,'nom').
```

Such a clause is easy to read: Case assignment from T to the specifier (case(TNull,Specifier)) succeeds if and only if T^0 carries the value *finite* (HasValue(TNull,'finit'))[5] and if the specifier receives the nominative feature (SetValue(Specifier,'nom'))[6].

In the above clause, the predicate **case** is the head of that clause; the body of the clause which follows the symbol for implication comprises in our example of two predicates.

Predicates have arguments: The predicate **case** for instance has two arguments, *TNull* and *Specifier*. Both these arguments have values (In Prolog terminology, both arguments are said to be instantiated: one must distinguish between the argument-value relationship and – as is the case above – the attribute-value relationship)

---

[3]For more details on the concept of inheritance cf. Khosafian & Abnous [15] and Reiser & Wirth [20]; for a very readable introduction to object orientation in Java cf. Goll et al. [12]. The concept as such is well known. It has its foundations in the Aristotelian theory of categorization and was elaborated in detail in scholastic philosophy.

[4]A Prolog interpreter was implemented as part of the LPS-system.

[5]the ':-' symbol expresses the implication.

[6]the comma between both predicates expresses logical conjunction.

In our case the value of *TNull* is the T^0 node of the tree structure, the value of *Specifier* is the D'' in the tree structure. *Has Value* checks if a node has a certain value: *SetValue* assigns a specific value to a node.

Yet there are three basic questions which have to be dealt with:

1. Where do the clauses come from?
2. How are the predicate arguments assigned their values? For instance, where does the *Specifier* argument obtain the D value from?)
3. What triggers the proof for the predicates?

These questions take us back to the components of a class, and as the case may be, to the components of an object derived from a class. So far only static components of a class have been presented here. The predicate `case` (the head of the clause) is a dynamic component of a class. This component appears in two classes, namely in class T^0 and class D''

```
CLASS T⁰;
case(Self,Other);
END;
CLASS D'' < D;
case(Other,Self);
END;
```

In terms of object orientation the predicate *case* is a method. *case* has two arguments; in the T^0 class these arguments are expressed by *Self* and by *Other*. The meaning of this expression becomes clear when one considers that a T^0 object is derived from a T^0 class. This object is the node of a tree (i.e. an object in the domain of *parole*). The object itself is the first argument of the *case* predicate. It is precisely this which *Self* is meant to express. This is also suggested in the *case* clause above. The name of the first argument in the clause is *TNull*. This refers to T^0 as an instance of the variable. Hence the first of the two questions formulated above has been partially answered: The clauses emerge as dynamic components from the classes and obtain an object passed on by means of the *Self* declaration, which is itself derived from a class.

However, the second argument, which is expressed through *Other* has not yet received a value – it has not yet been instantiated. In the following, we consider the class D''. This class also contains a predicate *case*. The order of its arguments are inversed. The first argument is *Other*, the second is *Self*. This *Self*, as was the case with T^0, is instantiated by the object derived from D''. In the process of derivation of an object from a class, methods of these classes emerge as dynamic components of the object. We thus can consider two objects containing case methods whose arguments are instantiated complementarily. In the light of this constellation, the third question formulated above can now be answered: the question as to where the proof for the *case* method is realized.

The *case* method can be proved when both arguments are instantiated. This is safeguarded by the following mechanism. The *case* method of the specifier D'' and the *case* method of the head T^0 are passed on in the phrase-structure tree till they meet. When two methods with the same name (e.g. the *case* method) encounter each other the method can be executed. We here call this process of "meeting" between methods and instantiated variables, *method unification*. If method unification succeeds, the method will be executed. The steps to be followed in the process are defined in the Prolog clause. If the method is executed, it follows that the given structure is licensed. If it fails then the structure is impermissible. The diagram on the following page summarizes the issue.

In order to generate a structure along the lines shown in the diagram, a mechanism has to be defined which selects the nodes to be projected. In our example the node concerned is the T node. The diagram interprets the *langue/parole* dichotomy in an object-oriented manner. The classes, the class components and the inheritance structure lie within the domain of *langue*. The nodes instantiated from classes as well as the tree-structures lie in the realm of *parole*.

The example shows how methods are passed on in structures and how they are executed as soon as method unification occurs. The range within which the passing of methods is permissable will be described in more detail in the following.

The *case* method of the T^0 class has to be passed upwards along the tree structure to T'' in order to be unified and executed precisely there. By default, the passing up of nodes is constrained to the maximum – methods in a tree structure can usually only be passed on from the instantiating nodes up to their parent nodes. This is the case with D'' in our example. The parent of D'' is T''. Since the D'' class does not specify how far up in the tree the `case (Other, Self)` method can be passed on, the default case comes into play. Upwards passing stops at the parent T''. We call the node which lies below the one where methods are unified and executed the *target node*. The target node for D'' is D'' itself. The default rule must be explicitly suspended for the method `case(Self, Other)` in the T^0 class since method unification can only take place two levels higher up. Therefore, the T^0 class method has to be expanded as follows:

```
CLASS T⁰;
case(Self,Other) T';
END;
```

The T' specification following the method defines where a method triggered by the T^0 node can move. The method has to be unified and executed in the parent of T^0. If this does not succeed, the suggested structure will not be licensed. One should note that methods in the parent of the target node must be unified and executed. If no target node is defined, it is the instantiated node itself which by default becomes the target node.

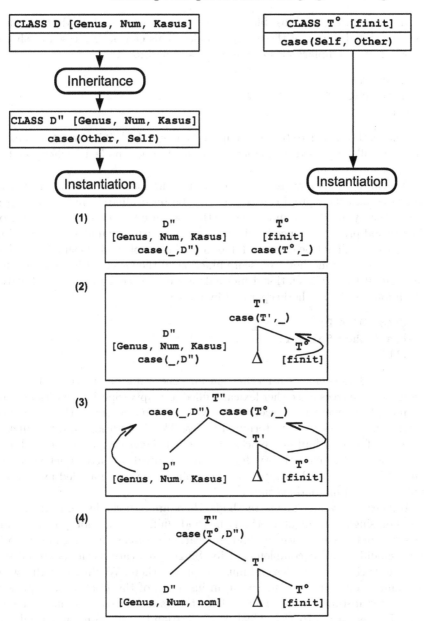

Fig. 4. LPS processes in linguistic object orientation.

In some cases it is possible to have several target nodes. Methods have to be executed in one of these target nodes. This can be expressed as shown below, using exemplary classes, methods and nodes.

```
CLASS X;
methodY(Self,Other) U',V';
END;
```

The methodY has to be passed up to the method U'. If it cannot be unified there, it will be passed on higher to the next target node V' where, at the latest, it must be unified.

Not all methods have to be proved. Some methods are optional. They can be proved and their proof contributes to a better evaluation in terms of optimality theory. If they do indeed fail, either because the method unification or the method proof is unsuccessful, the structure currently under consideration for licensing will not be discarded due to the failure of those methods. In other words facultative methods are score functions if they succeed but not penalty functions if they fail. Optional methods are expressed by an f (facultative) which follows the method separated by a comma.

```
CLASS D'' < D;
case(Other,Self),f;
END;
```

The X-Bar schema is a mechanism for generating binary trees. Terminal nodes in these trees are either lexically filled or empty. Special conditions constraining the generative power of the X-bar schema apply in the process of distributing filled or empty terminal nodes. These conditions are formulated in terms of special building blocks of so-called *linguistic knowledge*. In binding theory and in case theory, for instance, context sensitive phenomena of natural language can be captured. These phenomena are handled by method unification and method execution.

Besides the two-argument methods, there are also single-argument methods. For single argument methods method unification is not possible; this goes without saying since method unification assumes that argument positions should be filled complementarily. Single predicate methods are passed on up to their parent nodes and must be proved there. We illustrate this with an example on attribute-unification in instances of the D classes. D projections – for instance in German – share the values for gender, number and case. This is an approach typical of Unification-based grammar formalisms [23]. We express this form of unification by a method from which D^0, D' and D'' inherits. We thus state the method defined above more precisely in the following way:

```
CLASS D[Gen,Num,Kasus];
UnifyFeatures(Self);
END;
```

UnifyFeatures is specified in its associated *methodspecification*[7] file as follows:

```
UnifyFeatures(Daughter) :-
GetMother(Mother,Daughter),
Unify(Mother,Daughter,'Gen'),
Unify(Mother,Daughter,'Num'),
Unify(Mother,Daughter,'Kasus').
```

In the case of a D^0 node this method accesses the parent-node D' by way of the *BuiltIn*-Predicate *GetMother*, and unifies the attributes *gender*, *number* and *case* in the parent and sibling node by means of the *BuiltIn*-predicate *Unify*.

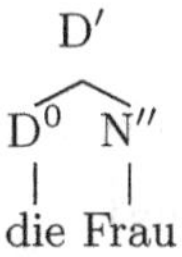

$$D'$$

$$D^0 \quad N''$$

die Frau

Fig. 5. A linguistic tree structure.

2.4 Instantiation and Late Binding

The term "instantiation" often came up in discussions concerning the nodes of a syntactic tree structure. In object-oriented programming, the instantiation of an object means the allocation of memory for that object and the typing of that reserved memory according to the description of a class. Here, the meaning of instantiation differs slightly from the original meaning. In our model, instantiation is the first typing of an already-present node object according to a class blueprint. Untyped node objects are generated by the LPS-system (cf. Section 3 on page 316). The system looks up a matching class for these untyped node objects. Even though the node objects are untyped, they still possess certain properties. The class that matches these properties assigns additional attribute values or methods to its node and thus expands the nodes with further components. For any given node it is necessary to define how it looks up a class from which it acquires further components. This process is here called instantiation. It could also be called *late binding* or more appropriately *late extension*. We however shall reserve this term for the typed extension of nodes, which have already received information from classes, i.e. in our terminology, which have already been instantiated.

---

[7]The methods defined in classes are specified in so called *method specification files* in order to keep class declarations and functions distinct.

2.5 Linguistic Object Orientation: The Details

While instantiation and late binding express the extension of node objects, inheritance expresses class extension. Consider the following class once again:

```
CLASS D⁰ < D;
END;
```

The D class notation following the extension symbol "<" marks the inheritance class. The D serves to identify the class. If there are further attributes, they may serve as data for inference, thus enriching the instantiated object. Attributes and values which this class can pass on, have to be explicitly defined. These attributes and values follow a "|" symbol in the declaration. Therefore the class described above:

```
CLASS D[Gen,Num,Kasus];
END
```

has to be redefined as follows:

```
CLASS D|[Gen,Num,Kasus];
END;
```

D^0, D' and D'' inherit the attributes Gen, Num, and $Kasus$ from this class. Gender, number, and case, however, are not only required by these classes, but also by the N and Adj classes. We can thus expand the above definition as follows:

```
CLASS NOUNPROPERTIES|[Gen,Num,Kasus];
END;
```

This example demonstrates why inherited attributes are declared after the "|" symbol. The class name NOUNPROPERTIES, a value of the attribute *Category*, is incompatible with the class names D or N, which are likewise values of the attribute *Category*. However since the class name is merely an identifier, what is of relevance here are the attributes or values to be inherited. Those are the ones that follow the vertical line "|".

One should note here that languages differ from each other with respect to their attributes and values for nominal features. There are languages which do not specify gender, while others do not distinguish in number. They have another, or a reduced set of nominal features at their disposal. Nonetheless they follow the same principle of feature distribution. This cannot be expressed by the abstract class NOUNPROPERTIES. Even there, modelling is far too language specific. We therefore refer to the principle of *interface classes*. Interface classes are surrogates for classes which are to be defined at a later stage. When defined they overwrite the interface classes. Such classes are located at the end of the module hierarchy. Consequently interface classes are initially empty and in course of a derivation process, implement the full functionality of a new class – by overwriting the empty interface-class – cf. polymorphism in modern object-oriented programming languages.

2.6 Modularization

Classes are grouped into modules by order of their functional aspects. This concept is well-known from object oriented languages – in Java for instance as packages. Inheritance is allowed beyond the borders of a module. For that purpose the module from which a class is inherited has to be specified. This is expressed in our notation using a prefix which follows the class name separated by a period. Let us assume that the features of Continental Germanic[8] languages were described in a module named KG, classes for the German language – grouped in a *Dts*[9] module – could inherit as follows:

```
(*Modul Dts.Clm; Klassen des Deutschen; Bearbeiter: JR; 2.02*)
....
CLASS D < KG.Nounproperties;
END;
```

Class Modules are files ending with a *.clm extension. For each class module we have a module defining the corresponding methods. The methods related to the Dts.clm module are defined in the method-specification file Dts.Msp.

Parent classes are always defined at the beginning of the declaration, be it the same module or a previously defined module. This way, class and module hierarchies of inheritance are created. They enable modelling linguistic knowledge by order of typological or genealogical criteria. Since hierarchies of inheritance express specialization it conversely is also true, that they imply generalization. This can be depicted as a directed graph, which originates in a module expressing universal grammatical properties – labelled appropriately by a *UG* prefix, as shown in the following diagram.[10]

The diagram shows how the verb is extended by its complement. For difference in word order it is crucial whether inheritance proceeds from *UG.Left* or *UG.Right*. It goes without saying that *UG.Left* expresses that complements stand to the left and *UG.Right*, that they stand to the right, and that verbs in specific languages either inherit from *UG.Left* or *UG.Right*. Continental Germanic languages inherit from *UG.Left* and the Romance languages from *UG.Right*. As said above, class hierarchies can either reflect genealogical or typological similarities. Classes for Continental Germanic or Romance languages express genealogical similarities: When languages such as Turkish or German fall back on to the common parent class *UG.Left*, this expresses a typological similarity.

---

[8]The term *Kontinentalgermanisch* is much more commonly used than its English equivalent *Continental Germanic*. As long as we speak of grammatical features – as we do here, the term is unproblematic.

[9]*Dts* for German. LPS notation has not yet been streamlined. Eventually all classes, methods and specifications will be abbreviated in English.

[10]I would like to thank Knud Möller for kindly allowing me to use this diagram from his thesis.

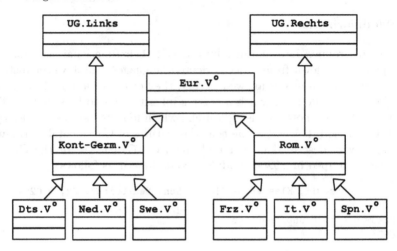

Fig. 6. LPS typological hierarchy in UML.

The example of case assignment shows how linguistic knowledge is deployed in the process of structure generation. As defined in the X-bar schema non-maximal nodes can project. Via projection, they constitute hypothetical parent nodes; such a hypothetical parent node can be considered as proven if at least one of its methods has been proven and when none of the methods returns false.

For computational linguistic processing it is however more suitable to carry out tests after any form of generation, i.e. after every projection stage. Such an approach enables hypotheses to be verified or disproved at an early stage. This is necessary, due to the unrestrictive nature of the X-bar schema, hence leading, as a consequence to overgeneration. Consequently, if they were not checked and rejected after every projection, numerous structures would be produced – this overgeneration is obviously neither efficient, nor is it cognitively plausible. In the model described here, the projection mechanism is embedded in a strongly-modified form of a Marcus parser [18].

Early local licensing of potentially over generated structures is also suitable for optimising language processing systems. This will be demonstrated in the following section.

3 Competition and Licensing

There are many causes of overgeneration. For instance, a projecting node can adopt various maximum projections as potential sibling nodes. It is also possible for the node to take an empty element as a sibling, i.e. a connector whose binding element is also part of another (partial) structure. Lexical entries with multiple meanings also contribute to overgeneration[11] It is important to remember that a projecting node can have a sibling, but does not necessarily have to. The processing system makes decisions in each of these cases. One should keep in mind, however, that the decisions made, based on considerations at the local level may be incorrect or sub-optimal on the global level.

A two-step strategy can be used to solve this problem. The first step of this strategy involves the principle of local licensing. The licensing is modified so that none of the applied methods can return false, and focusses instead on the sum of the individual results of each hypothetical configuration (from projecting nodes, from parent nodes, and in some cases, from selected sibling nodes). The configuration with the greatest number of methods that have returned true (and has not returned any false) wins the trial configuration competition.

This leads to local optimization: the best choice is selected from several positive hypotheses for the construction of a partial structure. Since not all information on structural relationships in trees is available at the local level, the local optimization cannot guarantee a global optimization. A more likely scenario is that a sub-optimal local hypothesis will create the conditions necessary for a better (global) solution at some later point. Yet not all sub-optimal local hypotheses should be considered at that later point. This would be at odds with efficient and principally deterministic linguistic processing.

Therefore sub-optimal alternatives come into play only when structure assembly later fails. Garden path sentences are good examples for this. Humans are able to analyze them, albeit with some difficulty and time spent in the process. Such sentences cannot be analyzed with strictly deterministic processing systems. We can however use the following strategy. Positive, but sub-optimal configuration hypotheses are not discarded, but rather kept in a memory stack. If it turns out that the analysis of an input sentence cannot be completed, the system resets the structure built up to that point. The system then selects the sub-optimal configuration for further processing. If it too is unsuccessful, then a possibly even worse configuration is next in line. If there is none, then the system resets itself one step further back. When resetting, the system always accesses the topmost element of the alternatives stack; accessing these naturally removes them from the stack.

4 Licensing and Performance

Formal grammars postulate idealized speakers. They are the basis for competence models. Variation is explained as a problem of performance and is neglected in linguistic description. Yet linguistic reality cannot be modelled in this manner. Remarkably, communication works even when the grammaticality of a linguistic utterance is doubtful or entirely lacking. It is precisely

---

[11]Lexical transfer is described in detail in [22].

on this point that formal and natural languages differ. Formal languages, e.g. programming languages, are encoded entirely syntactically and need only syntactic rules for their analysis, without exception. Relying on anything other than syntactic knowledge is not necessary. Formal languages' syntactic rigidity blocks any intrinsic evolution. In light of the parallel development of formalisms into programming languages and into natural languages, it is easy to assume that the abstraction required by a rigid notion of competence is much less a virtue as it is a necessity born of the notion of grammar sketched in section 2 on page 305.

For practical work in computational linguistics, determining strategies for performance-related licensing will prove to be useful. Two approaches should be considered for the system described here.

- The first approach operates within the sources of knowledge organized in modules and classes that have been hitherto described. The goal was local optimization via the summation of the method results of a node hypothesis configuration. A hypothesis is considered incorrect if a required method returns false. If it is not possible to assemble a syntactic structure due to the linguistic material being analyzed, then the method conditions can be relaxed. One possibility is to reduce a required method to an optional one. Required methods that return false cause a configuration hypothesis to fail; if these methods are marked as optional, then the hypothesis will not fail.

- The second approach in linguistic processing takes advantage of persistent knowledge. Here, natural communication serves as a model. Each communication participant possesses persistent linguistic and extralinguistic knowledge from which inferences are made and which is used to complete partial utterances. Linguistic knowledge and extralinguistic knowledge are encoded in tree structures. Together they form a knowledge base which may be consulted by the system. This knowledge base differs from translation memories in that we are not storing strings but tree structures. Extralinguistic knowledge is also described using language (language is already known to play an excellent role in the production, organization, and distribution of knowledge). Thus computers are enabled to know about what they communicate in an approximately similar way as humans.

The path we have followed here is simple to put into practice. Linguistic and extralinguistic information within language is encoded as tree-structures. These tree-structures are then stored as XML. This is useful because in this way the two-dimensional graphic representation of tree-structures is reduced by one dimension. Since one-dimensional structures can be treated as character strings, the problem before us can be dealt with as a character sequence comparison. The character sequences from persistent data are then compared to the character sequences that were extracted through linguistic processing and partial trees.

The Levenshtein algorithm [16] provides a proven method for soft matching when dealing with the comparison of character sequences (here we are dealing with tree-structures that have been transformed into character sequences). The Levenshtein algorithm provides a measurement of variance or similarity of character sequences. It can be used to find an optimal persistent structure for a given partial structure. If the variance remains within a predefined range, the matched structure can then be used as a model for the continued assembly of the partial structure.

Due to built-in predicates in the Prolog interpreter developed for LPS, soft matching with the Levenshtein algorithm is tied into the mechanism for processing methods described above.

Finally one must ask whether this approach makes the traditional generative concept of competence obsolete. The answer is negative. The application domain of the concept of competence is simply shifted: it no longer serves to license structures from the domain of performance, but rather it licences those from the domain of persistent knowledge. In principle, this involves a simple change in direction. Licensing structures on the basis of performance was a short circuit, mainly the fault of computational linguistics. Licensing deals with competence, a form of persistent knowledge. Grammaticality judgements are binary: structures are either acceptable or unacceptable. When applied to performance, the Levenshtein algorithm shows the extent of variance. This is how it assures communicative success for variant utterances.

References

[1] N. Chomsky. Three Models for the Description of Language. *IRE Transactions on Information Theory IT-2*, 2(3):113–124, 1956.

[2] N. Chomsky. *Syntactic Structures*. Mouton, The Hague, 1957.

[3] N. Chomsky. On Certain Formal Properties of Grammars. *Information and Control*, pages 137–167, 1959. (Reprinted in Readings in Mathematical Psychology, 2nd ed., edited by Luce, Bush, Galanter, 125-55. New York, Wiley and Sons, 1965.).

[4] N. Chomsky. Formal Properties of Grammars. *Handbook of Mathematical Psychology*, 1963.

[5] N. Chomsky. *Aspects of the Theory of Syntax*. MIT Press, Cambridge, 1965.

[6] N. Chomsky. *Some Concepts and Consequences of The Theory of Government and Binding*, volume 6. MIT Press, Cambridge, 1982.

[7] N. Chomsky. *Modular Approaches to The Study of The Mind*. State UP, San Diego, 1984.

[8] W. F. Clocksin and C. S. Mellish. *Programming in Prolog*. Springer, New York, 1981.

[9] W. Daelemans and K. de Smedt. Default Inheritance in an Object-Oriented Representation of Linguistic Categories. *International Journal of Human-Computer Studies*, 41:149–177, 1994.

[10] W. Daelemans, K. de Smedt, and G. Gazdar. Inheritance in Natural Language Processing. *Computational Linguistics*, 18(2):205–218, 1992.

[11] J. A. Fodor. *The Modularity of Mind. An Essay on Faculty Psychology.* MIT Press, Cambridge, 5 edition, 1983.

[12] J. Goll, C. Weiss, and F. Müller. *Java als erste Programmiersprache.* Teubner, Stuttgart, 3 edition, 2001.

[13] L. Haegeman. *Introduction to Government and Binding Theory.* Blackwell, Oxford, 1994.

[14] R. Jackendoff. *X-Bar Syntax.* MIT Press, Cambridge, 1977.

[15] S. Khoshafian and R. Abnous. *Object-Orientation: Concepts, Languages, Databases, User Interfaces.* John Wiley & Sons, New Jersey, 1990.

[16] T. Kohonen. *Self-Organizing Maps.* Springer, Heidelberg, 1 edition, 1985.

[17] J.-Y. Lalande. Verbstellung im Deutschen und Französischen unter Anwendung eines CAD-basierten Expertensystems. In *Linguistische Arbeitshefte.* Niemeyer, Tübingen, 1997.

[18] M. Marcus. *A Theory of Syntactic Recognition for Natural Language.* MIT Press, Cambridge, 1980.

[19] K. Möller. Die Codierung Linguistischen Wissens in Topic Maps. Master's thesis, University of Cologne, Cologne, 2003.

[20] M. Reiser and N. Wirth. *Programming in Oberon, Steps beyond Pascal and Modula.* Addison-Wesley, New York, 1992.

[21] J. Rolshoven. *Linguistische Objektorientierung.* Sprachliche Informationsverarbeitung, Köln, 2001.

[22] J. Rolshoven. Transfer in Machine Translation with OO:LPL. In *Computerlinguistik: Was geht, was kommt? Festschrift für Winfried Lenders*, pages 239–246. Gardez, St Augustin, 2002.

[23] E. Wehrli. *L'Analyse Syntaxique des Langues Naturelles. Problèmes et Méthodes.* Masson, Paris, 1 edition, 1997.

[24] N. Wirth. *Programming in MODULA-2.* Springer, Heidelberg, 3 edition, 1985.

[25] N. Wirth. *Compilerbau. Eine Einführung.* Teubner, Stuttgart, 4 edition, 1986.

Part V

Text Categorization and Classification

The Surface of Argumentation and the Role of Subordinating Conjunctions[*]

Winfried Lenders

University of Bonn
Lenders@uni-bonn.de

1 Introduction

In modern language processing systems like Machine Translation (MT), Information Extraction (IE) or Information Retrieval (IR) systems, the classification of texts and the *a priori* declaration of the domain of a particular text may help to optimize the system's performance.[1] However, in the future, in most of these systems the identification of text type and text classification may be done by statistical methods in combination with some structural analysis (tagging, parsing, lemmatization).[2] It may be interesting to see whether there are structural indicators for particular classes of texts as has been postulated by previous discourse theories. Statistical methods primarily use characteristics of the vocabulary of a text like the conditional probability that a certain element belongs to a certain class, neighbourhood measure or entropy etc., in order to identify the most probable class or domain of a text. These tools operate so much better when a text contains special terminology or special language. They do not use any structural criteria, which are – from the linguist's point of view – most interesting and relevant for a typological classification. It seems, however, that up to now we have relatively poor knowledge of these characteristic differences and structural similarities of texts and registers (cf. Biber et al. [2, p. 106]). This could only be explored by an exhaustive and multidimensional exploration of large text corpora [2]. The reason why up to now such explorations are not yet available is that we do not have enough really sufficiently working tools for automatic structural analysis.

Given this situation, the aim of this paper is to combine a vocabulary approach with a structural approach, i.e. to find out whether there are functional

---

[*]This work was supported by Korea Research Foundation Grant (KRF-2004-042-A00091)

[1]E.g. the 'semantic domain recognizer' of the Systran MT-system.

[2]For more details see Brückner [4], where the relevant statistical methods for text classification are briefly introduced.

W. Lenders: *The Surface of Argumentation and the Role of Subordinating Conjunctions*, StudFuzz **209**, 323–337 (2007)
www.springerlink.com   © Springer-Verlag Berlin Heidelberg 2007

words which may serve as structural operators and indicators for typical text structures and text types.

The more restricted question is whether there are elements which indicate particular argumentative structures or other sequences of speech acts which are characteristic for descriptive or narrative or argumentative texts (cf. van Dijk [10, p. 128ff.]). A preceding study exploring the German Limas Corpus (www.ikp.uni-bonn.de/Limas/index.htm) already concluded firstly that the frequency of single argumentative operators like *weil* and *denn* is not high enough to be used for the identification of arguments and secondly that it could be worthwhile to consider a greater number and more varied types of structural operators. Consequently it first seems to be necessary to consider a much bigger set of possible argumentative operators, e.g. all German conjunctions with a causal and consecutive function which have to be examined with respect to their distribution in a large text corpus, and, moreover, in different types of texts. As a further approach, in this paper we extended the exploration to a few more operators. A more exact and exhaustive elaboration must be reserved for a later occasion. The empirical basis again is the German Limas Corpus which was constructed as a representative corpus for the German of 1970 and which contains an elaborated classification of 33 text domains and branches.

As a result of this study we do not expect that the chance to use function words as indicators for particular text types really exists. We expect it to become clearer that only an exhaustive analysis of complex sentences and their interconnection in discourse can give sufficient information for their typological classification and comparison.

The paper is divided into three parts: First, in order to fix the position of this study we discuss several approaches of texts classification. In the second part we present the method of investigation, i.e. the corpus data and their characteristics and the tools. The third part presents some preliminary results. As a conclusion, we summarize what the benefits of this study and its consequences for further analysis are.

2 Aspects of Text Classification

The type of classification which is done by the afore mentioned automatic systems leads to a domain oriented characterization and could be called thematic classification. There are other more theoretical approaches which classify texts according to different dimensions, e.g. their function in a communicative context or its linguistic structure. There is no doubt that an outline of the state of the art in this area would require a volume of handbook size. This study therefore starts with a very simple assumption, that there are three general dimensions which are used to classify texts from different points of view:[3]

---

[3]In German linguistics these approaches are elaborated in detail e.g. by Brinker [3].

The *first* dimension concerns the thematic view and allows a classification of texts according to their particular *domain* or their specific purpose. This is the point of view which is predominant in currently existing computational linguistic systems (see above).

The *second* dimension is the *communicative* or *functional* dimension. From a functional perspective a text as a complex speech act has a particular function according to its predominant usage in communicative acts or communicative situations. E.g. Brinker distinguishes four types of texts according to four basic communicative functions: (a) informative texts (news, reports, specialized books, reviews), (b) appellative texts like advertising, commentaries, proposals, petitions etc., (c) texts which involve or constitute an obligation like a contract, a warranty etc., and (d) texts which serve to establish or care for contacts like thank-you letters or letters of condolence ([3], 107 ff).

The *third* dimension is concerned with the general internal text structure and allows the classifying of a text as a certain *type of discourse*. According to van Dijk [10] in this dimension are just three global textual structures, namely narrative, argumentative and descriptive structures. This distinction has been adopted by de Beaugrande & Dressler [5], who also distinguish between descriptive, narrative and argumentative texts (loc. cit. p. 190). According to this theory, the function of a descriptive text is to add knowledge about an isolated object or situation to that which is already known about objects and situations. Narrative texts on the other side describe certain events or actions which proceed in time, the different states of an event and the changes between these states. Argumentative texts, finally, are those texts which operate with certain types of logical relations between sentences like *claim, conclusion, condition* etc. in order to convince a person or to make a person sure about a certain fact. Of course, in reality none of these 'types' occurs in a pure form, but as a mixture with others [5, p. 191]. Brinker, in correspondence to his functional classification of texts, differentiates between descriptive, narrative, explicative and argumentative progress of topics in texts (loc. cit. p. 131f.).

A possible fourth approach to text classification comes from a sociological point of view which tries to classify texts into different 'registers' according to their characteristics in speech and writing coming from certain social conditions like profession, social status or age. For instance, [2, loc. cit.] distinguishes between *Academic Prose, Official Documents, Conversations, Prepared Speech* etc. This typology is similar in its intention as the above mentioned second dimension, the characterization of texts according to functional criteria.

A final remark concerns languages for specific purposes (in German *Fachsprachen*). These are languages which use a specific vocabulary or terminology, and the domain of these texts can easily be detected by looking up this particular vocabulary.

Since the aim of this paper is to study structural characteristics, it belongs to the third dimension. It should be pointed out whether there are particular

Table 1. Domains of the LIMAS-Corpus.

1) Allgemeines	14) Volkskunde, Völkerkunde	30) Hauswirtschaft
2) Religion	15) Geographie, Reisen	31) Sport
3) Philosophie	16) Naturwissenschaften	32) Spiel, Unterhaltung
4) Psychologie	17) Astronomie, Geodäsie	33) Belletristik
5) Kultur	18) Physik	
6) Recht	19) Chemie	
7) Politik, Verwaltung	20) Geologie, Mineralogie	
8) Gesellschaft	21) Meteorologie	
9) Wirtschaft	22) Biologie	
10) Sprache	23) Botanik	
11) Literatur	24) Botanik	
12) Bildende Kunst,	25) Anthropologie	
Musik, Theater,	26) Medizin	
Tanz, Film,	27) Mathematik	
Rundfunk	28) Technik	
13) Geschichte	29) Land- und Forstwissenschaft	

functional words which can be seen as operators indicating argumentative structures.

The following paragraphs will show how far the investigated corpus meets the requirements of such a study.

3 Method of Investigation

The description of a particular text type or register in comparison to other text types or registers requires as many linguistic features as possible. Biber et al. [2] point out, that "[...] register studies based on a few, selected linguistic features do not provide comprehensive register descriptions, and generalisations based on such studies are likely to be inaccurate" (136). Moreover, "comprehensive register studies have three important requirements: inclusion of a large number of texts, consideration of a wide range of linguistic characteristics, and comparison across registers" (136). Keeping this in mind, the aim of this paper is not to find this recommended complete set of 'linguistic characteristics' or to indicate what the structural differences between texts in law or literature are, but to demonstrate to what extend the statistical distribution of conjunctions (understood as argumentative operators) are able to separate text types, e.g. narrative texts on the one and argumentative texts on the other side.

3.1 The Corpus and its Text Types

The LIMAS-Corpus has been collected in the early 70th under very simple pragmatic conditions and considering the problem of text types in a very

Table 2. Domain 6: *Recht* (law) with subdivisions.

06.0	RECHT
06.1	PRESSE
06.2	BUECHER UND PERIODIKA
06.21	INLAENDISCHES RECHT
06.211	GESETZE UND ENTSCHEIDUNGEN
06.212	FALLSAMMLUNGEN
06.213	KOMMENTARE UND MONOGRAPHIEN
06.22	AUSLAENDISCHES RECHT UND VOELKERRECHT
06.31	KRIMINOLOGIE

trivial manner. Nevertheless we are able to identify the above described three typological dimensions:

- The first dimension is represented by a distinct relation of each of the 500 texts to one of 33 domains (table 1). Each of these domains represents a thematically homogeneous set of texts, each text has a standardized length of about 2000 running words. The 33 domains were taken from the classification system of the German National Bibliography.
- The second, i.e. the functional dimension can be found in some subdomains into which the domains are divided. Unfortunately this subdivision has been made in a very non-homogenous way. Some of the domains are subdivided in *'Presse'* and *'Bücher and Periodika'*, and these are again subdivided into different subdomains, others are subdivided according to some prominent subjects of the intended text type. Tables (2) and (3) show some of these subdivisions. Because of its in-homogeneity this characterization of subdomains is only of a limited value for this study (and for others as well). It may be that groups like *Presse and Bücher and Periodika* from a functional point of view are of a certain value but unfortunately this could not be elaborated in this study.
- Finally the third dimension is represented implicitly in those domains in which a particular type of discourse structure, narrative, argumentative or descriptive, is obviously predominant. This is the case with *Belles Lettres (Belletristik)*, Law (*Recht*) and Sciences (*Naturwissenschaften*). In the LIMAS-Corpus these fields are filled with by the following amounts of texts: domain 6, Law: 21 texts; domains 16 to 27, Sciences: 51 texts; domain 33, Belles-Lettres : 60 texts.

The leading hypothesis for the following is that in some of these sets a characteristic discourse structure can be found and verified from surface characteristics like frequency of particular elements and their co-occurrence. The immediate goal is to find those characteristics for argumentative structures.

From a methodical point of view, the first step of this study concerns the complete inspection of the corpus in order to obtain a general overview of the frequency of certain structural operators. As a second step, from the list,

Table 3. Domain 8: *Gesellschaft* (society) with subdivisions.

08.0 GESELLSCHAFT
08.1 PRESSE
08.11 SEX
08.12 VERBRECHEN
08.13 KLATSCH
08.14 JUGEND
08.15 WERKZEITUNG
08.16 HOROSKOPE
08.2 BUECHER UND PERIODIKA
08.21 ARBEITUNDSOZIALES
08.22 SOZIOLOGIE

those which do not have the potential for an interesting result are identified
and excluded from the study (see 4.3). In a third step the frequency and occurrences of the remaining operators were extracted from the three mentioned
domains (Law, Belles-Lettres, Sciences) in order to find out whether there are
typical indicators for their intended discourse structure.

3.2 Tools

Most of the textual explorations in this study have been done by the Word-
Smith-Tools which allow easy compilation of concordances, frequency lists,
co-occurrence patterns, type-token ratio, etc. In particular the conjunctions
consisting of more than one part (cf. 4.2) could be identified with the co-
occurrence tool. Several additional programming has been done with Perl-
Scripts on the basis of the completely XML-annotated corpus.[4]

4 Results

4.1 Conjunctions and their Distribution in the Corpus

As has been pointed out at the beginning, the aim of this paper is to explore
whether there are operators in texts which are typical for particular text types
and which allow a classification of a text according to a specific classification
system. What we have in mind are those operators which indicate argumentative structures, i.e. which combine sentences playing a certain role as parts
of an argument, like claims, conditions and concessive phrases.[5] It has often
been argued that these operators can be found in the vocabulary in the word

---

[4]I am grateful to Shu-Ju Lee who wrote several of these scripts and made some
statistical surveys.

[5]These according to the basic essay in argumentation theory by Toulmin [9] are
the main components of an argumentative structure.

Table 4. Coordinating conjunctions (Erben [7]).

copulative:	und (29023); sowohl (202) - als auch (145); nicht nur(460) - sondern auch (169); weder - noch (153)
succession:	und (29023); und zwar (179); und auch (11); sowie(456)
contrast:	oder (4072); oder auch (79 [oder ... auch (137)[8]]); entweder (124) - oder[9]; aber (3210); hingegen (56); indessen (33); als (6531); denn (791); jedoch (813); doch (1029); allein (369); nur (3569); allerdings (393), freilich (142); wenigstens (118); zumindest (104); sondern (1059); vielmehr (187)
causal:	denn (791); nämlich (304); daher (403); deshalb (411); darum (162); infolgedessen (18); also (960); folglich (16); mithin (36); (und) somit (130)

class 'conjunctions'. In accordance with the already mentioned preceding exploration for this study a list of German conjunctions is used which has been extracted from the grammar of [7].[6] This list is divided into two parts, the coordinating conjunctions and the subordinating conjunctions. The tables (4) and (5) list these conjunctions together with their absolute frequency in the corpus.[7]

4.2 Simple and Compound Conjunctions

As can be seen from these lists, several conjunctions consist of simply one word; we call them simple conjunctions. The others which contain more than one element, like *wenn auch, auch wenn, und wenn, angenommen (dass), im Falle dass* etc., are named compound conjunctions. With a few exceptions (*auch wenn, wenn auch, im Falle dass, vorausgesetzt dass*) they are discontinuous because several other words, even whole sentences, can be embedded within them. This fact has some consequences for the certainty of their identification by means of co-occurrence patterns (cf. 4.4). The indicated frequency

---

[6]A result of this study is that the German grammar by Engel [6] would be a more suitable basis for studies like this because this grammar is more oriented to the description of discourse types and structures. Engel, instead of one single but subdivided class of 'conjunctions' has two classes called 'subjunctors' and 'conjunctors', which both function for interconnecting sentences. Engel also presents not only a list of these types of conjunctions and of their individual readings, he also discusses their function in the construction of complex sentences and phrases (cf. Engel [6], § S 097 ff.).

[7]Frequencies of compound conjunctions have been extracted by the co-occurrence tool of WordSmith.

[8]In this case, just to show how it works, the numbers in square brackets show the occurrences of 'oder... auch' with up to 5 other words between them.

[9]*entweder* occurs in the corpus twice without a succeeding *oder* and again twice with succeeding *aber auch*.

Table 5. Subordinating conjunctions (Erben [7]).

local:	wo (479), woher (33), wohin (32), soweit (188)	
temporal:	als (6531); wie (4021); solange (81); während (709); nachdem (181); seitdem (35); sobald (64); sowie (456); wenn (2609); bis (1798); bevor (88); ehe (145)	
causal:	concessive:	obgleich (25); obschon (8); obwohl (180); trotzdem (103); ungeachtet (14); wenn auch (108); auch wenn (72); und wenn (72); wenngleich (21); wennschon (0); wiewohl (5)
	conditional:	falls (67); wenn (2609); so (3838); sofern 71); wo (479); wofern (0); angenommen (dass) (87); im Falle dass (0); vorausgesetzt (dass) (38)
	causal:	da (1271), deswegen (44), weil (781)
	final:	damit (992); (auf) dass (1); um - zu (> 700)
	consecutive:	(so) dass (272); weshalb (39); weswegen (1); ohne dass (78); als dass(16)
modal:	comparison:	wie (4021); als wenn (17); als ob (45)
	contrast (adversativ):	während (709), derweil (3); indes (35); indessen(35); wohingegen (0)
	degree:	so dass (272); seit (539); soviel (43); desto (40); je desto (24); je nachdem (9)
	circumstances:	indem (198), dadurch dass (121)

in this list is only a tentative one because it has been identified by simple automatic methods, through the exploration of co-occurrence lists. Only a few of them, namely some occurrences of *wenn* are discussed later.

4.3 Subclasses

From the list of coordinating conjunctions (table 4) it is easy to conclude that there are candidates (like *oder, aber, als, denn, nämlich, also* etc.) for successful further analysis of their argumentative function. Unfortunately some of these conjunctions are compound, others are ambiguous, and therefore this group requires a careful lexical and syntactical analysis, before their function in discourse can be investigated. Since up to now this analysis can not be done automatically, for the current study we leave this group aside.

Although this is also true for the subordinating conjunctions, this group (table 5) should be taken into further consideration. This group can be classified according to their frequency and ambiguity into the following three subclasses:

A first subclass contains those subordinating conjunctions which occur with a relative frequency less than one per text or which occur occasionally

Table 6. Rare conjunctions and their distribution in texts.

conj		Total freq.	> 4	4	3	2	1
conditional	falls	67	1	2	1	6	57×
	angenommen (verb:79) conj.: 8	8				2 (121)	6×
	vorausgesetzt (dass)	38			1 (121)	6	23
	weshalb	35			1 (496)	4	30×
concessive	obgleich obschon obwohl	214(26, 8, 180)		1	10	30	73×
	trotzdem	103; adv.: 101, conj.: 2					2
	deswegen	44				5	38

with a frequency of 1 or 2 in different texts. These are *woher* (33), *wohin* (32), *soweit* (188), *solange* (81), *seitdem* (35), *sobald* (64), *bevor* (88), *ungeachtet* (14), *wenngleich* (21), *wennschon* (0), *wiewohl* (5), *sofern* (71), *wofern* (0); *weswegen* (1), *während* (709), *derweil* (3), *indes* (35), *indessen* (35), *wohingegen* (0), *seit* (539), *soviel* (43), *desto* (40).

A second subclass contains those which have a very low total frequency but which occur in one or a few particular texts with a higher frequency and together with other conjunctions. In these cases they may be relevant as indicators for certain structural phenomena. Examples of these probably are *falls* with a total frequency of 67, occurring 5 times in one text, 4 times in two, 3 times in 1 and 2 times in 6 texts, *angenommen (dass)*, in its conditional sense, which has a total frequency of 8 (the other 79 occurrences are inflexional forms of the verb *nehmen*; two of these occur in one text, together with *vorausgesetzt (dass)*). This clearly characterizes the underlying text as formal argumentation which is usual in sciences like mathematics and formal logic. In this case the underlying text (no. 121) is from domain 27 (mathematics) and its topic is a very special mathematical problem (Tschebyscheff-Approximation) which is demonstrated by means of many formula (line 217/218: *"Angenommen, für (Formel) sei (Formel). Dann ist (Formel) für alle (Formel) und damit (Formel) für alle (Formel)"*). A few other conjunctions belonging to this group are *wo, ehe, nachdem, sowie, seit, indem, je desto, je nachdem, dadurch dass*. Concerning future studies this group becomes more relevant if conjunctions with the same or a similar function like *obgleich, obschon, obwohl* or *wenn* and *falls* are taken (in a tagged corpus) as one unique functionally defined element (cf. section 5 'conclusion').

The third group contains those conjunctions which occur with a relatively high frequency and a relatively in-homogeneous distribution across the corpus. These are *als, wie, während, bis, um-zu, so/so dass, da, wenn, weil* and *damit*. From this group five conjunctions, *so/so dass, da, wenn, weil* and *damit* have been selected to show their distribution over the corpus (cf. table 7). The

Table 7. Most frequent subordinating conjunctions and their distribution in texts.

Conj.	total freq.	per text	>5 [10]	>4	>3	>2	>1	<1
so/so dass	3838/272	7.7	57	49	92	124	122	47
da	1271	2.5	4	3	15	25	157	196
wenn	2611 [11]	3.0	24	20	57	88	174	109
weil	781	1.6	0	2	8	24	82	204
damit	952	1.9	0	0	4	18	126	254

Table 8. Remaining conjunctions and their ambiguity.

so			da		wenn 2611	weil	damit
Adv.			Conj.	causal	conditional 1518	Conj. causal	conj. final
Conj.	concessive			temporal	temporal 790		adv.
	conditional	= falls	Adv.	local	concessive 274		
	consecutive	= so dass, sodas		temporal	wish/desire 7		
Part.		confirming, expressing uncertainty		modal	comparisons 22		

others are not taken into account because obviously in most cases they show a temporal reading and their frequency exceeds the intended size of this study.

With the exception of *weil*, these five conjunctions are ambiguous as specified in table (8). From these only *wenn* will be taken into further consideration. Therefore this conjunction has been disambiguated manually into its main functional readings.

4.4 Distribution of Conjunctions in Different Text Domains

The next step is to show how these conjunctions are distributed in different types of texts. For this reason the texts of three domains have been selected which seem to be relatively homogeneous: Law, containing 21 texts; Sciences, containing 51 texts and Belles-Lettres, containing 60 texts.

---

[10]The numbers in table 7 represent the frequency of one conjunction per 1000 running words. E.g., the conjunction *so* occurs more than 5 times per 1000 running word in 57 texts, 4 times per 1000 running words in 49 texts etc.

[11]Total frequency in 500 texts including orthographical variants like *wenn's*, *wenns*.

Table 9. Frequencies of conjunctions in three domains.

	number of texts	wenn abs.	wenn per text	so abs.	so per text	da abs.	da per text	weil abs.	weil per text	damit abs.	damit per text
law	21	190	9.04	146	6.95	44	2.09	38	1.80	41	1.95
sciences	51	190	3.70	421	8.25	157	3.07	38	0.74	114	2.23
belles-lettres	60	447	7.45	598	9.96	324	5.4	150	2.5	74	1.23

The absolute and relative frequencies of the remaining conjunctions in these groups are shown in table (9).

These numbers are relatively less evident because of the remaining ambiguity which has to be resolved before further analysis. Only for *wenn* has this ambiguity resolution been done (cf. table 8) and leads to an interesting perspective: *wenn* occurs relatively often in legal texts (9.04 times), similarly in belles-lettres (7.45 times)[12], and with a relatively low frequency of 3.7 in Sciences. With regard to the ambiguity of *wenn*, the hypothesis is that *wenn* in legal texts in which the argumentative superstructure is predominant[13] occurs mainly in the conditional and concessive sense, and in belles-lettres mainly in its temporal sense. Even by a very rough disambiguation, this hypothesis could be confirmed.

This is not surprising, but it is interesting that the disambiguation of *wenn* in these two domains can be supported by characteristic collocations with other functional words like *dann, auch, aber, oder, nur, nicht, und* and *also* which can be used as reliable indicators for the particular reading of the conjunction and which possibly could facilitate its automatic disambiguation. Though concessive sense is dominant in the composition *und - wenn, auch - wenn* and conditional sense is indicated by occurrences of *dann - wenn, aber - wenn, oder - wenn, nur - wenn, nicht - wenn.*

Since the concessive and conditional readings are also characteristic for argumentation these compound conjunctions are qualified as indicators for (argumentative) legal texts. They express a strengthening or a restriction of the intended condition and in most cases they are in accordance with common language use. There is no significant indicator *wenn* for science. Of course, there are also occurrences of *wenn* in its temporal sense, as can be illustrated by an example from text 289: *"Ist es 'kühn', wenn Juristen endlich auf Uhr und Kalender blicken?"*

---

[12]This corresponds with *weil* which also seems to be more frequent in law and belles-lettres than in sciences.

[13]Legal texts often have been taken as outstanding instance of argumentative discourse (cf. Toulmin [9]).

Table 10. Absolute and relative frequencies for *wenn* and its collocates.

	Sciences (51 texts) 190 occ. = 3,7 per text				Law (21 texts) 190 occ. = 9,04 per text				Belles-lettres (60 texts) 447 occ. = 7,45 per text			
	5L/5R		10L/10R		5L/5R		10L/10R		5L/5R		10L/10R	
dann-wenn	22	0,43	29	0,57	32	1,52	37	1,76	14	0,23	20	0,33
wenn-dann	0	0,	5	0,1	0	0	1	0,05	4	0,11	28	0,47
auch-wenn	14	0,27	18	0,35	16	0,76	20	0,95	18	0,3	31	0,52
wenn-auch	6	0,12	9	0,18	10	0,48	12	0,57	15	0,1	12	0,20
aber-wenn	0	0	4	0,08	3	0,14	7	0,33	19	0,3	31	0,52
wenn-aber	0	0	3	0,06	2	0,09	2	0,09	3	0,07	7	0,12
oder-wenn	4	0,08	4	0,08	5	0,24	7	0,33	6	0,1	16	0,27
wenn-oder	4	0,08	9	0,18	0	0	7	0,33	4	0,08	16	0,27
nur-wenn	6	0,12	12	0,24	14	0,67	18	0,86	10	0,16	12	0,20
wenn-nur	7	0,14	8	0,16	2	0,09	3	0,14	8	0,15	15	0,25
nicht-wenn	9	0,18	10	0,20	12	0,57	21	1,0	22	0,33	35	0,58
wenn-nicht	8	0,16	12	0,24	18	0,86	26	1,24	34	0,55	58	0,97
und-wenn	11	0,21	28	0,55	5	0.24	20	0,95	47	0,71	97	1,62
wenn-und	19	0,37	32	0,63	5	0,24	23	1,10	23	0,47	67	1,12
also-wenn	3	0,06	5	0,10	0	0	0	0	6	0,1	12	0,20
wenn-also	3	0,06	3	0,06	0	0	0	0	1	0	1	0,02
	115		191		124		205		224		458	

On the other side, *und wenn* seems to be more characteristic for narrative texts, where of course also other compound conjunctions like *nicht - wenn, dann - wenn* etc. occur.

Finally, in many cases, including in argumentative and in narrative texts, *wenn* simultaneously involves the two main components, the temporal and the conditional.

The co-occurrence patterns of *wenn* for an environment of 5 resp. 10 words before and after the keyword have been compiled in table (10). Only those collocations with a min. occurrence of 5 in the environment are mentioned.

It may be interesting that only *auch - wenn* and *und - wenn* occurs in immediate environment ($1L/1R$) with a reasonable frequency (table 11).

With regard to the interpretation of these tables it seems that in the 5L/5R environment in sciences and law *wenn* co-occurs relatively often with other particles like *nur, nicht, dann, aber*, etc., between 60% and 70%, and in belles-lettres around 50%. But it is important to consider that this conclusion could be incorrect because even a span of 5 words between the collocations may exceed the boundaries of sentences or phrases. This becomes more obvious the more the span increases: For a span of ten words between collocates (10L/10R) the number of co-occurrences exceeds the number of occurrences of the keyword (see table 10). An illustrative example for this can be found in

Table 11. Co-occurrences of *wenn* in immediate environment $(1L/1R)$.

	Sciences		Law		Belles-lettres	
auch-wenn	7	0,14	4	0,19	11	0,18
wenn-auch	9	0,18	6	1,28	9	0,15
aber-wenn	0		0		13	0,21
wenn-aber	0		1		0	
und-wenn	0		0		22	0,37
wenn-und	0		1		1	

text No. 219 (belles-lettres), Z. 85ff.[14], where *wenn* co-occurs with *nicht* and *dann* in different sentences:

  | Z_083 .... Ich finde ja, wenn man alles so an sich
  | Z_083 vorbeipassieren läßt, man wird dann. Ich meine, im Laufe der
  | Z_084 Zeit. Die Zeit ist ja. Die Zeit. Also eine Minute zum
  | Z_085 Beispiel. Wenn man eine Uhr hat, nicht wahr, dann ist eine
  | Z_086 Minute ziemlich lang. Wenn Sie keine Uhr haben und auf einem
  | Z_087 Sockel stehen, dann. Auch die Menschen.

Future studies have to avoid this by using more sophisticated methods which restrict the 'horizons' of co-occurrences to the sentence level.

5 Conclusion

The basic idea of this study was that there are characteristic indicators on the surface level which allow the classification of a text according to its particular discourse type (argumentative, descriptive or narrative) and which are available by quantitative methods. Although it was clear from the beginning that, in order to answer this question, many linguistic features are to be considered, we restricted the study to subordinating conjunctions, to get ready in time.

With these – and some additional – restrictions the results are not more than only a certain kind of interim report. It should have become clear that the investigation of the functional role of conjunctions in concrete texts requires much more detailed linguistic analysis, including a careful and exact disambiguation. For argumentative texts it can be expected that the frequency of causal and conditional conjunctions is not that high as was hoped originally, and that means that other structural characteristics must be used, like thematic progression or the general temporal structure of a text (cf. Lenders [8]). But it seems to be plausible that particular senses of conjunctions correspond with particular discourse types.

As could be demonstrated by an exploration of the occurrences of *wenn* this conjunction has a relatively high total frequency, but distributed to at

---

[14]From: Martin Walser: Aus dem Wortschatz unserer Kämpfe. Szenen. Stierstadt/ Taunus: Verlag Eremiten-Presse, 1971.

least four different senses. *Wenn* in its conditional sense seems to be characteristic for law texts, in its temporal sense for narrative texts. With regard to the chances of an automatic disambiguation, it doesn't seem to be possible to find contextual criteria for all, but there are indicators that some conjunctional senses may be differentiated by means of certain regularities in their collocational environment.

These observations lead to the idea of a more elaborated study. A first recommendation for such a study is that it doesn't seem worthwhile investigating only single conjunctions, but sets of conjunctions which have the same or similar function. Therefore we first have to consider what sets of conjunctions according to their function, are candidates for indicators e.g. for argumentative (or narrative) superstructures. As can be easily extracted from the excellent German Grammar by Engel [6] a few functionally defined sets of conjunctions could be specified, like the concessive set (including *wenn-auch, wenn-schon, wiewohl, trotzdem, sowenig, so, obwohl, obgleich, obschon, obzwar*), the conditional set (including *falls, im Falle, wenn, so, sofern*), the adversativ (*während, wohingegen*), and the causal set (*weil, da*) etc. These sets should also include the relatively rare compound conjunctions like *vorausgesetzt dass, angenommen dass* etc.

A second obvious recommendation is to provide a complete functional tagging of these 'candidates'. The different readings have to be disambiguated before they can be used for the validation of any structural hypothesis, with the exception of a few unambiguous cases.

References

[1] D. Biber. *Dimensions of Register Variation: A Cross-Linguistic Comparison.* Cambridge University Press, Cambridge, 1995.

[2] D. Biber, S. Conrad, and R. Reppen. *Corpus Linguistics. Investigating Language Structure and Use.* Cambridge University Press, Cambridge, 1998.

[3] K. Brinker. *Linguistische Textanalyse. Eine Einführung in Grundbegriffe und Methoden.* Erich Schmidt, Berlin, 5th edition, 2001.

[4] T. Brückner. Textklassifikation. In K.-U. Carstensen, C. Ebert, and C. Endriss, editors, *Computerlinguistik und Sprachtechnologie: Eine Einführung.* Spektrum, Akademie Verlag, Heidelberg/Berlin, 2001.

[5] R. A. de Beaugrande and W. Dressler. *Einführung in die Textlinguistik.* Niemeyer, Tübingen, 1981.

[6] U. Engel. *Deutsche Grammatik.* Julius Gross Verlag, Heidelberg, 2nd edition, 1988.

[7] J. Erben. *Deutsche Grammatik. Ein Abriss.* Max Hueber, München, 1980.

[8] W. Lenders. *Argumentative Texte. Eine korpusanalytische Exploration zu weil und denn*. In L. Cyrus, H. Feddes, F. Schumacher, and P. Steiner, editors, *Sprache zwischen Theorie und Technologie*, pages 149–164. Deutscher Universitätsverlag, Wiesbaden, 2003.

[9] S. E. Toulmin. *The Uses of Argument*. Cambridge University Press, Cambridge, 1958.

[10] T. A. van Dijk. *Textwissenschaft. Eine interdisziplinäre Einführung*. Niemeyer, Tübingen, 1980.

[11] M. Wolters. *Towards Entity Status*. PhD thesis, Bonn, 2001.

Computing with Words for Text Categorization

Janusz Kacprzyk[1] and Slawomir Zadrożny[2]

[1] Polish Academy of Sciences
  kacprzyk@ibspan.waw.pl
[2] Warsaw School of Information Technology (WSISiZ)
  zadrozny@wsisiz.edu.pl

Summary. We discuss the use of some elements of Zadeh's computing with words and perceptions paradigms (cf. Zadeh and Kacprzyk [37, 38]) for the formulation and solution of automatic text document categorization. This problem is constantly gaining importance and popularity in view of a fast proliferation of textual information available on the Internet. The main issues addressed are the document representation and classification. The use of fuzzy logic for both problems has already been quite deeply studied though for the latter, i.e. classification, mainly in a more general context. Our approach is based mainly on the use of usuality qualification in the computing with words and perception paradigm that is technically handled by Zadeh's classic calculus of linguistically quantified propositions [36]. Moreover, we employ results related to fuzzy (linguistic) queries in information retrieval, in particular various interpretations of weights of query terms. The methods developed are illustrated by example of a well known text corpus.

1 Introduction

This paper presents an attempt at a more explicit use of Zadeh's computing with words and perceptions paradigm (cf. Zadeh and Kacprzyk's [37, 38] volumes) to formulate and solve the problem of text categorization viewed here as a specific problem class within information retrieval. Basically, the computing with words and perceptions paradigm tries to grasp the human ability to effectively represent and process vague and imprecise information. This leads to the representation of some variables, and relations between them by linguistic expressions rather than by strict numerical values and, e.g., functional relationships. Moreover, a further processing of such information, notably aggregation, fusion etc., may be done by using flexible operators, exemplified by linguistic quantifiers.

Problems encountered in broadly perceived information retrieval (IR, for short) may be viewed to be calling for such a paradigm as they are characterized by uncertainty (mostly of a probabilistic nature), partial matching,

J. Kacprzyk and S. Zadrożny: *Computing with Words for Text Categorization*, StudFuzz **209**, 339–362 (2007)
www.springerlink.com

incompleteness of queries, a vague concept of relevance, etc. – cf. van Rijsbergen [28].

Much research has been done in the application of fuzzy logic in information retrieval, cf. [4, 5, 6, 11, 12, 18, 19]. The main issues of text document representation and their querying have been addressed within this framework. The use of fuzzy logic related concepts for query structure and interpretation is especially promising. This is due to the fact that some elements of the classical IR system interface may be artificially precise and too rigid for a human user. In addition to the main task of an IR system, i.e., the retrieval of documents relevant for the user needs, there are many other related tasks. Among them, quite an important task is that of text document categorization. Basically, it boils down to assigning some thematic categories to the documents. This may be, and often is, done manually which is both costly and time consuming. However, in case of huge document sets, exemplified by those available through the Internet, this becomes ineffective and inefficient. Thus, automatic approaches are more and more often applied. They are usually based on machine learning techniques.

The problem of text categorization exhibits some imprecision. Even a human being may be unsure as to a clear-cut classification of a document to just one category. Moreover, it is quite natural to consider a degree of belongingness to a category. This becomes even more apparent in case of an automatic classification procedure. We may easily expect that the results of classification may be ambiguous. Fuzzy logic approaches have proven to be useful in such a context. We have implemented (cf. Zadrożny et al. [39]) a pilot version of an Internet oriented IR system featuring some elements of fuzzy logic built into the user interface and making it more human consistent. Here, we investigate a possible application of some fuzzy logic related concepts to the very classification process. Our approach is mainly based on the use of linguistically quantified propositions in the sense of Zadeh to model some intuitively appealing rules of classification. Moreover, we adopt results on fuzzy extensions to the querying language of an IR system proposed by other authors – cf. [5, 6, 4, 11, 12, 18, 19] since, viewed from a certain perspective, classification may be treated as a specific querying task in the space other than the original space of text documents of an IR system.

In Section 2 we overview the main concepts of information retrieval. Section 3 presents the essence of Zadeh's computing with words and perceptions paradigm, mainly Zadeh's calculus of linguistically quantified propositions and the concept of a linguistic variable. Section 4 discusses some known extensions to the Boolean model of IR. Section 5 discusses text document categorization and possible approaches to its representation and solution using tools proposed in preceding sections, and describes some computational results.

2 Information Retrieval: Document Representation and Query Matching

The most important issues in any IR project are: document representation and query language. Usually, it is assumed that there exist three main approaches (models):

- Boolean,
- vector space, and
- probabilistic.

Fuzzy logic based concepts have been so far primarily discussed within the framework of the two first models, and this is also adopted in this paper. For a brief description of these models we assume the following notation:

- $D = \{d_i\}$, $i = 1, \ldots, N$ – a set of text documents,
- $T = \{t_j\}$, $j = 1, \ldots, M$ – a set of index terms.

The Boolean model represents a document as a set of terms assigned to it:

$$d_i = \{t_k\}, \quad k = 1, \ldots, K, \quad d_i \in D, \quad t_k \in T \tag{1}$$

and the representation of documents may be formally expressed via the following function:

$$F : D \times T \longrightarrow \{0, 1\} \tag{2}$$

In the vector space model a document is represented as a vector:

$$d_i = [w_1, \ldots, w_M], \quad d_i \in A^M, \quad A \subseteq R \tag{3}$$

where each dimension corresponds to an index term and the value of w_j (weight) determines to which extent a term $t_j \in T$ is essential for the description of the content of the document. Most often, $A = [0, \ 1]$ is assumed, and therefore function F is:

$$F : D \times T \longrightarrow [0, 1] \tag{4}$$

The index terms may be some general concepts describing the content of documents. In librarian IR systems these are usually carefully selected terms indicating, e.g., in case of scientific library, the discipline of a book/journal or keywords for a journal article. In such a case, usually an expensive and time consuming work of an expert is required to assign index terms. An alternative approach, popular in case when automatic indexing is needed, is to select some words that actually appear in a document as its indexing terms. In the literature there are proposed many forms of function F (4) for this alternative. One of the most popular is based on the requirement that such a function should assign to a term (word) t_j in a document d_i a weight directly proportional to its frequency in this document and inversely proportional to the number of documents in which it appears. It may be formalized as a so-called $tf \times idf$ function:

$$F(d_i, t_j) = (f_{ij}/\arg \max_j f_{ij}) \star [\log(\frac{N}{n_j})/\arg \max_j \log(\frac{N}{n_j})] \tag{5}$$

where f_{ij} is the frequency of term t_j in document d_i, N is the number of all documents in set (collection) D and n_j is a number of documents from D where term t_j appears (document frequency). Thus, the first factor is the normalized frequency of term (tf, term frequency) t_j in document d_i, while the second factor is the normalized inverted frequency (idf, inverted document frequency) in the collection D of documents in which term t_j appears at least once. Other normalization schemes may be employed, too.

In addition to document representation, each classical model also offers a querying formalism. In the Boolean model the query is a formula in the sense of propositional calculus. Each term t_j is identified with an atomic formula (proposition), z_j. These may be combined using logical connectives, notably the conjunction and disjunction. Then, such a formula/query is evaluated for each document as it is done in model theory of propositional calculus. More precisely, each proposition z_j of a query is assigned, with respect to document d_i, a truth value *true* if $t_j \in d_i$ and a truth value *false* in the opposite case. Then, a document is treated as relevant, i.e., matching the query, if the whole formula corresponding to the query is true after such an assignment of truth values to its atomic formulae (propositions). Thus, in the classical Boolean model the relevance (matching) is a binary concept: a document is relevant or not, no intermediate situation is possible.

In the vector space model the query takes the form of a vector as in (3), securing a unified representation of documents and queries. The relevance of a document has here a gradual character. The matching degree of a query q and a document d is computed as a similarity $sim(d, q)$ of vectors that represent them. The most popular similarity measure used is the cosine of both vectors:

$$sim(d, q) = \frac{\sum_{i=1}^{l} w_i q_i}{\sqrt{\sum_{i=1}^{l} w_i^2} \cdot \sqrt{\sum_{i=1}^{l} q_i^2}} \tag{6}$$

where $d = [w_1, \ldots, w_l]$ and $q = [q_1, \ldots, q_l]$. Thus, the matching degree is a number from $[0, 1]$.

Another popular family of similarity measures is based on the notion of distance in the space of documents, notably the Euclidean distance.

In both models we can distinguish elementary queries, i.e., those referring to just one term. Formally, they are expressed as atomic formulae (single propositions) and one-dimensional vectors in the Boolean and vector space models, respectively. This concept will be useful for our further considerations.

Here we just briefly describe the primary task studied in IR, i.e., the organization of text documents and their retrieval by matching queries against documents. There are many other related tasks considered within IR (for their discussion as well as for an extensive presentation of the whole domain of IR cf., e.g., [1, 2, 17, 25]). One of them, text categorization, being a primary topic of this paper, is discussed more thoroughly in Section 5.

3 The Computing with Words and Perception Paradigm, and Linguistic Approaches to Information Aggregation and Representation

Fuzzy logic makes it possible to represent and quantify imprecise information. This provides for a more flexible information representation and processing with words and perceptions (cf. Zadeh and Kacprzyk's [37, 38] volumes). Its idea may be briefly summarized as follows.

Information processing comprises, among others, an aggregation of pieces of information. Basic classical aggregation operators related to logical connectives (AND, OR) and quantifiers ('for all', 'there exists') are often too strict – for more on that, see Section 4. In many practical situations a human being would express a rule as "Most of the pieces of information should be taken into account in the process of aggregation". The word "most" may be replaced here with some other linguistic quantifier: "almost all", "much more than 50%", etc. As it often happens that all/some conditions quantified are of a gradual type, both the conditions and quantifier are best modeled within the framework of fuzzy logic.

Zadeh [36] introduced two types of linguistically quantified propositions:

$$QX\text{ 's are }G\text{ 's \quad (type I)} \tag{7}$$

$$QB\text{ 's are }G\text{ 's \quad (type II)} \tag{8}$$

where Q is a linguistic quantifier, and G and B are fuzzy sets in the universe X. Fuzzy linguistic quantifiers are represented by fuzzy sets defined in an appropriate universe. The absolute linguistic quantifiers such as "approximately 3", "several", etc. are represented as fuzzy subsets on domain of positive real numbers, R^+; proportional linguistic quantifiers such as "most", "almost all", etc. are represented by fuzzy subsets, Q, of the unit interval $[0, 1]$:

$$\mu_Q : [0, 1] \longrightarrow [0, 1] \tag{9}$$

Zadeh proposed an interpretation for the proportional linguistic quantifiers such that the degree of truth T of proposition (7) is computed by:

$$T = \mu_Q \left(\frac{\text{card}(G)}{\text{card}(X)} \right) = \mu_Q \left(\frac{\sum_i \mu_G(x_i)}{n} \right) \tag{10}$$

where $\mu_Q(.)$ is the membership function of quantifier Q and n is the cardinality of the universe X.

For propositions of type (8) we have:

$$T = \mu_Q \left(\frac{\text{card}(G \cap B)}{\text{card}(B)} \right) = \mu_Q \left(\frac{\sum_i [\mu_G(x_i) \wedge \mu_B(x_i)]}{\sum_i \mu_B(x_i)} \right) \quad (11)$$

Thus, the truth of a proposition of type (7) is proportional to the fraction of elements of the universe X that belong to its subset G. An exact form of this relationship is determined by the membership function of Q which may be, for "most" of the following, piece-wise linear, form:

$$\mu_{\text{"most"}}(y) = \begin{cases} 1 & \text{for } y \geq 0.8 \\ 2y - 0.6 & \text{for } 0.3 \leq y \leq 0.8 \\ 0 & \text{for } y \leq 0.3 \end{cases} \quad (12)$$

On the other hand, the truth of a proposition of type (8) is proportional to the fraction of elements of a (fuzzy) set $B \subseteq X$ that at the same time belong to $G \subseteq X$. Thus, B plays here a role similar to the scope in case of the classical quantifiers. However, due to the nature of a linguistic quantifier, the type II proposition is not equivalent to the type I propositions with B connected with G using the implication or conjunction as it is true for the classic general and existential quantifier, respectively.

This interpretation of fuzzy linguistic quantifiers is very attractive due to its simplicity both in the definition and use of a quantifier. The type II propositions offer a capability of a weighted aggregation. However, it may become inconvenient in some applications, mainly due to the use of a simple cardinality of fuzzy sets, the so-called $\sum$Counts calculated for fuzzy sets as $\sum_i \mu_A(x_i)$. This makes the number of elements with a low membership degree to count as one element with a high degree.

A convenient approach to handle fuzzy linguistic quantifiers is to use the ordered weighted averaging (OWA) operators, introduced by Yager [30]; see also Yager and Kacprzyk's book [31].

Linguistic quantifiers provide for a flexible processing of fuzzy information. However, information to be aggregated has still to be provided in a strict, numerical form. It is argued that for some applications it may be counter-intuitive, and not human consistent. Thus, also in IR related fuzzy logic applications, Zadeh's [35] concept of a linguistic variable is often employed. It may be briefly described as follows. A linguistic variable is a variable taking on the values in the form of linguistic terms (labels). Formally, a linguistic variable is a 5-tuple $(H, T(H), U, G, M)$, where H is a name of the variable, $T(H)$ is a set of its values (linguistic terms); $U = \{u\}$ denotes the universe under consideration, [i.e., fuzzy sets defined over U provide the interpretation for particular terms belonging to $T(H)$], G is a rule that generates values for the linguistic variable [if $T(H)$ is finite, then G may be just a simple enumeration of linguistic terms]; M is a semantic rule providing for each value $l \in T(H)$ its meaning $M(l) \in U$. For example, treating age as a linguistic variable, one may assume: T("age")={"very young", "young", "middle aged", "old", "very old"}, $U = [1, 100]$, M associates with particular values of T ("age") fuzzy numbers defined over the interval $[0, 100]$ and intuitively corresponding to individual descriptions of the age. For example, with the term "young" a trapezoidal fuzzy number $(0; 0; 25; 35)$ may be associated. Thus, basically, the meaning of linguistic terms is provided by fuzzy numbers (usually triangular or trapezoidal) associated via M and then all operations on linguistic terms, are done on these fuzzy numbers, see, e.g., [7]. This may lead to some problems when the results of such an aggregation are to be again expressed in terms of the original linguistic terms (e.g., in case of the averaging of values of a linguistic variable). Then, linguistic approximation has to be applied and the results may not be fully reliable.

In another approach, the set of linguistic terms is assumed to be finite and ordered. Thus, the semantics of a term is provided just by its position in the order imposed – no fuzzy numbers are associated. In such a case all operations on the linguistic terms have to be specifically defined. For details, see [8, 11, 12]. Such an ordinal (symbolic) linguistic approach has been proposed for IR systems in [11, 12].

4 Fuzzy Extensions of the Boolean Model

The vector space model has been widely accepted as an effective and efficient way of dealing with the tasks addressed within IR. On the other hand, the query language of this model is rather limited. Practically, a query matches such documents that are represented with the terms weighted similarly as in the query. This corresponds, more or less, to the logical conjunction of the elementary queries. That is in contrast with the query language of the Boolean model in which the user may freely combine elementary queries using logical connectives.

The classical Boolean model suffers from an oversimplified representation of documents as sets of terms. It has been observed (cf., e.g., Baeza-Yates & Ribeiro-Neto [1]) that a combination of flexibility of the Boolean querying language and the vector representation of documents may be worthwhile. This has led to many proposals for extensions of the Boolean model. The extended Boolean model has also become a starting point for many proposals for the use of fuzzy logic concepts in IR. A vector representing a document via function F such as in (5) may be easily interpreted as a fuzzy set of terms.

Extensions to the Boolean model may modify only the document's representation or both documents' representations and queries. In the former case, the documents are represented like in the vector space model and the query language remains unchanged. The evaluation of classic Boolean queries against documents represented with weighted terms may take various forms.

First of all, queries may be interpreted as formulae of a fuzzy propositional calculus. Thus, they may be true to a degree from the interval $[0, 1]$. As in the classical case (cf. Section 2), the matching of a query against a document is computed as the truth degree of the query/formula under the evaluation of propositions/terms provided by the document, or more precisely, by weights of the query terms in the document. Classical fuzzy operators/connectives of min and max are used in place of AND and OR, respectively. Obviously, variants of this approach may be obtained by using some triangular norms and conorms to represent logical connectives.

It is widely acknowledged that classical aggregation operators corresponding to the AND and OR connectives often fail to represent real requirements of the user (not only in the context of querying but in broadly meant decision making). The present authors were among the first to propose a solution for that problem in the context of querying of databases. We discuss it briefly later in this paper and now let us look at a more precise statement of that problem and a solution proposed to it even earlier by Salton et al. [1, 24].

Let us assume that a query is the conjunction of elementary queries:

$$q = t_1 \wedge \ldots \wedge t_l \tag{13}$$

and notice that we equate here propositions z_i with terms t_i, for readability.

Under the classical interpretation of the AND connective, if just one term of $t_1, \ldots, t_l$ is absent in a document, then it makes the document completely irrelevant (non-matching) to the query. Even under a fuzzy interpretation the document is relevant only to a degree determined by a term t_j to which the lowest weight is assigned in the document, thus possibly also to degree 0.

It seems to be fully rational to expect that such a matching degree should vary depending on the number of terms of the query that are well/poorly matched in the document. Salton et al. [24] observed that the relevance of a document should be inversely proportional to the distance between two l-dimensional vectors $[w_1, \ldots, w_l]$ and $[1, \ldots, 1]$, where the former gathers weights in the document of terms used in the query, $t_1, \ldots, t_l$. Analogously, for a query that is a disjunction of elementary queries $q = z_1 \vee \ldots \vee z_l$, its matching degree should be proportional to the distance between vectors $[w_1, \ldots, w_l]$ and $[0, \ldots, 0]$.

This idea has been adopted in an extension to the Boolean model, a so-called p-norm model. The distance between the vectors is computed using a p-norm (more often referred to as an l-norm) for a selected value of parameter p. For $p = 1$ we obtain the classical vector space model, and for $p = \infty$ we obtain a simple fuzzy model described above which employs the min and max operators. The very same problem of some deficiencies of the classical AND and OR, as well as their fuzzy counterparts the min and max were addressed by Kacprzyk et al. [14, 15, 16] in the context of fuzzy querying of a classical relational database (thus the setting assumed there is somehow dual to the one considered here: queries are fuzzy while the content of a database is crisp; however the idea of a linguistic quantifier guided aggregation applies in both cases). They proposed to aggregate elementary queries using linguistic quantifiers, such as, e.g., "most". Thus, instead of insisting on the fulfillment of all elementary queries as it is required by the AND connective (and the general quantifier, $\forall$, corresponding to it) or just one elementary query as it is allowed by the OR connective (and the existential quantifier, $\exists$, corresponding to it) the user may require to have "most", "almost all", "many" etc. of elementary queries satisfied. As the above observation of Salton et al. [24] illustrates, the use of such a flexible quantifier may be convenient even in case when we have in mind a strict conjunction of the elementary queries. If there are no documents fully meeting our requirements, we get an empty answer with the classical aggregation operator AND, while using a linguistic quantifier, we get a list of documents almost meeting the query, not perfectly but at least to a certain degree satisfying the query.

The ordered weighted min (OWmin), cf. Dubois et al. [9], operator provides another scheme for the evaluation of a query (13). The motivation here is exactly the same as in case of a linguistic quantifier guided aggregation, i.e., instead of requiring that all elementary queries in (13) are matched, we are satisfied with most of them being matched. This is formalized in a way slightly different to that of linguistic quantifiers. Namely, the concept of a requested majority of matched elementary queries, e.g., most, is modeled as a fuzzy set I in the space $\{0, 1, 2, \ldots, l\}$, such that $\mu_I(0) = 1$; $\mu_I(i) \geq \mu_I(i + 1)$, cf. (9). Thus (cf. Dubois et al. [9]), if we require that "at least k elementary queries be matched", then we set $\mu_I(i) = 1$ for all $0 \leq i \leq k$ and $\mu_I(i) = 0$ for all $i > k$.

Moreover, let us assume that $t_i(d_j)$ denotes the matching of document d_j to the elementary query t_i. Then, we sort vectors $[t_1(d_j), t_2(d_j), \ldots, t_l(d_j)]$ in a non-increasing order to obtain $t_{1*}(d_j) \geq t_{2*}(d_j) \geq \ldots \geq t_{l*}(d_j)$ where $t_{1*}(d_j)$ is the greatest value from among $t_1(d_j), t_2(d_j), \ldots, t_l(d_j)$; $t_{2*}(d_j)$ is the second greatest value, etc. Then, to obtain a matching degree of document d_j to the overall query (13) we compute:

$$\min_{i=1,\ldots,l} \max(1 - \mu_I(i), t_{i*}(d_j)) \qquad (14)$$

In this approach the concept of a linguistic quantifier in the sense of Zadeh may be directly employed to provide a definition of a fuzzy set I. This may be interpreted in a more general setting as the Sugeno measure; for details cf. Dubois et al. [9].

All the above min (or, more generally, a t-norm), p-norm and linguistic quantifier guided models produce an aggregated evaluation (matching degree) of the conjunction of elementary queries (13). Sometimes this is more than needed and we may be quite satisfied with just an ordering of the documents from the best matching the query to the least matching one. For such a purpose many more methods are available.

For example, LEXIMIN [9] compares two documents in terms of their matching to the query (13) in the following way. Let us assume the same notation as in case of OWmin discussed above, however this time the vector $[t_1(d_j), t_2(d_j), \ldots, t_l(d_j)]$ is sorted in the non-decreasing order. Then, d_1 is said to better match the overall query (13) than d_2 if there exists such a k

that $t_{i*}(d_1) = t_{i*}(d_2)$ for all $i < k$ and $t_{k*}(d_1) > t_{k*}(d_2)$. Thus, cf. Dubois et al. [9], LEXIMIN favors documents failing to match as few elementary queries as possible.

LEXIMAX [9], on the other hand, favors documents matching as many elementary queries as possible. Assuming the same notation as for OWmin, this boils down to declaring d_1 as better matching the overall query (13) than d_2 if there exists such a k that $t_{i*}(d_1) = t_{i*}(d_2)$ for all $i < k$ and $t_{k*}(d_1) > t_{k*}(d_2)$.

Other fuzzy extensions to the Boolean model assume weights assigned to terms of the query. In the classical vector space model the interpretation of such weights in queries is quite simple: we seek documents that have terms weighted similarly as in the query. However, in the extended Boolean model more interpretations are possible, and to sketch them, we assume a weighted query in the disjunctive normal form:

$$q = ((t_{11}, w_{11}) \wedge \ldots \wedge (t_{1u}, w_{1u})) \vee \ldots \vee ((t_{d1}, w_{d1}) \wedge \ldots \wedge (t_{dw}, w_{dw})) \quad (15)$$

where t_{ij} denotes a term and w_{ij} denotes its weight in the query.

Assuming such a canonical form, we focus on the matching degree of a single conjunction:

$$(t_{11}, w_{11}) \wedge \ldots \wedge (t_{1u}, w_{1u}) \quad (16)$$

referred to as a *disjunct.*

The matching degree of the whole query is obtained via an aggregation, for instance by using the max operator, of matching degrees of all disjuncts. In the literature three interpretations of the query weights w_{ij} are considered [5]:

- relative importance,
- thresholds of importance, and
- ideal weights.

According to the first interpretation (i.e. relative importance), weight w_{ij} of term t_{ij} in a query indicates to which extent the appearance of term t_{ij} in a document is important for the document to satisfy the query. If the weight is low (close to 0), then the absence of term t_{ij} in a document (i.e., a low, possibly equal 0, weight of this term in the document) does not exclude the matching of this document against the query. If the weight of a term in a query is high (close to 1), then the document has to contain the term (i.e. to have a high weight assigned to this term) to qualify for matching the query.

According to the second interpretation (i.e. thresholds of importance), the weights of particular terms in the documents sought have to be higher than threshold values w_{ij} given in the query. There are further possible interpretations depending on how the undersatisfaction of query terms is treated – a further discussion is given below. Herrera-Viedma [12] proposed a modified interpretation of query weights in this interpretation. Namely, high query weights are treated as mentioned above but low weights require that the corresponding weights in the documents be lower (i.e., set a lower bound).

The third interpretation (i.e. ideal weights) is somehow analogous to that assumed in the vector space model: documents sought should be characterized by weights of terms similar to those specified in the query.

The existence of these various interpretations of query term weights poses some theoretical difficulty. Fortunately, their analysis may be to some extent unified due to results obtained in the area of using fuzzy logic in multicriteria decision making, fuzzy querying of databases and fuzzy information retrieval, cf. Dubois et al. [10]. These may be summarized in terms of IR as follows. Let the matching degree of document d and query q of the form (16) be denoted by $\nu(q, d)$. Moreover, assume the matching degree of an elementary query $q_i = t_i$ (without a weight) and a document d be equal to the weight of term t_i in document d (4), that is:

$$\nu(q_i = t_i, d) = F(d, t_i) \tag{17}$$

The matching degree of the whole query (16) is calculated as:

$$\mu(q, d) = \min_i(q_i, d) \tag{18}$$

Dubois and Prade [9, 10] analyzed several interpretations of importance weights assigned to elementary queries (cf. Bookstein [3] and Yager [29]). They observed that a number of them may be treated as a special case of the following general scheme:

$$\nu(q_i, d) = w_i \longrightarrow F(d, t_i) \tag{19}$$

where "$\longrightarrow$" is a fuzzy implication operator.

Then, using different implication operators we recover various interpretations of importance weights of the terms.

For the Kleene–Dienes implication:

$$x \longrightarrow y = \max(1 - x, y) \tag{20}$$

we obtain

$$\nu(q_i, d) = \max[F(d, t_i), 1 - w_i] \tag{21}$$

which is Yager's [29] interpretation that may be expressed as follows. If an elementary query is completely unimportant ($w_i = 0$), then it does not pose any constraints on the form of the document that has to meet it (the matching degree is always equal 1). Otherwise, a document to satisfy the query has to contain term t_i with a high weight w_i. Thus, this interpretation corresponds to the concept of a relative importance.

For the Gödel implication:

$$x \longrightarrow y = \begin{cases} 1 \text{ if } x \leq y \\ y \text{ otherwise} \end{cases} \tag{22}$$

we get

$$\nu(q_i, d) = \begin{cases} 1 & \text{if } F(d, t_i) \geq w_i \\ F(d, t_i) & \text{otherwise} \end{cases} \qquad (23)$$

which is another Yager's [10] interpretation, and requires that term t_i has in a document a weight higher than that indicated in the query. Thus, this directly refers to the concept of an importance threshold. If the weight of the term in the document, $F(d, t_i)$, does not reach the query weight of this term w_i, such a document satisfies this elementary query to a degree equal to $F(d, t_i)$.

Another, continuous treatment of such an undersatisfaction is obtained while using the Goguen implication:

$$x \longrightarrow y = \begin{cases} 1 & \text{if } x \leq y \\ \frac{y}{x} & \text{otherwise} \end{cases} \qquad (24)$$

and we obtain:

$$\nu(q_i, d) = \begin{cases} 1 & \text{if } F(d, t_i) \geq w_i \\ \frac{F(d, t_i)}{w_i} & \text{otherwise} \end{cases} \qquad (25)$$

Yet another characterizations of importance thresholds has been provided in the literature, cf. Radecki [22].

A similar analysis may be provided for the disjunction of elementary queries, and we refer the reader for details to Bordogna et al. [4, 6]. Here we add the following observation to the characterization of ideal weights provided in [4, 6], namely:

$$\nu(q_i, d) = w_i \longleftrightarrow F(d, t_i) \qquad (26)$$

where "$\longleftrightarrow$" is a fuzzy equivalence operator.

Then, using the definition of a fuzzy equivalence based on the Goguen implication:

$$x \longleftrightarrow y = \min(x \longrightarrow y, y \longrightarrow x) = \begin{cases} 1 & \text{if } F(d, t_i) = w_i \\ \frac{w_i}{F(d, t_i)} & \text{if } F(d, t_i) \geq w_i \\ \frac{F(d, t_i)}{w_i} & \text{if } F(d, t_i) \leq w_i \end{cases} \qquad (27)$$

we obtain a reasonable interpretation of ideal weights.

5 Fuzzy Concepts in Document Categorization

The primary problem in IR is to retrieve documents which are relevant to the user. This task decomposes into the representation of documents and queries, and their matching. In Sections 2 and 4 we briefly reviewed fuzzy logic based approaches proposed for this purpose. Now we briefly discuss a related problem of automated text categorization. Next, we describe our idea of using fuzzy logic based models outlined in the previous section for this purposes. Then, we present our computational experiments and their results.

5.1 The Concept and Purpose of Automatic Text Categorization

In order to make clearer the concept and purpose of automatic text categorization, let us consider some scenarios in which it may be applicable and useful. The first one is that of a Web Spider, a software agent "traversing" the Web and automatically classifying documents found with the purpose to provide the user only with the documents of interest (i.e., belonging to a pre-specified category/categories). The second example may be a "translation agency". In this case, the purpose of the system is to automatically assign to interpreters documents sent by customers for translation. Interpreters prefer certain categories of documents and the aim is to match their preferences so as to secure a high efficiency of the whole translation process. In both cases the classification may be done manually. However, it may be not as good a solution as it may seem. Firstly, in particular in the former case, it is unreasonable to expect that all documents are classified by their authors or some other bodies (see, e.g., Yahoo!). Secondly, the classification provided by the author may be useless for, or inconsistent with, the very purpose or expectations of the document "consumer". Both scenarios assume a set of pre-specified categories of documents.

Basically, we can distinguish two classes of approaches to automatic text categorization. The first consists in a manual construction of a set of explicit classification rules that are then automatically applied to classify the documents. Thus, this methodology belongs to the field of expert systems. The second approach consists in using techniques of machine learning to automatically produce a classifier. We follow the latter and try to use some elements of fuzzy logic, notably those mentioned in Sections 3 and 4. Another dimension along which the text categorization tasks may be distinguished is that of how many classes are considered and how many categories may be assigned to one document. The most general approach (adopted in this paper) assumes a multiclass multilabel task, i.e., there are more than two categories and more than one may be assigned to a document. Still another dimension that is possible is to distinguish the two classes of categorization tasks depending on whether the documents to be classified are available one at a time ("on-line categorization") or in larger portions ("batch categorization"). This distinction is to some extent formal but is important from the point of view of thresholding strategies considered later.

Thus, text categorization as discussed here is a typical example of classification. More precisely, the process consists of two phases:

1. learning of classification rules (explicit or implicit; i.e. building a classifier) from examples of documents with known class assignments (supervised learning),
2. classification of documents unseen earlier using rules derived in Phase 1.

We start with a numerical representation of documents as discussed in Section 2 and formalized by (4). Then, any one of numerous classifier construction algorithms may be applied, including rule-based systems, decision trees, artificial neural networks, etc.

One of classical algorithms developed in the area of IR is that of Rocchio [13, 23]. The learning phase consists in computing a centroid vector for each category of documents. Then, in the classification phase, a document is classified to a category whose centroid is the most similar to this document. The similarity may be meant in several ways – in the original Rocchio's approach it corresponds to the Euclidean distance. In the next subsection we propose to apply some fuzzy logic related concept to build such a classifier. Here, we further precisiate the classification task that is addressed and steps that have to be taken to develop a classifier of the type considered.

In our computational experiments, cf. Section 5.3, we use the Reuters corpus [20] that is widely accepted as a testbed for text categorization algorithms. This is a collection of newswire stories that are usually classified to a number of categories. Thus, this calls for methods dealing with a multiclass and multilabel case. The multilabel categorization requires the solution of an additional problem while building a classifier. Namely, a classifier such as the one considered here produces for a document a list of categories to which it possibly belongs. These categories are ordered non-increasingly according to their matching with the document. Then, a decision has to be made which of them, or more precisely, how many of those from the top of the list are to be assigned to a document under consideration. This is referred to as a *thresholding strategy* [26, 32, 33]. Usually [33], the following strategies are considered:

- rank-based thresholding (RCut),
- proportion based assignment (PCut), and
- score-based local optimization (SCut).

The first strategy (RCut) consists in choosing r top categories for each document. Parameter r may be set by the user or automatically tuned (learned) using a part of the training set of documents. The second strategy (PCut) works for "batch categorization" and assigns to each category such a number of documents from a batch of documents to be classified so as to preserve a proportion of the cardinalities of particular categories in the training set. The third method (SCut) assigns a document to a category only if a matching score of this category and document is higher than a certain threshold. Thresholds are tuned using a part of the training set of documents, separately for each category. In the next section we propose other strategies using the concept of a linguistic quantifier.

To summarize, decisions related to the following issues have to be made while building a classifier of the type considered here:

1. representation of documents and queries, of which an integral part is feature selection,
2. definition of a distance (matching degree) of a centroid and a document, and
3. choice of a thresholding strategy.

The next subsection describes our experiments applying the elements of fuzzy logic mainly for the purposes of Type 2 and 3, but also to some extent of Type 1. The computational experiments require some measures of effectiveness. In our tests we use standard measures as discussed in Subsection 5.3.

5.2 Fuzzy and Linguistic Approaches to the Construction of a Rocchio Type Classifier

The Rocchio type classifier fits into a more general scheme of *profile-based classifiers* [26]. The idea is to compute a *profile* (referred to elsewhere in the text as a *centroid*) for each category and then to base the categorization of a document on its distance (more generally, some measure of similarity) from centroids of all categories. In this way, we can order all categories from the best to the worst by matching the document content. The origin of the Rocchio style classifier is related to the formula for relevance feedback in the vector space model. It is a way of modifying an original user query so as to take into account his or her feedback as to the relevance of particular documents retrieved according to this original query. More precisely, the user picks out relevant documents, then the system computes their centroid (average), possibly taking into account irrelevant documents as well, by subtracting their centroid from the centroid of relevant documents. Next, such a centroid of the class of relevant documents is used as a modified query to once again retrieve documents from the whole collection. Thus, in terms of the categorization task it corresponds to a binary (only two classes of relevant and irrelevant documents are considered), one-class (each document is classified either as relevant or irrelevant, but not both), and "batch" rather than on-line problem. It is, then, quite different in comparison to multiclass multilabel on-line categorization task addressed here.

Our approach assumes, classically, the computation of centroids (see the next subsection for details) for all categories. Then, a document to be classified (more precisely, its representation) is treated as a query against the set of centroids. Within the extended Boolean framework such a query may be treated as the conjunction of the form (13) or (16). As we pointed out, a simple interpretation of the conjunction as the min operator does not work well for classical queries considered in IR, i.e., queries constructed by a human user. This is true even to a further extent in case of the categorization task considered here. Thus, in our experiments we try various flexible schemes of aggregation presented in Section 4 as well as, in case of (16), different weight interpretations discussed in the same section.

For our further discussion let us assume the following notation:

$$C = \{c_p\}_{p \in [1,P]} \tag{28}$$

is a set of centroids, one for each of P categories. Each centroid is represented by a vector:

$$c_p = [c_{p1}, \ldots, c_{pM}] \tag{29}$$

where M denotes, as previously, the number of terms used to index the documents.

These centroids are constructed in a different way than in the typical Rocchio type classifier. Namely, weights c_{pi} are not calculated directly as the averages of the weights of all training documents belonging to a given category but according to the following formula:

$$c_{pj} = \frac{f_{pj} \star \log(\frac{P}{n_j})}{\arg \ \max_j [\log(\frac{P}{n_j}) \star f_{pj}]} \tag{30}$$

where f_{pj} is a frequency of term j in all documents belonging to category p and n_j is the number of categories in documents of which term j appears (category frequency). By analogy to (5) it may be called a $tf \times icf$ representation where icf stands for an inverted category frequency. A document to be classified, d, which is here treated as query q against a set of centroids, is represented, as previously, by a vector, or more precisely as in (16):

$$d = q = [w_1, \ldots, w_M] \equiv (t_1, w_1) \wedge \ldots \wedge (t_M, w_M) \tag{31}$$

where

$$w_j = \frac{f_j \star \log(\frac{P}{n_j})}{\arg \ \max_j [\log(\frac{P}{n_j}) \star f_j]} \tag{32}$$

where f_j is a frequency of term t_j in document d and n_j is the category frequency of this term, cf. (30). Thus, this setting most naturally fits the extended Boolean model where both queries and documents are represented using weights.

Now, we base our decision on the classification of document d as pertinent to category p on its similarity to a corresponding centroid c_p, i.e., on the matching of query q (31) and this centroid. In order to compute this matching degree we are going to employ our discussion of Section 4, in particular various interpretations of the query weights.

Let us observe that, in the given context, the similarity of a query and a centroid intuitively means that terms representing them have weights that are comparable, relatively or absolutely. Thus, the relative importance and ideal weights interpretations seems to be more suitable than the thresholds of importance. In order to justify the latter claim let us assume that a document is represented by means of 2 terms with high weights and 10 terms with very low weights. Then, the category whose centroid is represented by all 12 terms with high weights that perfectly fits the document/query according to the threshold logic, does not seem to be a good fit.

Below we list and briefly comment upon the matching schemes we tested in our experiments with an automatic text categorization. Their main goal is to overcome a deficiency of the min operator which is reinforced by a high dimensionality of the representation space of both the query (31) and documents (29).

We group the tested schemes according to the query weight interpretation adopted.

I. Relative importance

M.I.1. Original weight interpretation via the Kleene-Dienes implication

This concept of matching (i.e., document classification), cf. (18),(19), may be linguistically expressed as:

"A document matches a category if *all of the important terms* present in the document are also present in the centroid of the category."

The matching of the elementary and overall queries are here computed using (19) and (18), respectively.

M.I.2. Linguistic majority

This concept of matching (i.e., document classification) may be linguistically expressed as:

"A document matches a category if *most of the important terms* present in the document are also present in the centroid of the category."

The idea refers directly to our previous experiences with a fuzzy database querying (cf. Kacprzyk and Zadrożny [14]–[16]). The above linguistic expression is formalized using Zadeh's calculus of linguistically quantified propositions by, cf. (8):

$$QB\text{'s are }G\text{'s}$$

where X, the universe considered, is a set T of all index terms, B is a fuzzy set of terms important for the document d, i.e.,

$$\mu_B(t_j) = w_j = F(d, t_j)$$

and G is a fuzzy set of terms present in centroid c_p of category p, i.e.,

$$\mu_G(t_j) = c_{pj}$$

Due to a high dimensionality of the considered space and known deficiencies of linguistic quantifiers in the sense of Zadeh (cf. Section 3) we also tested a modified version where only terms weighted in the query higher than a certain threshold are considered.

II. Ideal weights

M.II.1. Cosine

The classical vector space model formula (6) has been employed.

M.II.2. Fuzzy equivalence based approach

The ideal weight logic is here represented using fuzzy equivalence, as expressed by the formulae (26) and (27). The overall matching degree computes using the min, as in (18) or OWmin, as in (14). Also a modified version where only terms weighted in the query higher than certain threshold are considered, has been tested.

III. **Thresholding strategies**

Concerning the thresholding strategy we also propose an approach based on fuzzy linguistic concepts. The underlying idea may be expressed as follows:

T.I. "Select such a threshold r that *most of the important categories had a number of sibling categories similar to r in the training data set.*"

Thus, for each $r \in [1, R]$ we compute the degree of truth of the above clause in italics (R is a parameter). This is again formalized using Zadeh's calculus of linguistically quantified propositions as:

$$QB\text{'s are } G\text{'s} \tag{33}$$

where X, the universe considered, is a set C of 10 categories with the highest matching score, B is a fuzzy set of important categories for a given document d, i.e.,

$$\mu_B(c_p) = \nu(q, c_p)$$

where $\nu(\cdot, \cdot)$ is the matching function (18) used and q is a query/document to be classified. G is a fuzzy set of categories, that, on the average, had in the training set the number of sibling categories similar to r for which the truth value of (33) is calculated. This similarity is modeled by a similarity relation which is another parameter of the method. For the purposes of this strategy (and others, as given below), for each category the number of average sibling categories in the training data set is first computed. By the sibling category for a category c_p we mean a category that is assigned to the same document as the category c_p.

T.II. Another approach exploiting the concept of sibling categories works as follows. Only categories whose matching score is higher than a certain parameter (in our experiments usually 0.2 is assumed) are taken into account. Their scores are normalized (divided by the sum of their scores) and then the weighted sum of the average number of siblings is taken as a threshold cut (rounded to the nearest integer value).

T.III. For the comparison we also tested the simple RCut with a threshold rank equal 2, i.e., two top scored categories are assigned to each document.

The whole classification procedure proceeds then in the following steps:

- Training phase
    1. The training documents are read and data on their frequency in documents and categories are gathered. Also the average

number of sibling categories is computed for each category (for the thresholding strategy purposes),

2. The training documents are read once again and the centroids of particular categories are calculated according to formula (30).

- Testing phase

   1. A test document is read and its representation is calculated using formula (32) and normalized (in the experiments all coordinates are divided by the highest one),

   2. For each category the matching degree (score) of its centroid and the document is computed using one of the approaches outlined in the previous subsections and this vector of scores is sorted in the non-increasing order,

   3. One of the threshold strategies as outlined above is used to decide which categories assign to the document.

5.3 Computational Results

In our general setting for computational experiments we use Yang and Liu's work [34]. The text corpus used is Reuters-21578 as made available over the Internet by Lewis [20]. More precisely we are using the Modified Apte ("ModApte") split of data, i.e. for the training phase a subset of news that are characterized by the attributes LEWISSPLIT="TRAIN" and TOPICS="YES" and for testing phase a subset LEWISSPLIT="TEST"; TOPICS="YES". In both cases, we use only news that actually contains topics and body of the text or at least the title. This gives rise to 7,728 training, 3,005 test documents and 114 categories. The title of the document and its body are concatenated to produce the document. The documents are preprocessed by removing stop words [27] and numbers. Stemming is done using the standard Porter's [21] algorithm. The terms space dimensionality reduction is done using a simple approach based on document and category frequencies of terms. Namely, only terms with a document frequency higher than 3 and a category frequency lower than 75% are used. This rule yields 5,565 index terms.

The evaluation of particular approaches tested has been carried out by using standard measures of recall, precision, F1 measure and 11-point average precision. Both micro- and macro-averaging results are presented. These measures are expressed by the following formulae:

micro-averaging

$$\text{precision} = \frac{\text{number of correct classifications made by the system}}{\text{total number of all classifications made by the system}} \quad (34)$$

$$\text{recall} = \frac{\text{numer of correct classifications made by the system}}{\text{total number of all categories indicated in test documents}} \quad (35)$$

Table 1. Comparison of matching schemes for T.II. thresholding strategy.

Matching scheme	micro-averaging			macro-averaging			11-pt AVP*****
	precision	recall	F1	precision	recall	F1	
M.I.1.	0.3763	0.5521	0.4475	0.2212	0.3704	0.2770	0.6793
M.I.2.	0.3914	0.8215	0.5302	0.4038	0.5322	0.4592	0.8311
M.I.2a.*	0.4226	0.6765	0.5203	0.3416	0.6174	0.4398	0.7673
M.II.1.	0.2226	0.6462	0.3311	0.1235	0.4943	0.1976	0.6511
M.II.2a.**	0.5597	0.4597	0.5048	0.5601	0.0934	0.1601	0.5926
M.II.2b.***	0.5847	0.5015	0.5399	0.5231	0.1349	0.2145	0.6356
M.II.2b.****	0.3809	0.6961	0.4923	0.2978	0.4287	0.3515	0.7397

* – only terms weighted above 0.2 are considered in matching degree computation
** – aggregation via min
*** – aggregation via min; only terms weighted above 0.2 are considered in matching degree computation
**** – aggregation via OWmin; only terms weighted above 0.2 are considered in matching degree computation
***** – 11-point average precision

$$F1 = 2 \star \text{precision} \star \frac{\text{recall}}{\text{precision} + \text{recall}} \qquad (36)$$

Note that the number of classifications is higher than the number of test documents as more than one category may be assigned to a document. By a system we mean an automatic classifier based on a given approach.

macro-averaging

First, precision, recall and F1 measure are calculated separately for each category using formulae (34)– (36) and then the arithmetic mean of them is calculated.

Below in the tables we present some of the results of our experiments. First, in Table 1 we show a comparison of results obtained for the T.II. thresholding strategy.

In Table 2 we test various thresholding strategies for the linguistic guided aggregation used in M.I.2.

Table 2. Comparison of different thresholding strategies for the M.I.2. matching scheme.

Matching scheme	micro-averaging			macro-averaging			11-pt AVP*****
	precision	recall	F1	precision	recall	F1	
T.I.	0.5531	0.7765	0.6460	0.4891	0.4785	0.4837	0.8311
T.II	0.3914	0.8215	0.5302	0.4038	0.5322	0.4592	0.8311
T.III.*	0.4642	0.7478	0.5728	0.4776	0.4309	0.4530	0.8311

The results reported for the state-of-the-art approaches, cf. Yang & Liu [34], are better, especially for micro-averaging. The best reported F1 measure for macro-averaging reaches 0.5242 and our result places more or less in the middle of the five methods tested in [34]. For micro-averaging our precision score is essentially worse than the best reported (0.8507). However, two comments on that should be made. First of all, our 11-point average precision is pretty good (high) – there are no scores reported on that in [34] – which may suggest that the main source of our relatively poor results in this respect is a weak thresholding strategy. This will be the subject of our further research. It seems that in this respect, there is still a space for an essential improvement in the framework of fuzzy logic based approaches. The results for particular matching schemes are essentially worse than the one reported for the state-of-the-art approaches, cf. Yang & Liu [34]. Moreover, our approach offers, typical for Rocchio type classifiers, both fast learning and training phases which may be important in some applications.

To summarize, however, there is still a number of various factors that influence the effectiveness of an automatic categorization system, including the representation of documents, tuning the parameters, and a thresholding strategy. The very nature of fuzzy linguistic approaches makes it possible to tune parameters that possess an interpretation that is easier to grasp by the human user involved. In this short study we only try to check a general applicability of fuzzy logic based concepts for text categorization.

6 Concluding Remarks and Further Research

In the paper we discussed the automatic text document categorization that has recently attracted a lot of attention and interest. In our approach we tried to use results obtained by other authors proposing some fuzzy logic based extensions to classical IR models, notably the Boolean model. Although they mainly address the primary task of retrieval of the documents relevant to the human user, their underlying ideas are also applicable to text categorization.

Our starting point was fuzzy querying of crisp (nonfuzzy) databases. We tried to adapt some of the ideas we proposed earlier in this respect, notably of a linguistically guided aggregation of partial matching degrees, for the purposes of text document categorization. We illustrated the preliminary results on a standard document corpus used to test most sophisticated and successful classifiers. Further research will focus on tuning various parameters that are included in our proposed approaches. The most important factor for the potential improvement is a better choice and a more sophisticated tuning of the thresholding strategy. This will be a subject of our next efforts. Another important factor is a particular document representation scheme. We plan to compare the results obtained with our classifier for other traditional representation schemes, notably the $tf \times idf$.

References

[1] R. Baeza-Yates and B. Ribeiro-Neto, editors. *Modern Information Retrieval*. Addison-Wesley, Reading, Massachusetts, 1999.

[2] R. K. Belew and C. J. van Rijsbergen. *Finding Out About: A Cognitive Perspective on Search Engine Technology and the WWW*. Cambridge University Press, New York, NY, USA, 2000.

[3] A. Bookstein. Fuzzy Requests: An Approach to Weighted Boolean Searches. *Journal of the American Society for Information Sciences*, 31:240–247, 1980.

[4] G. Bordogna, P. Bosc, and G. Pasi. Extended Boolean Information Retrieval in Terms of Fuzzy Inclusion. In O. Pons, M. A. Vila, and J. Kacprzyk, editors, *Knowledge Management in Fuzzy Databases*, pages 234–246. Physica, Heidelberg, New York, 2000.

[5] G. Bordogna, P. Carrara, and G. Pasi. Fuzzy Approaches to Extend Boolean Information Retrieval. In P. Bosc and J. Kacprzyk, editors, *Fuzziness in Database Management Systems*, pages 231–274. Physica, Heidelberg, 1995.

[6] G. Bordogna and G. Pasi. Application of Fuzzy Sets Theory to Extend Boolean Information Retrieval. In F. Crestani and G. Pasi, editors, *Soft Computing in Information Retrieval*, pages 21–47. Physica, Heidelberg, New York, 2000.

[7] C. Carlsson and R. Fuller. A New Look at Linguistic Importance Weighted Aggregation. In *Proceedings of the Fourteenth European Meeting on Cybernetics and Systems Research*, pages 169–174, Vienna, 1998. Austrian Society for Cybernetic Studies.

[8] M. Delgado, J. L. Verdegay, and M. A. Vila. On Aggregation Operations of Linguistic Labels. *International Journal of Intelligent System*, 8:351–370, 1993.

[9] D. Dubois, H. Fargier, and H. Prade. Beyond Min Aggregation in Multicriteria Decision: (Ordered) Weighted Min, Discri-min, Leximin. In R. R. Yager and J. Kacprzyk, editors, *The Ordered Weighted Averaging Operators. Theory and Applications*, pages 181–192. Kluwer Academic Publishers, Boston, Dordrecht, London, 1997.

[10] D. Dubois and H. Prade. Using Fuzzy Sets in Flexible Querying: Why and How? In T. Andreasen, H. Christiansen, and H. L. Larsen, editors, *Flexible Querying Answering Systems*, pages 45–60. Kluwer Academic Publishers, Boston, Dordrecht, 1997.

[11] E. Herrera-Viedma. An Information Retrieval System with Ordinal Linguistic Weighted Queries Based on Two Weighting Elements. *International Journal of Uncertainty, Fuzziness and Knowledge-Based Systems*, 9:77–88, 2001.

[12] E. Herrera-Viedma. Modeling the Retrieval Process of an Information Retrieval System Using an Ordinal Fuzzy Linguistic Approach. *Journal of*

the *American Society for Information Science and Technology (JASIST)*, 52(6):460–475, 2001.

[13] T. Joachims. A Probabilistic Analysis of the Rocchio Algorithm with TFIDF for Text Categorization. In *Proceedings of ICML-97, 14th International Conference on Machine Learning*, pages 143–151, Nashville, US, 1997. Morgan Kaufmann.

[14] J. Kacprzyk and S. Zadrożny. Computing with Words in Intelligent Database Querying: Standalone and Internet-Based Applications. *Information Sciences*, 134:71–109, 2001.

[15] J. Kacprzyk, S. Zadrożny, and A. Ziółkowski. FQUERY III+: a "human-consistent" database querying system based on fuzzy logic with linguistic quantifiers. *Information Systems*, 14:443–453, 1989.

[16] J. Kacprzyk and A. Ziółkowski. Database Queries with Fuzzy Linguistic Quantifiers. *IEEE Transactions on Systems, Man and Cybernetics*, 16:474–479, 1986.

[17] R. R. Korfhage. *Information Storage and Retrieval.* John Wiley and Sons, New York, 1997.

[18] D. H. Kraft, G. Bordogna, and G. Pasi. An Extended Fuzzy Linguistic Approach to Generalize Boolean Information Retrieval. *Journal of Information Sciences*, 2(3):119–134, 1994.

[19] D. H. Kraft, G. Bordogna, and G. Pasi. Fuzzy Set Techniques in Information Retrieval. In J. C. Bezdek, D. Dubois, and H. Prade, editors, *Fuzzy Sets in Approximate Reasoning and Information Systems (The Handbook of Fuzzy Sets Vol. 3)*, pages 469–510. Kluwer Academic Publishers, Norwell, 1999.

[20] D. D. Lewis. Reuters-21578, Dist. 1.0. online. http://www.research. att.com/~lewis.

[21] M. F. Porter. An Algorithm for Suffix Stripping. *Program*, 14(3):130–137, 1980.

[22] T. Radecki. Fuzzy Set Theoretical Approach to Document Retrieval. *Information Processing and Management*, 15:247–260, 1979.

[23] J. Rocchio. Relevance Feedback in Information Retrieval. In G. Salton, editor, *The SMART Retrieval System: Experiments in Automatic Document Processing*, pages 313–323. Prentice-Hall Inc., 1971.

[24] G. Salton, E. A. Fox, and H. Wu. Extended Boolean Information Retrieval. *Communications of ACM*, 26(11):1022–1036, 1983.

[25] G. Salton and M. J. McGill. *Introduction to Modern Information Retrieval.* McGraw Hill, New York, 1983.

[26] F. Sebastiani. A Tutorial on Automated Text Categorisation. In *Proceedings of ASAI-99, 1st Argentinian Symposium on Artificial Intelligence*, pages 7–35, Buenos Aires, 1999.

[27] Stop Words list. http://www.indiana.edu/cgi-bin-local/ doIsearch.pl?Stopwords.

[28] C. J. van Rijsbergen. *Information Retrieval.* Butterworths, London, Boston, 1979.

[29] R. R. Yager. A Note on Weighted Queries in Information Retrieval Systems. *Journal of the American Society for Information Science*, 38:23–24, 1987.

[30] R. R. Yager. On Ordered Weighted Averaging Aggregation Operators in Multi-Criteria Decision Making. *IEEE Transactions on Systems, Man and Cybernetics*, 18:183–190, 1988.

[31] R. R. Yager and J. Kacprzyk, editors. *The Ordered Weighted Averaging Operators: Theory and Applications*. Kluwer Academic Publishers, Boston, 1997.

[32] Y. Yang. An Evaluation of Statistical Approaches to Text Categorization. *Journal of Information Retrieval*, 1(1/2):67–88, 1999.

[33] Y. Yang. A Study on Thresholding Strategies for Text Categorization. In W. B. Croft, D. J. Harper, D. H. Kraft, and J. Zobel, editors, *Proceedings of ACM SIGIR Conference on Research and Development in Information Retrieval (SIGIR'01)*, pages 137–145, New Orleans, US, 2001. ACM.

[34] Y. Yang and X. Liu. A Re-examination of Text Categorization Methods. In M. A. Hearst, F. Gey, and R. Tong, editors, *Proceedings of ACM SIGIR Conference on Research and Development in Information Retrieval (SIGIR'99)*, pages 42–49, Berkeley, US, 1999. ACM.

[35] L. A. Zadeh. The Concept of Linguistic Variable and its Applications to Approximate Reasoning. Parts I,II,III. *Information Sciences*, 8, 9:199–251 (8), 301–357 (8), 43–80 (9), 1975.

[36] L. A. Zadeh. A Computational Approach to Fuzzy Quantifiers in Natural Languages. *Computers and Mathematics*, 9:149–184, 1983.

[37] L. A. Zadeh and J. Kacprzyk, editors. *Computing with Words in Information/Intelligent Systems. Part 1: Foundations*. Physica, Heidelberg, New York, 1999.

[38] L. A. Zadeh and J. Kacprzyk, editors. *Computing with Words in Information/Intelligent Systems. Part 2: Applications*. Physica, Heidelberg, New York, 1999.

[39] S. Zadrożny, K. Ławcewicz, and J. Kacprzyk. Intelligent Linguistic Characterization and Retrieval of Textual Documents: An Internet-Based Application. In B. Bouchon-Meunier, L. Foulloy, and R. R. Yager, editors, *Intelligent Systems for Information Processing – From Representation to Applications*, pages 153–164. Elsevier, Amsterdam, 2003.

Neural Networks, Fuzzy Models and Dynamic Logic

Leonid I. Perlovsky

Air Force Research Lab
Leonid.Perlovsky@hanscom.af.mil

Summary. The paper discusses possible relationships between computational intelligence, known mechanisms of the mind, semiotics, and computational linguistics. Mathematical mechanisms of concepts, emotions, and goals are described as a part of information processing in the mind and are related to language and thought processes in which an event (signals from surrounding world, text corpus, or inside the mind) is understood as a concept. Previous attempts in artificial intelligence at describing thought processes are briefly reviewed and their fundamental (mathematical) limitations are analyzed. The role of emotional signals in overcoming these past limitations is emphasized. The paper describes mathematical mechanisms of concepts applicable to sensory signals and linguistics; they are based on measures of similarities between models and signals. Linguistic similarities are discussed that can utilize various structures and rules proposed in computational linguistic literature. A hierarchical structure of the proposed method is capable of learning and recognizing concepts from textual data, from the level of words and up to sentences, groups of sentences, and towards large bodies of text. I briefly discuss a role of concepts as a mechanism unifying thinking and language and their possible role in language acquisition. A thought process is related to semiotic notions of signs and symbols. It is further related to understanding, imagination, intuition, and other processes in the mind. The paper briefly discusses relationships between the mind and brain and applications to understanding-based search engines.

1 Language and the Mind

Language and thinking are distinctly human abilities. Even if one prefers to consider the difference between human and animal minds in terms of degrees, the difference is formidable. Close relationships between language and thinking encouraged equating these abilities in the past. Rule-based systems, using the mathematics of logic, implied significant similarities between the two. The situation has changed, in part due to the fact that logic-rule systems have not been sufficiently powerful to explain thinking, nor language abilities, and in part due to improved scientific understanding (psychological,

L.I. Perlovsky: *Neural Networks, Fuzzy Models and Dynamic Logic*, StudFuzz **209**, 363–386 (2007)
www.springerlink.com © Springer-Verlag Berlin Heidelberg 2007

cognitive, neural, linguistic) of the mechanisms involved. Among contemporary linguists there is a growing appreciation of a possibility that language and thinking could be distinct and different abilities of mind (see [43] for further references). Comparing apes with parrots, the first having significant intellectual capabilities and the second having significant linguistic capabilities, one may conclude that language and thinking might have evolved along separate evolutionary paths; and some researchers believe there are reasons for this conclusion.

Human language mechanisms include abilities to acquire a large vocabulary, rules of grammar, and to use the finite set of words and rules to generate virtually infinite number of phrases and sentences [16, 44]. Human thinking includes abilities to understand the surrounding world in terms of objects, their relationships (scenes and situations), relationships among relationships, and so on [40]. Researchers in computational linguistics, mathematics of intelligence and neural networks, cognitive science, neuro-physiology and psychology during the last twenty years significantly advanced understanding of the mechanisms of the mind involved in learning and using language, mechanisms of perception and cognition – for the discussions and further references see [10, 16, 25, 28, 40, 44, 46] Much less advance was achieved toward deciphering mechanisms relating linguistic competence to understanding and thinking. Although it seems clear that language and thinking are closely related abilities, intertwined in evolution, ontogenesis, and everyday use, still the currently understood mechanisms of language are mainly limited to relations of words to other words and phrases, but not to the objects in the surrounding world, not to cognition and thinking. Possible mathematical approaches toward integrating language and thinking, words and objects, phrases and situations are discussed in this paper.

The paper starts with a mathematical description of thinking, which still is an issue of much controversy. Among researchers in mathematical intelligence it has become appreciated, especially during the last decades that thinking is not just a chain of logical inferences [10, 28, 40]. Yet, mathematical methods describing thinking as processes involving concepts, instincts, emotions, memory, imagination are not well known, although significant progress in this direction was achieved [10, 28, 40]. A brief historical overview of this area including difficulties and controversies is given in the next two sections from mathematical, psychological and neural standpoints; it is followed by a mathematical description of thinking processes. Then the paper discusses the ways in which the mathematical description of thinking can be combined with language, taking advantage of recent progress in computational linguistics. It touches upon novel ideas of computational semiotics relating language and thinking through signs and symbols. In conclusion, I briefly discuss relationships between mathematical, psychological, and neural descriptions of thinking processes and language as parts of the mind.

Words like *mind, thought, imagination, emotion, concept* are often used colloquially in many ways, but their use in science and especially in mathematics of intelligence has not been uniquely defined and is a subject of active research and ongoing debates [10, 28]. According to a dictionary [42], mind includes conscious and unconscious processes, especially thought, perception, emotion, will, memory, and imagination, and it originates in brain. These constituent notions will be discussed throughout the paper.

A broad range of opinions exists on the mathematical methods suitable for the description of the mind. Founders of artificial intelligence thought that formal logic was sufficient [31] and no specific mathematical techniques would be needed to describe the mind [29]. An opposite point of view is that there are few specific mathematical constructs, "the first principles" of the mind organization. Among researchers taking this view is Grossberg, who suggested that the first principles include a resonant matching between lower-level signals [10] and higher-level representations and emotional evaluation of conceptual contents [12]; several researchers suggested specific principles of the mind organization [18, 27, 40, 51]. Hameroff, Penrose, and the author (among others) considered quantum computational processes that might take place in the brain [14, 33, 36]. Although, it was suggested that new unknown yet physical phenomena will have to be accounted for explaining the working of the mind [33]. This paper describes mechanisms of the mind that can be "implemented" by classical physics mechanisms of the brain neural networks and, alternatively, by using existing computers.

2 Theories of the Mind and Combinatorial Complexity

Understanding signals coming from sensory organs involves associating subsets of signals corresponding to particular objects with internal representations of these objects. This leads to recognition of the objects and activates internal brain signals leading to mental and behavioral responses, which constitute the understanding of the meaning (of the objects).

Developing mathematical descriptions of the very first *recognition* step of this seemingly simple association-recognition-understanding process has not been easy, a number of difficulties have been encountered during the past fifty years. These difficulties have been summarized under the notion of combinatorial complexity (CC) [37]. The problem was first identified in pattern recognition and classification problems in the 1960s and was named "the curse of dimensionality" [2]. The following thirty years of developing adaptive statistical pattern recognition and neural network algorithms designed for self-learning led to a conclusion that these approaches often encountered *CC of learning requirements*: recognition of any object, it seemed, could be learned if "enough" training examples were used for an algorithm self-learning. The required examples had to account for all possible variations of "an object", in all possible geometric positions and in *combinations* with other objects, sources of light, etc., leading to astronomical (and worse) numbers of required examples.

By the end of the 1960s a different paradigm became popular: logic-rule-based systems (or expert systems) were proposed to solve the problem of learning complexity. An initial idea was that rules would capture the required knowledge and eliminate a need for learning. The first Chomskian ideas concerning mechanisms of language grammar related to deep structure [4] were also based on a similar idea of logical rules. Rule systems work well when all aspects of the problem can be predetermined. However, rule systems and expert systems in the presence of unexpected variability, encountered *CC of rules*: more and more detailed sub-rules and sub-sub-rules, one contingent on another, had to be specified.

In the 1980s model-based systems became popular, which were proposed to combine advantages of adaptivity and rules by utilizing adaptive models. Existing knowledge was to be encapsulated in models and unknown aspects of concrete situations were to be described by adaptive parameters. Along similar lines were *rules and parameters* ideas of Chomsky [5]. Model-based systems encountered *computational CC* (N and NP complete algorithms). The reason was that considered algorithms had to evaluate multiple combinations of elements of data and rules (models). CC is prohibitive because the number of combinations is very large: for example, consider 100 elements (not too large a number) whose combinations had to be evaluated; the number of combinations of 100 elements is 100^{100}, a number comparable to the number of elementary particles in a Universe; no computer would ever be able to compute that many combinations. The CC became a ubiquitous feature of intelligent algorithms and seemingly, a fundamental mathematical limitation.

Combinatorial complexity has been related to the type of logic, underlying various algorithms and neural networks [37]. Formal logic is based on the "law of excluded third", according to which every statement is either true or false and nothing in between. Therefore, algorithms based on formal logic have to evaluate every little variation in data or internal representations as a separate logical statement; a large number of combinations of these variations cause combinatorial complexity. In fact, combinatorial complexity of algorithms based on logic has been related to the Gödel theory: it is a finite system manifestation of the incompleteness of logic [34]. Multivalued logic and fuzzy logic were proposed to overcome limitations related to the law of excluded third [17]. Yet the mathematics of multivalued logic is no different in principle from formal logic. Fuzzy logic encountered a difficulty related to the degree of fuzziness: if too much fuzziness is specified, the solution does not achieve a needed accuracy, if too little, it becomes similar to formal logic.

3 Mind: Concepts and Emotions

Seemingly fundamental nature of mathematical difficulties discussed above led many to believe that classical physics cannot explain the working of the mind. Yet, I would like to emphasize another aspect of the problem: often mathematical theories of the mind where proposed before the necessary physical intuition of how the mind works was developed. Newton, as often mentioned, did not consider himself as evaluating various hypotheses about the working of the material world, he felt that he possesses what we call today a physical intuition about the world [50]. An intuition about the mind points to mechanisms of concepts, emotions, instincts, imagination, behavior generation, consciousness and unconscious. An essential role of emotions in the working of the mind was analyzed from the psychological and neural perspective by Grossberg [13], from the neuro-physiological perspective by Damasio [6], and from the learning and control perspective by the author [8, 38, 39]. One reason for engineering community being slow in adopting these results is the cultural bias against emotions as a part of thinking processes. Plato and Aristotle thought that emotions are "bad" for intelligence, this is a part of our cultural heritage ("one has to be cool to be smart"), and the founders of Artificial Intelligence repeated this truism about emotions [31]. Yet, as discussed in the next section, combining conceptual understanding with emotional evaluations is crucial for overcoming the combinatorial complexity as well as the related difficulties of logic.

Let me summarize briefly and in a much simplified way several aspects of the working of the mind, which seem essential to the development of the mathematical descriptions of the mind mechanisms: instincts, concepts, emotions, behavior generation. The mind has evolved for the purpose of survival and therefore it serves for a better satisfaction of the basic instincts, which have emerged as survival mechanisms even before the mind. Instincts operate like internal sensors: for example, when a sugar level in our blood goes below a certain level an instinct "tells us" to eat. The most accessible to our consciousness mechanism of the mind is concepts: the mind operates with concepts. Concepts are like internal models of the objects and situations; this analogy is quite literal, e.g., during visual perception of an object, an internal concept model projects an image onto the visual cortex, which is matched there to an image projected from retina (this simplified description will be refined later).

An ability for concepts evolved for instinct satisfaction, and the mechanism linking concepts and instincts involves emotions. Emotions are neural signals connecting instinctual and conceptual brain regions. Whereas in colloquial usage, emotions are often understood as facial expressions, higher voice pitch, exaggerated gesticulation, these are the outward signs of emotions, serving for communication. A more fundamental role of emotions within the mind system is that emotional signals evaluate concepts for the purpose of instinct satisfaction. This evaluation is not according to rules or concepts (like in rule systems of artificial intelligence), but according to a different instinctual-emotional mechanism described in the next section. This emotional mechanism is crucial for breaking out of the "vicious circle" of combinatorial complexity.

The results of conceptual-emotional understanding of the world are actions (or behavior) in the outside world or within the mind. In this paper we touch on only one type of behavior, the behavior of improving understanding and knowledge of the language and world (including self). In the next section we describe a mathematical theory of a "simple" conceptual-emotional recognition and understanding. As we will discuss, in addition to concepts and emotions, it involves with necessity mechanisms of intuition, imagination, conscious, unconscious, and aesthetic emotion. And this process is intimately connected to an ability of mind to form symbols and interpret signs.

The mind involves a hierarchy of multiple levels of concept-models, from simple perceptual elements (like edges, or moving dots), to concept-models of objects, to complex scenes, and up the hierarchy ... toward the concept-models of the meaning of life and purpose of our existence. Hence the tremendous complexity of the mind, yet relatively few basic principles of the mind organization go a long way explaining this system.

4 Modeling Field Theory (MFT)

Modeling field theory [40], summarized below, associates lower-level signals with higher-level concept-models (or internal representations), resulting in understanding of signals, while overcoming the difficulties of CC described in Section 2. It is achieved by using measures of similarity between the concept models and the input signals combined with a new type of logic, i.e. the fuzzy dynamic logic. Modeling field theory is a multi-level, hetero-hierarchical system. This section describes a basic mechanism of interaction between two adjacent hierarchical levels of signals (fields of neural activation); sometimes, it will be more convenient to talk about these two signal levels as an input to and output from a (single) processing level.

At each level, the output signals are concepts recognized (or formed) in input signals. Input signals $\mathbf{X}$ are associated with (or recognized, or grouped into) concepts according to the representation models and similarity measures at this level. In the process of association-recognition, models are adapted for better representation of the input signals; and similarity measures are adapted so that their fuzziness is matched to the model uncertainty. The initial uncertainty of models is high and so is the fuzziness of the similarity measure; in the process of learning models become more accurate and the similarity more crisp, the value of the similarity measure increases. I call this mechanism *fuzzy dynamic logic*.

4.1 Internal Models, Learning, and Similarity

During the learning process, new associations of input signals are formed resulting in evolution of new concepts. Input signal $\{\mathbf{X}(n)\}$, $n = 1, \ldots, N$, a field of input neuronal synapse activation levels, enumerates the input neurons. $\mathbf{X}(n)$ are the activation levels. A set of concept-models $h = 1, \ldots, H$ is characterized by the models (or representations) $\{\mathbf{M}_h(n)\}$ of the signals $\mathbf{X}(n)$.

Each model depends on its parameters $\{\mathbf{S}_h\}, \mathbf{M}_h(\mathbf{S}_h, n)$. In a highly simplified description of a visual cortex, n enumerates the visual cortex neurons, $\mathbf{X}(n)$ are the "bottom-up" activation levels of these neurons coming from the retina through visual nerve, and $\mathbf{M}_h(n)$ are the "top-down" activation levels (or priming) of the visual cortex neurons from previously learned object-models[1]. Learning process attempts to "match" these top-down and bottom-up activations by selecting "best" models and their parameters. Mathematically, learning increases a similarity measure between the sets of models and signals, $L(\{\mathbf{X}(n)\}, \{\mathbf{M}_h(n)\})$. The similarity measure is a function of model parameters and associations between the input synapses and concept-models. It is constructed in such a way that any of a large number of objects can be recognized, no matter if they appear on the left or on the right. Correspondingly, a similarity measure is designed so that it treats each concept model as an alternative for each subset of signals

$$L(\{\mathbf{X}\}, \{\mathbf{M}\}) = \prod_{n \in N} \sum_{h \in H} r(h) l(\mathbf{X}(n)|\mathbf{M}_h(n)); \qquad (1)$$

$l(\mathbf{X}(n)|\mathbf{M}_h(n))$ (or simply $l(n|h)$) is a conditional partial similarity between one signal $\mathbf{X}(n)$ and one model $\mathbf{M}_h(n)$ – all possible combinations of signals and models are accounted for in this expression. Parameters $r(h)$ are proportional to the number of signals $\{n\}$ associated with the model h.

In the process of learning, concept-models are constantly modified. From time to time a system forms a new concept, while retaining an old one as well; alternatively, old concepts are sometimes merged. (Formation of new concepts and merging of old ones require a modification of the similarity measure (1); the reason is that more models always result in a better fit between the models and data. This is a well-known problem, it can be addressed by reducing the r.h.s. of equation (1) using a "penalty function", $p(N, M)$ that grows with the number of models M, and this growth is steeper for a smaller amount of data N. For example, an asymptotically unbiased maximum likelihood estimation leads to multiplicative $p(N, M) = \exp(-N_{par}/2)$, where N_{par} is the total number of adaptive parameters in all models (this penalty function is known as *Akaike Information Criterion*, see [40] for further discussion and references).

4.2 Fuzzy Dynamic Logic and MFT

The learning process consists in estimating model parameters $\mathbf{S}_h$ and associating subsets of signals with concepts by maximizing the similarity (1). Note, that equation (1) contains a large number of combinations of models and signals, a total of H^N items; this was a reason for the combinatorial complexity of the past algorithms discussed in section 2. Modeling field theory (MFT)

---

[1]In fact, there are many levels between the retina, visual cortex, and object models.

solves this problem by fuzzy dynamic logic [35, 40]. MFT introduces fuzzy association variables $f(h|n)$:

$$f(h|n) = \frac{r(h)l(n|h)}{\sum_{h' \in H} r(h')l(n|h')} \tag{2}$$

These variables give a measure of correspondence between signal $\mathbf{X}(n)$ and model $\mathbf{M}_h$ relative to all other models, h'. A mechanism of concept formation and learning, an internal dynamics of the modeling fields (MF) is defined as follows,

$$\mathbf{S}_h = \mathbf{S}_h + \alpha \sum_n f(h|n) \left(\frac{\partial \ln l(n|h)}{\partial \mathbf{M}_h} \right) \frac{\partial \mathbf{M}_h}{\partial \mathbf{S}_h} \tag{3}$$

$$r(h) = \frac{N_h}{N}; N_h = \sum_n f(h|n); \tag{4}$$

Parameter α determines the iteration step and speed of convergence of the MF system; N_h can be interpreted as a number of signals $\mathbf{X}(n)$ associated with or coming from a concept object n. As already mentioned, in the MF internal dynamics, similarity measures are adapted so that their fuzziness is matched to the model uncertainty. Mathematically, this can be accomplished in several ways, depending on the specific parameterization of the conditional partial similarity measures, $l(n|h)$; for example, they can be defined as Gaussian functions,

$$l(n|h) = (2\pi)^{-\frac{d}{2}} (det\mathbf{C}_h)^{-\frac{1}{2}} \exp\{-0.5(\mathbf{X}(n) - \mathbf{M}_h(n))^T \\ \mathbf{C}_h^{-1}(\mathbf{X}(n) - \mathbf{M}_h(n))\} \tag{5}$$

In this formula, d is the dimensionality of the vectors $\mathbf{X}$ and $\mathbf{M}$, and $\mathbf{C}_h$ is a covariance. The dynamics of fuzziness of the MF similarity measures is defined as

$$\mathbf{C}_h = \sum_n f(h|n)(\mathbf{X}(n) - \mathbf{M}_h(n))(\mathbf{X}(n) - \mathbf{M}_h(n))^T / N_h \tag{6}$$

Initially, models do not match data, covariances are large, and association variables, $f(h|n)$, take homogeneous values across the data, associating all concept-models h with all input signals n. As matching improves, covariances become smaller, and the association variables, $f(h|n)$, tend to high values 1 for some subsets of signals and models and zero for others; thus certain concepts get associated with certain subsets of signals (objects are recognized and concepts formed). The following theorem was proven in [40]:

Theorem 1. *Equations (2) through (6) define a convergent dynamic system MF with stationary states given by* $\max_{\{\mathbf{S}_h\}} L$.

In plain language this means that the above equations indeed result in concept-models in the "mind" of the MFT system, which are most similar – in terms of similarity measure (1) – to the sensory data. Despite a combinatorially large number of items in equation (1), a computational complexity of the MF method is relatively low, it is linear in N and could be implemented by a physical system (like a computer or a brain). These equations describe a closed loop system, which is illustrated in figure 1. A reference to the *closed* loop emphasizes that the loop sustains its operations on its own, the loop is not closed in that there are input signals into the loop and output concepts from the loop.

Comment. Equation (5) of conditional partial similarities using Gaussian functions can be considered a basis for the following probabilistic interpretation: A model $M_h(S_h, n)$ is a conditional statistical expectation of signals from object h described by parameters S_h. A similarity measure (1) is a total likelihood. Let me emphasize that such an interpretation could be valid if for some values of the parameters, the models are accurate (that is, models actually *are* conditional statistical expectation). If models are approximate in a non-statistical sense, other similarity measures could be more preferable mathematically, like mutual information in the models about the data [40]. I would also like to emphasize that unlike usual "Gaussian assumption" this model is quite general, it *does not assume that the signal distribution is Gaussian*, but only the deviations between the models and signals are, this model can represent any statistical distribution [40].

Summary of the MF convergence: during an adaptation process initial fuzzy and uncertain models (internal structures of the MF system) are associated with structures in the input signals, fuzzy models are getting more definite and crisp. The type, shape and number of models are selected so that the internal representation within the system is similar to input signals: The MF concept-models represent structure-objects in the input signals. Mathematical equations which describe this process are called *fuzzy dynamic logic* [40] which in terms of mind-internal processes describes an elementary thinking process involving instincts, imagination, emotions and concepts. But before discussing this cognitive-psychological interpretations, lets us briefly look into integrating this process with language.

4.3 Integrating Language and Thinking

During visual perception, when internal representation-models are matched in the visual cortex to retinal signals, cortex representations maintain their spatial topology and continuity. A number of MFT models have been developed for visual perception, for other sensor modalities, and for cognition of simple situations [40]. By using concept-models with multiple sensor modalities, a MFT system can integrate signals from multiple sensors, while adapting and improving internal concept-models. Similarly, MFT can be used to integrate

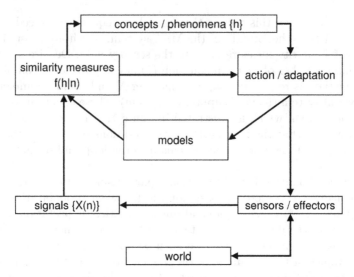

Fig. 1. For a single level of MFT, input signals are unstructured data $\{\mathbf{X}(n)\}$ and output signals are recognized or formed concepts $\{h\}$ with high values of similarity measures. The MFT equations (2) through (6) describe a continuous loop operation involving input signals, similarity measures, models, and actions of the model adaptation (the inner loop in this figure). Psychologically, a similarity measure corresponds to the knowledge instinct and its changes to aesthetic emotions.

language and thinking. This requires the development of linguistic MFT models. Here, I briefly outline an approach to the development of MFT linguistic models. Like MFT, language is a hierarchical system. Among other things, it involves sounds, phonemes, words, phrases, sentences, and texts, where each level operates with its own models. Like other models of the mind, these models are a result of evolution; for computational intelligent systems we have to develop them, and this development at each level is a research project, which is added by a number of already described linguistic models [16, 26, 44, 46].

In order to give an illustration, I discuss an approach to the development of models of phrases from words in the context of text understanding which, for example, could be used for an understanding-based search engine. The input data, $\mathbf{X}(n)$, in this "phrase-level" MF system are word strings of a fixed length S. Thus: $\mathbf{X}(n) = \{w_{n+1}, w_{n+2}, \ldots, w_{n+S}\}$. w_n are words of a given dictionary $W = \{w_1, w_2, \ldots, w_K\}$ of size K, and n is the word position in a body of texts. A simple phrase model is "a bag of words", that is, a model is a subset of words from a dictionary, without any order or rules of grammar,

$$\mathbf{M}_h(\mathbf{S}_h, n) = \{w_{h,1}, w_{h,2}, \ldots, w_{h,S}\} . \tag{7}$$

The parameters of this model are its words, $\mathbf{M}_h(\mathbf{S}_h, n) = \mathbf{S}_h = \{w_{h,1}, w_{h,2}, \ldots, w_{h,S}\}$. The language acquisition project in this simplified context consists in defining models-concepts-phrases best characterizing the given body of texts in terms of a similarity measure.

Conditional partial similarities between a string of text, $\mathbf{X}(n)$, and a model $\mathbf{M}_h$ could be defined by a proportion of the matches between the two sets, $\mathbf{X}(n)$ and $\mathbf{M}_h$, $l(n|h) = |\mathbf{X}(n) \cap \mathbf{M}_h|/S$. Thus, similarity in terms of equation (1) would be defined and could be maximized over the unknown parameters of the system, $\{\mathbf{S}_h\}$, that is, over the word contents of phrases. This would result in learning models-concepts-phrases, accomplishing the goal of the language acquisition project. The difficulty of the above approach is that the dynamics of MFT cannot be used for the similarity maximization, in particular, equation (3) requires evaluating derivatives, which requires a smooth dependence of models on their parameters. Without the fuzzy dynamic logic of MFT, the computational complexity of this language acquisition project becomes combinatorial, i.e. $\sim K^{(H^* N^* S)}$, which is a prohibitively large number.

The combinatorial complexity of the above solution is related to a "logic-type" similarity measure, which treats every potential phrase model (every combination of words) as a separate logical statement. The problem can be solved by using dynamic fuzzy phrase contents as follows. First, define fuzzy conditional partial similarity measures:

$$l(n|h) = (2\pi\sigma_h^2)^{-\frac{S}{2}} \exp\{-0.5 \sum_s e(n, h, s)^2/\sigma_h^2\}, \tag{8}$$

where $e(n, h, s)$ is a distance (measured in the numbers of words) between the middle of the word sequence $\mathbf{X}(n)$, that is $n + S/2$, and the closest occurrence of the word $w_{h,s}$; the sum here is over words belonging to the phrase model h. In practical implementations, the search for the nearest word can be limited by $\pm 3\sigma_h$ words, and $e(n, h, s)$ falling outside this range can be substituted by a $(3\sigma_h + 1)$. The dynamics of fuzziness of this similarity measure is given by a modification of equation (6),

$$\sigma_h^2 = \sum_n f(h|n) \sum_s e(n, h, s)^2/N_h \ . \tag{9}$$

Second, define fuzzy phrase contents, that is a degree of the word $w_{h,s}$ "belonging" to a model-phrase h, $\phi(s|h)$; this is a function of the average distance of the word $w_{h,s}$ from the phrase model $\varepsilon(s, h)$:

$$\varepsilon(h, s) = \sum_n f(h|n) e(n, h, s)^2/N_h; \tag{10}$$

$$\phi(s|h) = p(h|s)/\sum_{s' \in h} p(h|s');$$

$$p(h|s) = (2\pi\sigma_h^2)^{-1/2} \exp\{-0.5 \sum_s \varepsilon(h, s)/\sigma_h^2\}, \tag{11}$$

The dynamics of the word contents of the phrase models is given by modifying S (the number of words in phrases) in the iteration process, say, by defining $S_h \sim S\sigma_h$, or by requiring $\phi(s|h)$ to be above a threshold value, and keeping in each phrase model the words satisfying this criteria. The dynamics defined in this way results in learning phrase models (concepts) and accomplishes the goal of the language acquisition project without combinatorial complexity, the computational complexity is moderate, $\sim H^*K^*S^2$.

The "bag-of-words" phrase models considered above are much simpler than tree-like dependencies or known structures of natural languages [16, 25, 26, 44, 46, 48]. These more complicated "real" linguistic models can be used in place of a simple distance measure $e(n, h, s)$ in equation (8). In this way the models of noun and verb phrases and tree structures can be incorporated into the above formalism of MFT.

Integration of language and cognition in MFT is attained by characterizing objects and situations in the world with two types of models, linguistic models considered above and cognitive models considered in section 4.2 and in [40]. Such integrated MFT system learns – similarly to human – in parallel in three realms: (1) linguistic models can be learned to some extent independently from cognition, when linguistic data are encountered for the first time with limited or no association with perception and cognition (like in a newborn baby); (2) similarly, cognitive models can be learned to some extent independently from language, when perception signal data are encountered for the first time in limited or no association with linguistic data; and (3) linguistic and cognitive models are learned jointly, when linguistic data are present in some association with perception signals, like during mother talking to a baby: "this is a car" (perception models and word models), and like during more complicated conversations: "Look at Peter and Ann, they are in love" (cognitive models and phrase models).

A Constructed Example. A real-life example of this approach would be too voluminous and boring to follow. Here is a simplified constructed example to illustrate some of the main points of learning phrase models. It starts with a large text data base (hundreds of millions of words) and partitions it into 10-word chunks. Four of these chunks are shown here containing a word *chair*:

fifth chair foundation not-for-profit organization devoted fostering online bridge
    education hickory chair furniture catalog register wish list store locator contact
    fork picnic table set choice chairs benches sets fork table site give information
    university course provide software engineering chair involved

After several iterations the algorithm learned phrase models limited to 6-word length; one hundred thousand of the mostly often used phrase models were retained, among them the following four were connected to the previous chunks with appreciable probabilities:

chair foundation nonprofit organization online education
    online furniture catalog store brand discount

ascii table hexadecimal octal set character
university chair professor appointment invitation name

After several more iterations ten thousands of most useful 4 word phrase models were retained, among them the following five were connected to the previous chunks with appreciable probabilities:

organization nonprofit community service
online furniture catalog website
brand name furniture discount
ascii table hexadecimal octal
university chair professor appointment

Higher levels of generalization (fewer word sentences with broader meanings) require moving to a higher level of a multi-level hierarchical system.

4.4 MFT Hierarchical Organization

The previous subsections described a single processing layer in a hierarchical MFT system. Inputs to each layer are signals $\mathbf{X}(n)$, or in neural terminology, an input field of neuronal activations. Outputs are the activated models $\mathbf{M}_h(\mathbf{S}_h, n)$; it is a set of models or concepts recognized in the input signals. Equations (2-6) and (8-11), as shown in figure 1, can be interpreted as a loop process: at each iteration the equations contain association variables $f(h|n)$ and model parameters computed at the previous iteration. In other words, the output models "act" upon the input to produce a "refined" output model (at the next iteration). This process is directed at increasing the similarity between the models and signals. It can be described as an internal behavior of model adaptation.

The output models initiate other actions as well. First, activated models (neuronal axons) serve as input signals to the next processing layer, where more general concept-models are recognized or created. Second, concept-models along with the corresponding instinctual signals and emotions may activate behavioral models and generate behavior directed into the outside world (a process not contained within the above equations). In general, a higher level in a hierarchical system provides a feedback input into a lower level. For example, sensitivities of retinal ganglion cells depend on the objects and situations recognized higher up in the visual cortex; or, a gaze is directed based on which objects are recognized in the field of view. These interactions within this hierarchical organization are illustrated in figure 2.

Concept objects identified at the output of the lower level of MFT system in figure 2 become input signals to the next MFT level which identifies more general concepts of relationships among objects and situations; at the same time more general concepts of understanding identified at a higher level activate behavioral concept-models that affect processes at a lower level. The

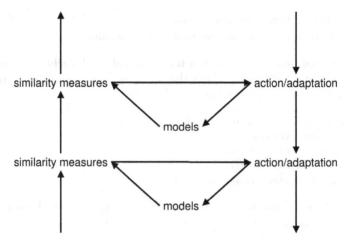

Fig. 2. Hierarchical organization of the MFT system. High levels of similarity measures correspond to concepts recognized at a given level in the hierarchy; these are the input signals to the next, higher level. Also concepts affect behavior (actions). Models at a higher level are more general than models at a lower level.

agent processes, or the loop processes of model concept adaptation, understanding and behavior generation continue up and down the hierarchy of the MFT levels.

The loop of operations of MFT can also be described as multiple loops each involving a single model $h, h = 1, \ldots, H$. To some extent these multiple loops are independent, yet some models interact when they are associated with the same input signals. Each model along with its adaptation mechanism is an intelligent agent, which possesses a degree of autonomy and is interacting with other agents. Thus MFT is an intelligent system composed of multiple adaptive intelligent agents. Each agent, including its concept model along with the similarity measure and behavioral response, is a continuous loop of operations, interacting with other agents from time to time; an agent is "dormant" until activated by a high similarity value. When activated, it is adapted to the signals and other agents, so that the similarity increases. A subset of data in input signals may activate several concept agents, in this way data provide evidence for the presence of various objects (or concepts). Agents compete with each other for evidence (matching to signals), while adapting to the new signals.

A multi-level hierarchical linguistic MFT system can be developed by adding more levels similar to a word phrase level described in section 4.3. A relatively simple system can use similar "bag" models for each layer, like "bag of phrases" model for the next level of concepts (say, sentence), and so on. Alternatively, more realistic linguistic models of sentences, paragraphs and large bodies of texts can be utilized (cf. Rieger [47] and Mehler [24]). Among many possible commercial applications of such systems could be understanding-based search engines; everybody familiar with the frustration of the web searches would appreciate a search engine that even remotely understands user queries and contents of the web pages.

5 MFT Theory of Mind

5.1 MFT Dynamics

Equations (2-6) and (8-11) describe elementary processes of perception or cognition, in which a number of model concepts compete for incoming signals, model concepts are modified and new ones are formed, and eventually, more or less definite connections (high values of $f(h|n)$, close to 1) are established among signal subsets on the one hand, and some model concepts on the other, accomplishing perception and cognition.

A salient mathematical property of this processes ensuring a smooth convergence is a correspondence between uncertainty in models (that is, in the knowledge of model parameters) and uncertainty in associations $f(h|n)$. In perception, as long as model parameters do not correspond to actual objects, there is no match between models and signals; many models poorly match many objects, and associations remain fuzzy (between 0 and 1). Eventually, one model (h') wins a competition for a subset $\{n'\}$ of input signals $\mathbf{X}(n)$, when parameter values match object properties, and $f(h'|n)$ values become close to 1 for $n \in \{n'\}$ and 0 for $n \notin \{n'\}$. This means that this subset of data is recognized as a specific object (concept). Upon the convergence, the entire set of input signals $\{n\}$ is divided into subsets, each associated with one model object, uncertainties become small, and fuzzy a priori concepts become crisp concepts. Cognition is different from perception in that models are more general, more abstracts, and input signals are the activation signals from concepts identified (cognized) at a lower hierarchical level; the general mathematical laws of cognition and perception are similar and constitute a basic principle of the mind organization. Kant was the first one to propose that the mind functioning involves three basic abilities: Pure Reason (concept-models), Judgment (emotional measure of correspondence between models and input signals), and Practical Reason (behavior; we only considered here the behavior of adaptation and learning) [21, 20, 22]. Let us discuss relationships between the MFT theory and concepts of mind originated in psychology, philosophy, linguistics, aesthetics, neuro-physiology, neural networks, artificial intelligence, pattern recognition, and intelligent systems.

5.2 Elementary Thought Process, Conscious, and Unconscious

A thought process or thinking involves a number of sub-processes and attributes, including internal representations and their manipulation, attention, memory, concept formation, knowledge, generalization, recognition, understanding, meaning, prediction, imagination, intuition, emotion, decisions, reasoning, goals, behavior, conscious and unconscious [25, 27, 40]. Here and in the following subsections we discuss how these processes are described by MFT.

A "minimal" subset of these processes, *an elementary thought process*, has to involve mechanisms for afferent and efferent signals [10], in other words, bottom-up and top-down signals coming from outside (external sensor signals) and from inside (internal representation signals). According to Carpenter and Grossberg [3] every recognition and concept formation process involves a "resonance" between these two types of signals. In MFT, at every level in a hierarchy the afferent signals are represented by the input signal field $\mathbf{X}$, and the efferent signals are represented by the modeling fields $\mathbf{M}_h$; resonances correspond to high similarity values $l(n|h)$ for some subsets of $\{n\}$ that are "recognized" as concepts (or objects). The mechanism leading to the resonances between incoming signals and internal representations is given by equations in sections 4.2 and 4.3. The elementary thought process also involves elements of conscious and unconscious processes, imagination, memory, concepts, instincts, emotions, understanding and behavior as described later.

A description of working of the mind as given by the MFT dynamics was first provided by Aristotle [1], describing thinking as a learning process in which an a priori form-as-potentiality (fuzzy model) meets matter (sensory signals) and becomes a form-as-actuality (a concept). Jung suggested that conscious concepts are developed by the mind based on genetically inherited structures, archetypes, which are inaccessible to consciousness [19], and Grossberg [10] suggested that only signals and models attaining a resonant state (that is signals matching models) reach consciousness. Fuzzy uncertain models are less accessible to consciousness, whereas more crisp and certain models are better accessible to consciousness.

5.3 Understanding

In the elementary thought process, subsets in the incoming signals are associated with recognized model objects, creating *phenomena* (in the MFT-mind) which are *understood* as objects, in other words *signal subsets* acquire *meaning* (e.g., a subset of retinal signals acquires a meaning of a chair). There are several aspects to understanding and meaning. First, object-models are connected (by emotional signals [8, 12, 38, 39, 40]) to instincts that they might satisfy, and also to behavioral models that can make use of them for instinct satisfaction. Only two instincts and types of behavior are described within equations of section 4: (1) the knowledge instinct and behavior of learning of perception and cognition models (that is improving and adapting these models for better correspondence to the world), and (2) the language instinct and behavior of learning linguistic models (that is improving and adapting these models for better correspondence to the language data, like words and grammar). A formulation is proposed where these two instincts are closely related and can be considered as two aspects of the same instinct.

The second aspects of understanding and meaning is that an object, situation, or phrase is understood in the context of a more general situation in the next layer, consisting of more general concept-models, which accepts as input signals the results of object recognition. That is, each recognized object model (phenomenon) sends (in neural terminology, activates) an output signal; and a set of these signals comprises input signals for the next layer models, which 'cognize' more general concept-models. And this process continues up and up the hierarchy of the models and mind toward the most general models a system could come up with, such as models of universe (scientific theories), models of self (psychological concepts), models of meaning of existence (philosophical concepts), models of a priori transcendent intelligent subject (theological concepts).

5.4 Imagination

Visual imagination involves excitation of a neural pattern in a visual cortex in absence of an actual sensory stimulation (say, with closed eyes) [10]. Imagination was often considered to be a part of thinking processes; Kant [20] emphasized the role of imagination in the thought process, he called thinking "a play of cognitive functions of imagination and understanding". Whereas pattern recognition and artificial intelligence algorithms of recent past would not know how to relate to this [29, 31], Carpenter and Grossberg's resonance model [3] and the MFT dynamics both describe imagination as an inseparable part of thinking: imagined patterns are top-down signals that *prime* the perception cortex areas (*priming* is a neural terminology for making neural cells to be more readily excited). In MFT, models M_h give the imagined neural patterns. MFT (in agreement with neural data) just adds details to Kantian description: thinking is a play of *higher-hierarchical level* imagination and *lower-level* understanding. Kant identified this "play" (described by equations (2-6) or (8-11)) as a source of aesthetic emotions discussed later.

5.5 Mind vs. Brain

Historically, the mind is described in psychological and philosophical terms, whereas the brain is described in terms of neurobiology and medicine. Within scientific exploration the mind and brain are different description levels of the same system. Establishing relationships between these description is of great scientific interest. Today we approach solutions to this challenge [11], which eluded Newton in his attempt to establish physics of "spiritual substance" [50]. General neural mechanisms of the elementary thought process (which are similar in MFT and ART [3]) have been confirmed by neural and psychological experiments, this includes neural mechanisms for bottom-up (sensory) signals, top-down "imagination" model signals, and the resonant matching between the two [9, 10, 52]. Adaptive modeling abilities are well studied and adaptive parameters identified with synaptic connections [15, 23]; instinctual learning mechanisms have been studied in psychology and linguistics [4, 7, 41, 43]; identifying neural structures responsible for knowledge and language instincts is a next challenge for the neural sciences.

5.6 Instincts and Emotions

Functioning of the mind and brain cannot be understood in isolation from the system's "bodily needs". For example, a biological system (and any autonomous system) needs to replenish its energy resources (*eat*); this and other fundamental unconditional needs are indicated to the system by instincts, which could be described as internal sensors. Emotional signals, generated by this instinct are perceived by consciousness as "hunger", and they activate behavioral models related to food searching and eating. In this paper we are concerned primarily with the behavior of recognition: instinctual influence on recognition modify the object perception process (3) - (6) in such a way that desired objects "get" enhanced recognition. It can be accomplished by modifying priors, $r(h)$, according to the degree to which an object of type h can satisfy a particular instinct. Details of these mechanisms are not considered here, except for the two instincts considered in this paper.

5.7 Aesthetic Emotions and Instinct for Knowledge

Recognizing objects in the environment and understanding their meaning is so important for human evolutionary success that there has evolved an instinct for learning and improving concept models. This instinct (for knowledge and learning) is described in MFT by maximization of similarity between the models and the world according to equation (1). Emotions related to satisfaction/dissatisfaction of this instinct we perceive as harmony/disharmony (between our understanding of how things ought to be and how they actually are in the surrounding world). According to Kant [20] these are aesthetic emotions (emotions that are not related directly to satisfaction or dissatisfaction of bodily needs). Aesthetic emotions in MFT correspond to changes in the knowledge instinct (1). The aesthetic emotion is negative, when new input signals do not correspond well to existing models. The mathematical basis for the theorem stated after equation (6) can be interpreted psychologically: During iterations defined by the equations (2-6) the aesthetic emotion is always positive.

In sections 4.2 we considered perception and cognition concept models and similarity measures; using them in equation (1) yields an instinct driving the MFT system to improve the knowledge about the world. Similarly, using linguistic models in equation (1) and the similarity measures considered in section 4.3, yields the MFT system improving the knowledge of language, or language instinct. Combining cognitive and linguistic models results in a system with combined linguistic and thinking abilities: language and sensory information together help adapting both, linguistic and cognitive models. A specific mathematical mechanism combining language and cognition described in section 4 associates both types of models with every object and situation. We do not know if the mind works this way. Finding out actual neural mechanisms combining language and cognition is a future challenge.

5.8 Beauty and Intuition

Harmony is an elementary aesthetic emotion related to improvement of object-models. Higher aesthetic emotions are related to the development of more complex "higher" models: we perceive an object or situation as aesthetically pleasing if it satisfies our learning instinct, that is the need for improving the models and increasing similarity (1). The highest forms of aesthetic emotion are related to the most general and most important models. According to Kantian analysis [20], among the highest models are models of the meaning of our existence, of our purposiveness or intentionality, and beauty is related to improving these models: we perceive an object or a situation as beautiful, when it stimulates improvement of these highest models of meaning. Beautiful is what "reminds" us of our purposiveness.

Intuition includes an intuitive perception (imagination) of object models and their relationships with objects in the world, higher-level models of relationships among simpler models, and behavioral models. Intuition involves fuzzy unconscious concept models, which are in a state of being learned and being adapted toward crisp and conscious models (a "thought" or a theory); such models may satisfy or dissatisfy the knowledge instinct in varying degrees before they are accessible to consciousness, hence the complex emotional feel of an intuition. The beauty of a physical theory discussed often by physicists is related to satisfying our feeling of purpose in the world that is satisfying our need to improve the models of the meaning through understanding of the universe.

5.9 Theory Testing and Future Directions

The general neural mechanisms of the elementary thought process, which include neural mechanisms for bottom-up (sensory) signals, top-down "imagination" model signals, and the resonant matching between the two have been confirmed by neural and psychological experiments (these mechanisms are similar in MFT and ART [9, 10, 40, 52]). Adaptive modeling abilities are well studied and adaptive parameters have been identified with synaptic connections [15, 23]; instinctual learning mechanisms have been studied in psychology and linguistics [4, 7, 41, 43]. Ongoing and future research will confirm, disprove, or suggest modifications to specific mechanisms of parameter adaptation (equation 2-5), reduction of fuzziness during learning (equation 6), similarity measure (equation 1) as a foundation of aesthetic instinct for knowledge, relationships between psychological and neural mechanisms of learning on the one hand and, on the other, aesthetic feelings of harmony and emotion of beauty. Differentiated forms of (1) need to be developed for various forms of the knowledge instinct (differentiation between cognition and language instincts in this paper is a step in this direction). Future experimental research needs to study in details the nature of hierarchical interactions: to what extent the hierarchy is "hardwired" vs. adaptively emerging in ontogenesis and throughout life; theory of emerging hierarchical models will have to be developed. For a combined theory of language and cognition, future experimental research ought to identify neural mechanisms combining linguistic and cognitive concepts, prove or disprove the mechanisms proposed in this paper, and also study the ontogenesis of these mechanisms in child development processes.

5.10 Thinking Process and Semiotics

Semiotics studies symbol content of culture [49]. For example, consider a written word "chair". It can be interpreted by a mind to refer to something else: an entity in the world, a specific chair, or the concept "chair" in the mind. In this process, the mind, or an intelligent system is called *an interpreter*, the written word is called *a sign*, the real-world chair is called *a designatum*, and the concept in the interpreter's mind, the internal representation of the results of interpretation is called *an interpretant* of the sign. The essence of a sign is that it can be interpreted by an interpreter to refer to something else, a designatum. This process of sign interpretation is an element of a more general process called *semiosis*, which consists of multiple processes of sign interpretation at multiple levels of the mind hierarchy.

In mathematics and in "Symbolic AI" there is no difference between signs and symbols. Both are considered as notations, arbitrary non-adaptive entities with axiomatically fixed meaning. This non-differentiation is a "hangover" from an old superstition that logic describes mind, a direction in mathematics and logical philosophy that can be traced through the works of Frege, Hilbert, Russell, to its bitter end in Gödel theory, and its revival during the 1960s and 1970s in artificial intelligence. In general culture, symbols are understood also as psychological processes of sign interpretation. Jung emphasized that symbol processes connect conscious and unconscious [19], Pribram wrote of symbols as adaptive, context-sensitive signals in the brain, whereas signs he identified with less adaptive and relatively context-insensitive neural signals [45].

In classical semiotics [30, 32] words *sign* and *symbol* were not used consistently; in the context of the mathematical description in this paper, a sign means something that can be interpreted to mean something else (like a mathematical notation, or a word), and the process of interpretation is called a symbol process, or symbol. Interpretation, or understanding of a sign by the mind according to MFT is due to the fact that a sign (e.g., a word) is a part of an object model (or a situation model at higher levels of the mind hierarchy). The mechanism of a sign interpretation therefore involves first an activation of an object model, which is connected to instincts that the object might satisfy, and also to behavioral models that can make use of this object for instinct satisfaction. Second, a sign is understood in the context of a more general situation in the next layer consisting of more general concept-models, which accepts as input signals the results of lower-level sign recognition. That is, recognized signs comprise input signals for the next layer models, which 'cognize' more general concept-models.

A symbol process of a sign interpretation coincides with an elementary thought process. Each sign interpretation or elementary thought process, a symbol, involves conscious and unconscious, emotions, concepts, and behavior; this definition connecting symbols to archetypes (fuzzy unconscious model concepts) corresponds to a usage in general culture and psychology. As described previously, this process continues up and up the hierarchy of models and mind toward the most general models. In semiotics this process is called *semiosis*, a continuous process of creating and interpreting the world outside (and inside our mind) as an infinite hierarchical stream of signs and symbol processes.

References

[1] Aristotle. Metaphysics. In J. Barnes, editor, *Complete Works of Aristotle*. Princeton University Press, Princeton, NJ, 1995.

[2] R. E. Bellman. *Adaptive Control Processes*. Princeton University Press, Princeton, NJ, 1961.

[3] G. A. Carpenter and S. Grossberg. A Massively Parallel Architecture for a Self-organizing Neural Pattern Recognition Machine. *Computer Vision, Graphics and Image Processing*, 37:54–115, 1987.

[4] N. Chomsky. *Language and Mind*. Harcourt Brace Javanovich, New York, 1972.

[5] N. Chomsky. Principles and Parameters in Syntactic Theory. In N. Hornstein and D. Lightfoot, editors, *Explanation in Linguistics. The Logical Problem of Language Acquisition*. Longman, London, 1981.

[6] A. R. Damasio. *Descartes' Error: Emotion, Reason, and the Human Brain*. Avon, New York, 1995.

[7] T. W. Deacon. *The Symbolic Species: The Co-Evolution of Language and the Brain*. W. W. Norton & Company, 1998.

[8] V. A. Dmitriev and L. I. Perlovsky. Art Form as an Object of Cognitive Modeling (Towards Development of Vygotsky's Semiotics Model). In *Proceedings of the 1996 Conference on Intelligent Systems and Semiotics*, volume 2, pages 385–389, Gaithersburg, 1996.

[9] W. J. Freeman. *Mass Action in the Nervous System*. Academic Press, New York, 1975.

[10] S. Grossberg. *Neural Networks and Natural Intelligence*. MIT Press, Cambridge, MA, 1988.

[11] S. Grossberg. Linking Mind to Brain: The Mathematics of Biological Intelligence. *Notices of the American Mathematical Society*, 47:1361–1372, 2000.

[12] S. Grossberg and D. S. Levine. Neural Dynamics of Attentionally Modulated Pavlovian Conditioning: Blocking, Inter-stimulus Interval, and Secondary Reinforcement. *Psychobiology*, 15(3):195–240, 1987.

[13] S. Grossberg and N. A. Schmajuk. Neural Dynamics of Attentionally Modulated Pavlovian Conditioning: Conditioned Reinforcement, Inhibition, and Opponent Processing. *Psychobiology*, 15(3):195–240, 1987.

[14] S. R. Hameroff. *Toward a Scientific Basis for Consciousness*. MIT Press, Cambridge, MA, 1994.

[15] D. Hebb. *Organization of Behavior*. J. Wiley & Sons, New York, 1949.

[16] R. Jackendoff. *Foundations of Language: Brain, Meaning, Grammar, Evolution*. Oxford University Press, New York, 2002.

[17] J.-S. R. Jang, C.-T. Sun, and E. Mizutani. *Neuro-Fuzzy and Soft Computing: A Computational Approach to Learning and Machine Intelligence*. Prentice Hall, Upper Saddle River, 1996.

[18] B. D. Josephson. An Integrated Theory of Nervous System Functioning Embracing Nativism and Constructivism. In *International Complex Systems Conference*, Nashua, NH, September 21-26 1997.

[19] C. G. Jung. Archetypes of the Collective Unconscious. In *The Collected Works*, volume 9.II of *Bollingen Series XX, 1969*. Princeton University Press, Princeton, NJ, 1934.

[20] I. Kant. *Critique of Judgment*. Macmillan & Co., London, 2nd edition, 1914.

[21] I. Kant. *Critique of Pure Reason*. Wiley Book, New York, 1943.

[22] I. Kant. *Critique of Practical Reason*. Hafner, 1986.

[23] C. Koch and I. Segev, editors. *Methods in Neuronal Modeling: From Ions to Networks*. MIT Press, Cambridge, MA, 1998.

[24] A. Mehler. A Multiresolutional Approach to Fuzzy Text Meaning. A First Attempt. In A. Meystel, editor, *Proceedings of the 1996 International Multidisciplinary Conference on Intelligent Systems: A Semiotic Perspective*, volume I, pages 261–273, Gaithersburg, 1996. National Institute of Standards and Technology.

[25] A. Mehler. Components of a Model of Context-Sensitive Hypertexts. *Journal of Universal Computer Science*, 8(10):924–943, 2002.

[26] A. Mehler. Hierarchical Orderings of Textual Units. In *Proceedings of the 19th International Conference on Computational Linguistics, COLING'02, Taipei*, pages 646–652, San Francisco, 2002. Morgan Kaufmann.

[27] A. Meystel. *Semiotic Modeling and Situational Analysis*. AdRem, Bala Cynwyd, PA, 1995.

[28] A. M. Meystel and J. S. Albus. *Intelligent Systems: Architecture, Design, and Control*. Wiley, New York, 2001.

[29] M. Minsky. *The Society of Mind.* MIT Press, Cambridge, MA, 1988.

[30] C. Morris. *Writings on the General Theory of Signs.* Mouton, The Hague, 1971.

[31] A. Newell. Intellectual Issues in the History of Artificial Intelligence. In F. Machlup and U. Mansfield, editors, *The Study of Information.* J. Wiley, New York, 1983.

[32] C. S. Peirce. *Collected Papers of Charles Sanders Peirce.* Harvard University Press, Cambridge, MA, 1935-66.

[33] R. Penrose. *Shadows of the Mind.* Oxford University Press, Oxford, 1994.

[34] L. I. Perlovsky. Gödel Theorem and Semiotics. In *Proceedings of the 1996 Conference on Intelligent Systems and Semiotics,* volume 2, pages 14–18, Gaithersburg, 1996.

[35] L. I. Perlovsky. Mathematical Concepts of Intellect. In *Proceedings of the World Congress on Neural Networks,* pages 1013–1016, San Diego, 1996. Lawrence Erlbaum Associates.

[36] L. I. Perlovsky. Towards Quantum Field Theory of Symbol. In *Proceedings of the 1997 Conference on Intelligent Systems and Semiotics,* pages 295–300, Gaithersburg, 1997.

[37] L. I. Perlovsky. Conundrum of Combinatorial Complexity. *IEEE Transactions on Pattern Analysis and Machine Intelligence,* 20(6):666–670, 1998.

[38] L. I. Perlovsky. Cyberaesthetics: Aesthetics, learning, and control. In *STIS'98,* Gaithersburg, 1998.

[39] L. I. Perlovsky. Emotions, Learning, and Control. In *Proceedings of the International Symposium on Intelligent Control, Intelligent Systems & Semiotics,* pages 131–137, Cambridge, MA, 1999.

[40] L. I. Perlovsky. *Neural Networks and Intellect: Using Model-based Concepts.* Oxford University Press, New York, 2001.

[41] J. Piaget. *The Psychology of the Child.* Basic Books, 2000.

[42] J. P. e. Pickett, editor. *The American Heritage College Dictionary.* Houghton Mifflin, Boston, MA, 3rd edition, 2000.

[43] S. Pinker. *The Language Instinct: How the Mind Creates Language.* Harper Perennial, 2000.

[44] S. Pinker. *Words and Rules: The Ingredients of Language.* Harper Perennial, 2000.

[45] K. Pribram. *Languages of the Brain.* Prentice Hall, 1971.

[46] B. B. Rieger. Empirical Semantics II. A Collection of New Approaches in the Field. In *Quantitative Linguistics,* volume 13. Brockmeyer, Bochum, 1981.

[47] B. B. Rieger. Situation Semantics and Computational Linguistics: Towards Informational Ecology. In K. Kornwachs and K. Jacoby, editors, *Information. New Questions to a Multidisciplinary Concept,* pages 285–315. Akademie-Verlag, Berlin, 1995.

[48] B. B. Rieger. Tree-like Dispositional Dependency Structures for Non-propositional Semantic Inferencing: A SCIP Approach to Natural Language Understanding by Machine. In B. Bouchon-Meunier and R. Yager, editors, *Proceedings of the 7th International Conference on Information Processing and Management of Uncertainty in Knowledge-based Systems (IPMU-198)*, pages 351–358, Paris, 1998.

[49] T. A. Sebeok. *Sign: An Introduction to Semiotics.* University of Toronto Press, Toronto, 1995.

[50] R. S. Westfall. *Never at Rest: A Biography of Isaac Newton.* Cambridge University Press, Cambridge, 1983.

[51] L. A. Zadeh. Information Granulation and its Centrality in Human and Machine Intelligence. In *Proceedings of the 1997 Conference on Intelligent Systems and Semiotics*, pages 26–30, Gaithersburg, 1997.

[52] S. Zeki. *A Vision of the Brain.* Blackwell, Oxford, 1993.

Part VI

Cognitive Modeling

A Cognitive Systems Approach to Automatic Text Analysis

Gert Rickheit and Hans Strohner

Bielefeld University
{Gert.Rickheit,Hans.Strohner}@uni-bielefeld.de

Summary. By regarding cognitive aspects, some shortcomings of traditional accounts of automatic text analysis can be avoided. In particular, at least the aspects of world knowledge, the interaction between text and reader and the impact of the communicative situation should be included. With regard to verbal information, a cognitive system is able to process a text by relating the text information to world knowledge and situational demands. On the basis of this interaction, the system produces inferences, which may lead to text analysis, text evaluation and communicative responses. As a core component of automatic text analysis, we present a cognitive theory of inference building. According to this theory, textual inferences are the product of an intimate interaction of verbal input and world knowledge in certain contexts. Without such inferential abilities, automatic text analysis is severely restricted. In order to prove this claim, we present some examples from various research projects.

1 Introduction

In his ambitious enterprise to build up a comprehensive theory of cognitive text processing, Burghard Rieger has demonstrated that we need the close collaboration of all cognitive disciplines to reach this goal. Not only linguistics, but also philosophy, psychology, computer science and neuroscience must contribute their specific perspectives and experiences. As an integrative framework, Rieger uses the dynamic systems theory. On this basis, he is able to incorporate the important cognitive aspects of situatedness, grounding and embodiment into his theory. In empirical demonstrations, Rieger was able to show how the various components of a cognitive system work together in order to achieve language understanding. Since cognitive representations are central to language understanding, Rieger starts his theorising by clarifying the concept of representation. To put it in his own words:

"Modeling *semiotic cognitive information processing* (SCIP) systems' performances, the concept of *representation* is considered fundamental. To realize – instead of

G. Rickheit and H. Strohner: *A Cognitive Systems Approach to Automatic Text Analysis*, StudFuzz **209**, 389–399 (2007)
www.springerlink.com   © Springer-Verlag Berlin Heidelberg 2007

simulating – the experimental distinction of semiotic *processes* (of cognition) from their *results* (as representational structures) is – due to the traces these processes leave behind – a process of emergence of discernible forms of (interpreted) structures as *acquisition* of *knowledge*. Computational semiotics embarks on the venture to (re-) construct algorithmically these emergent structures from natural language discourse which lie at the base of cognitive processes and are representational for them." [7, p. 398].

We agree with Burghard Rieger on many of his theoretical and empirical points. Specifically, we support his interdisciplinary approach to the complex problem of text understanding. In Rieger's and our opinion, systems theory is a proper conceptual basis for the integration of cognitive science. Our contribution to this volume can be seen as an attempt to strengthen the ecological part of Rieger's theory. First, we will give an overview of our approach to an ecosystemic view on text understanding. Then we will present two empirical examples in order to support our arguments. Finally, we conclude that the cognitive systems approach is helpful in building up a cognitive theory of text analysis.

2 An Ecosystemic View on Text Understanding

When people try to solve a problem, they usually use their knowledge of already familiar events. Often they use metaphors as a basis for the analysis. Scientific metaphors are certainly very helpful, but not without risk. Like all metaphors, they only partially reveal the characteristics of the unknown subject. Other characteristics may be more or less concealed or even deformed.

In his book *Metaphors of mind*, Robert Sternberg [9] starts with the thesis that scientists are sometimes unaware of the exact nature of the metaphor underlying the research, and may even be unclear about the particular set of questions that their metaphor generates. In order to help scientists to recognise their own type of approach, Sternberg analyses the following internal and external metaphors of mind on the basis of their major motivating questions:

- *Internal metaphors:*
  - *Geographic:* What form does a map of the mind take?
  - *Computational:* What are the information-processing routines (programs) underlying thought?
  - *Biological:* How do the anatomy and physiology of the brain and the central nervous system account for intelligent thought?
  - *Epistemological:* What are the structures of the mind through which knowledge and mental processes are organised?

- *External metaphors:*
  - *Anthropological:* Which forms does intelligence take as a cultural invention?
  - *Sociological:* How are social processes in development internalised?

After discussing and criticising these approaches, Sternberg presents his triarchic theory of human intelligence as a possible way that crosscuts metaphors. He calls the triarchic theory a systemic approach which combines the internal and external worlds of the individual. In addition, a specific subtheory relates intelligence to the experience of the individual with tasks and situations. The components of intelligence are manifested at different levels of situational tasks which may vary in relevance to a person's life.

Some of the most popular metaphors for cognitive text understanding are the computer metaphor, the brain metaphor and the ecosystem metaphor. The traditional computer metaphor refers to a classical computer system with a central processing unit and a separate memory. According to the brain metaphor, human understanding is modelled as a neural network with many simple units, which combine processing and memory functions. However, human understanding is more than activation flow in the computer or in the brain. A basic feature of human understanding is its relation to the physical and social environment. Here is the starting point for the ecosystem metaphor [5].

In cognitive science more and more researchers agree to the integration of inner and outer views on cognition [7]. In order to emphasise this integrative function of systems theory we use the term *ecosystem* [5]. The notion of ecosystem refers not only to the processes inside the cognitive system but also to the external processes in its environment. Biological ecosystems are units of organisms and their natural environments. If only a single organism with its environment is focussed, the ecological analysis results in a description and explanation of the interactions of this single organism with its environment.

This individual level of analysis is highly relevant for a cognitive approach. In addition to biological processes such as nutrition, metabolism and reproduction, cognitive processes such as perception, thinking and learning are included. From an ecosystemic point of view, cognitive processes are not only internal events at the representational level, but also interactions between the system and its physical and social environment. Thus, not only are perception, thinking and learning the topics of cognitive science, but situated perception, situated thinking and situated learning also are.

The ecosystem metaphor of text understanding refers not only to the processes within the brain, but also to the sensory-motor connections of living systems to their environment and to the impact of the environment on the systems. From an ecological point of view, text understanding is the product of evolutionary processes, which aim at a better adaptation to the environment. On the basis of this position, much of the research into text understanding has to be criticized.

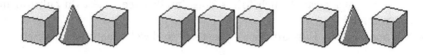

Fig. 1. A sample array of experimental objects.

With respect to text understanding, the cognitive system has to combine structural and functional aspects of the overall communicative situation including the text. In the course of discourse processing, the state of the cognitive system is subject to rapid change. Bottom-up and top-down processes interact quickly. In text understanding, processing often is initially data driven. Knowledge-based processes will be activated selectively when required. Central to knowledge-based processes in text understanding are inferences [4, 6]. In the following empirical examples, we focus on two aspects of discourse inferences. First, we have a look at referential inferences. Then we discuss how compositional inferences are built during text understanding.

3 Referential Inferences

Without referential understanding communicative text processing is impossible. Therefore, research into referential inferences is central to any comprehensive theory of cognitive text processing. However, the majority of current empirical approaches are theories of coreference accounting of the way people use words to refer to other words. This dominance of theories on coreference is probably due to methodological factors. Experimental studies on coreference need only control linguistic material, whereas studies on external reference also have to take into account the presentation of pictorial stimuli and, in addition, the relationship between linguistic and pictorial information in a certain communicative setting.

In order to take a closer look at the effect of pictorial, verbal and situational information on referential inferences, we carried out some simple experiments [10]. Since we were going to study inferential processes, we used referentially ambiguous materials. Reference resolution was possible by recourse to the discourse focus. In the experiment repeated here, subjects were asked to mark a potential referent presented pictorially. Every picture contained seven objects from one category (e.g. cubes) and two of a different kind (e.g. pyramids). The nine objects were arranged horizontally in three groups of three. An example is given in figure (1).

As the goal of the experiment was to gather initial information on the role of some relevant linguistic, communicative and cognitive factors in the resolution of ambiguous reference, the following factors were systematically varied:

- *Description type:* A syntactic-semantic factor that has attained some prominence in the linguistic discussion of reference is *indefiniteness*. Conceivably, a definite description causes a reader to employ a strategy aimed at processing one single referent entity. However, this strategy will not suffice in the case of referential ambiguity. Hence, we expected to find a difference in processing between definite descriptions (*Please mark the cube*) and indefinite descriptions (*Please mark a cube*). In the case of ambiguity, reference should be less successful with definite than with indefinite descriptions.

- *Goal orientation:* From a pragmatic point of view, ambiguity constitutes an obstacle to the mutual establishment of reference. Recipients may take referential ambiguity as a minor or a major flaw, depending on their orientation toward the communicative goal. With a cooperative attitude, violation of the singularity constraint could be more easily pardonable than with a critical attitude. Accordingly, we induced different levels of goal orientation by varying instructions: Under tolerant conditions, subjects were asked to mark the intended referent "even if the expression seemed inappropriate". Under critical conditions, they were asked "to indicate an inadequacy in the expressions". We expected goal orientation to interact with description type: Given definite descriptions, tolerant subjects should be more successful in reference than critical subjects.

- *Focus:* Being an integral part of cognitive activity, reference will also be influenced by the outcome of preceding cognitive processes. Prior experience could be a crucial factor in the resolution of referential ambiguity, since some of the potential referents might be preactivated by virtue of having been focused before. Thus, a cognitive account of reference will have to consider the individual's chain of foci. In order to disentangle primacy from recency effects, we systematically varied individual "focusing histories" by bundling reference tasks into trials of three. During the initial focusing task of each trial, we made subjects attend to one particular side of the picture by asking them to, e.g., *mark a cube on the left side* (or, alternatively, *on the right side*). Immediately afterward, during the second focusing task, again a particular part of the picture (the left side, the centre, or the right side) was put into focus by means of an appropriate instruction. According to the primacy hypothesis, the initial focus should exert the greater influence on the subject's behaviour when subsequently encountering an ambiguous situation in a target task. According to the recency hypothesis, the second focus should exert the greater influence.

- *Materials and procedure:* Pictorial and verbal materials were compiled in booklets, each one containing 36 reference tasks. The 36 tasks were grouped into 12 trials of three consecutive tasks each. Trials were separated by an extra sheet asking the subjects to turn the page over to proceed. In every trial, two focusing tasks were followed by one target task. In each task, the subject had to mark one of nine pictured objects in accordance with the corresponding verbal instruction. Pictorial stimuli were varied between trials, but were held constant within a trial; verbal stimuli, however, were varied between tasks in a trial. Thus, for example, with the accompanying picture of figure (1), one trial comprised three tasks such as the following (verbal materials translated from German):

1) First focusing task: *Please mark a cube on the left side.*
2) Second focusing task: *Please mark a cube on the right side.*
3) Target task: *Please mark the pyramid.*

The main results of the experiment were the following:

- Indefinite descriptions lead to more referential reactions than definite descriptions.
- With definite descriptions, subjects with a tolerant attitude show a higher percentage of referential reactions than subjects with a critical attitude.

Thus, there is evidence in favour of the hypothesis that, in an ambiguous situation, a definite description constitutes an obstacle to processing. Conceivably, recipients take definite descriptions as indicating that either the singularity constraint is met, or, at least, it is possible to single out a referent by taking additional information into account. Obviously, this inferential process agrees with an increase in processing load.

On the basis of the results at hand, the resolution of referential ambiguity should be viewed not as an all-or-nothing phenomenon, but as a matter of degree. In the establishment of reference, the cognitive system uses focused information for referential disambiguation depending on the specific communicative situation. As a consequence, focus adjustment could be modelled in a probabilistic manner.

Moreover, focus in referential ambiguity resolution is not a unidimensional, but a multidimensional phenomenon. Since multiple foci, such as the pictorial and verbal focus, may have an effect, one will have to consider the possibility of focus competition as well as mutual reinforcement of foci.

Finally, reference is not a stable state but a dynamic phenomenon. Reference changes with time, partially determined by contextual influences. Since every new input may alter the actual focus, it makes sense to study chains of foci. The dynamic processing and development of reference could be modelled, for instance, by the flow of activation in a system theoretical framework [1].

4 Compositional Inferences

From a cognitive systems perspective, research into compositional inferences is especially interesting because of its extreme flexibility. This flexibility is due to the fact that, on the one hand, words generally occur together with other words and thus illustrate the context dependency of language processing and, on the other hand, they often refer to actions and events and thus give evidence of the situatedness of language processing. These two aspects also seem to be the reason why, in the past, many experimental researchers have refrained from tackling compositional inferences more intensely. The variability induced by verbal context and social situations makes experimental testing difficult because an enormous number of factors has to be controlled. As a consequence, a great part of experimental research dealt with semantics of isolated nouns, thus following a research strategy which starts with the simple and already, at least, partially known, and only then turns to the more complex and unknown.

Recently, the situatedness of semantic composition has become more and more important to researchers. This specific change is due to a general move towards a broader range of issues in cognitive science, as for instance human-machine interaction, or the growing interest in various aspects of communication in natural settings.

One of the background issues for a cognitive theory of compositional semantics is the number of levels which contributes to language processing. While some researchers differentiate between a level closely related to the syntactic structure of the sentence and a conceptual level, others feel that the conception of a special linguistic structure in the processing of semantics is unnecessary and even misleading. From an empirical point of view, the immediacy of language processing, which has been shown in many studies, points to a highly integrated semantics. However, only precisely controlled experiments can give an answer to this theoretically motivated question.

The present study focuses on the semantic processes which occur during the combination of two concepts, as in the noun phrase *peeled apple*. During the comprehension of this phrase, the concepts of the participle and the noun may first be fully activated and then combined to a near conceptual entity. Alternatively, the combination process may be achieved in one step. If this second alternative is true, it can be an argument for the one-level theory of semantics, in which quick inferential processes play a central role [11].

When a participle is combined with a noun, as in *peeled apple*, a new cognitive complex emerges. In order to achieve the new conceptual structure, several processes must occur very quickly. Two theoretical views of the combination process have been discussed recently:

- The traditional *autonomous theory* views the composition of two concepts as a two-step process. During the first step, the concepts are constructed separately from each other, e.g. the isolated meaning of *peeled* on the one hand, and the isolated meaning of *apple* on the other are activated. During the second step, the two concepts are combined and some semantic features are changed, e.g. the colour of the apple turns from red to white.
- The *interactive theory* maintains that the combination of the two concepts starts immediately after the second concept has been encountered. According to this view, the concept of *apple* is activated with respect to

the concept *peeled* from the very first moment. Therefore, its colour is white and not red.

Springer & Murphy [8] reported empirical evidence in favour of the interactive theory. Working with a verification task, they presented their subjects with assertions such as *Peeled apples are white* and *Peeled apples are round.* Immediately after presentation, subjects had to decide as quickly as possible whether the assertions were true or false. The reaction times showed that *Peeled apples are white* was verified faster than *Peeled apples are round.* This means that an emergent feature of conceptual combination was activated earlier than a permanent feature of the noun concept. Springer and Murphy interpreted this result as a confirmation of the interactive theory. In a partial replication of the Springer and Murphy study, Strohner & Stoet [11] obtained results with German-speaking subjects, which are in good agreement with those of the original study.

According to this empirical evidence, the interactive theory comes off better than the autonomous theory. However, as Strohner and Stoet criticised, it remained unclear what exactly the interaction processes look like. Since verification reactions took over 1500 ms, different types of processes may occur during this time interval. Firstly, in a study with event-related brain potentials Kounios & Holcomb [3], besides activation processes, also discuss processes which have the function of evaluating the coherence of the emerging semantic structures. Secondly, one might think of inhibition processes which deactivate those semantic features not in the actual focus of attention. Thirdly, with regard to the autonomous and interactive models of semantic compositionality, there could be a first phase of composition, during which the autonomous model is true, and a second one, during which the autonomous processes are replaced by more interactive processes.

In sum, the verification method is too slow to be able to provide an insight into the composition processes, which would be differentiated enough for a process theory of semantic composition. These standards can only be met by observing the processing states at different times and by taking into account different types of semantic features. In order to get closer to the composition dynamics, Strohner & Stoet [11] used the lexical decision task and the naming task in two further experiments. Both tasks are well-established methods in psycholinguistic experiments and are able to show subtle activation processes.

In both experiments, eighteen German participle-noun combinations (e.g. *geschälter Apfel*, English: *peeled apple*) are presented word by word (after 100, 500 or 1000 ms) on a computer screen. Shortly after the onset of the noun, one of three different target words were presented: For the example of *geschälter Apfel* these target words were *rund* (*round*), *rot* (*red*) and *weiß* (*white*). The three target items represented permanent features (e.g. *rund*), canceled features (e.g. *rot*) or emergent features (e.g. *weiß*) of the critical object (e.g. *Apfel*). In a control study, the permanent, canceled and emergent features were proven to bear a clear relationship to the respective object. In addition, the three types of target words did not differ in their word length.

The results were strikingly similar in the lexical decision task and the naming task. After an interval of 100 ms between noun onset and presentation of the target word, the canceled features were activated faster than the permanent and emergent features, whereas no significant differences could be found between the three feature types for the intervals of 500 ms and 1000 ms. These results suggest that in forming concept combinations a lot of additional knowledge and inferences are used. However, the resulting structure of the combined concept seems to be directed by interaction with actual task demands. Without such a selective function of the environment, the network of conceptual units will remain more or less unstructured.

From these results we conclude that the interactive theory, which was confirmed by the verification studies, should be modified. According to our results, the entire composition process comprises several cognitive dimensions. Verification of the compositional features may be one dimension, and activation of these features another. Depending on the observation methods used, researchers will focus more on one or the other of these two crucial dimensions. In a similar vein, Khalidi [2] distinguished various concept theories, which were developed by using different empirical procedures. If the activation and evaluation aspects are combined, it may be possible to construct an integrative theory of semantic composition.

5 Conclusion

As already mentioned, some mentalistic shortcomings of traditional approaches may be avoided by analysing language processing in the framework of cognitive systems [5, 7]. According to this theory, a cognitive system comprises not only the mental processor but also the situation, which the processor perceives and to which it reacts.

Since the semantic knowledge of a cognitive system relates the linguistic to the object information, its structure consists of the following three complex relations:

- the *verbal concept*, which relates the linguistic knowledge to the conceptual knowledge,
- the *reference*, which connects the concept with certain entities in the situation models, and
- the *semantic composition*, which connects these models with other models with which they form higher-order semantic units.

Only when all three semantic dimensions are taken into account is a complete description of the semantic knowledge possible from a cognitive point of view. Concepts, references and semantic compositions are not abstract entities which only appear in mathematical formulae, they are concrete systems with concrete inputs and outputs. As concrete systems they have specific architectures, they change their states of activity and they are created by learning experiences.

Some of the main processing principles of such systems are the following:

- *Immediacy:* The strategy of immediacy ensures that all knowledge relevant to a linguistic unit is activated as quickly as possible. The cognitive processor does not wait until the end of the phrase or sentence before starting the inferential machinery.
- *Situatedness:* Cognitive systems do not only consider verbal input information but also the impact of the context and the communicative situation. Only those knowledge-driven process are activated, which are necessary to reach a coherent model of the referential information.
- *Sense constitution:* From a communicative point of view, language processing aims at integrating verbal knowledge not only into the already existing world knowledge but also into the social knowledge of the communication partners. Usually, a text is regarded as making sense if it fits into the social knowledge and the function of the particular communication.

By means of these processing principles the cognitive system is well equipped to deal with complex linguistic information, despite its limited working memory. It is also able to handle many of the semantic problems occurring in everyday language. The research into text understanding shows that theories in this domain have to take into consideration the specific architecture of the dynamics of the human cognitive system. Theories of cognitive semantics must, therefore, include notions of semantic processing in all three dimensions of the semantic system.

The favoured cognitive strategy for dealing with the questions of semantic text analysis is the close cooperation of linguists with researchers of other disciplines in cognitive science. Semantics is one of the major topics also in cognitive psychology, philosophy of mind and artificial intelligence. In a cognitive view, linguistic semantics is just a special type of general information processing. It is an important task for future research to describe its general characteristics as well as its special characteristics in the domains of referential and compositional inferences.

References

[1] K. Kessler, I. Duwe, and H. Strohner. Grounding Mental Models: Subconceptual Dynamics in the Resolution of Reference in Discourse. In G. Rickheit and C. Habel, editors, *Mental Models in Discourse and Reasoning*, pages 169–193. Elsevier, Amsterdam, 1999.

[2] M. A. Khalidi. Two Concepts of Concept. *Mind and Language*, 10:402–422, 1995.

[3] J. Kounios and P. J. Holcomb. Structure and Processes in Semantic Memory: Evidence from Event-related Brain Potentials and Reaction Times. *Journal of Experimental Psychology. General*, 121:459–479, 1992.

[4] G. Rickheit and H. Strohner, editors. *Inferences in Text Processing*, volume 29 of *Advances in Psychology*. North Holland, Amsterdam, 1985.

[5] G. Rickheit and H. Strohner. Cognitive Systems Theory: A Discussion of the Leading Metaphors. In G. Altmann and W. A. Koch, editors, *Systems: New Paradigms for the Human Sciences*, pages 404–419. De Gruyter, Berlin, 1998.

[6] G. Rickheit and H. Strohner. Inferenzen. In G. Rickheit, T. Herrmann, and W. Deutsch, editors, *Psycholinguistik*, pages 566–577. De Gruyter, Berlin, 2003.

[7] B. B. Rieger. Semiotic Cognitive Information Processing: Learning to Understand Discourse. A Systemic Model of Meaning Constitution. In R. Kühn, R. Menzel, W. Menzel, U. Ratsch, M. M. Richter, and I. O. Stamatescu, editors, *Perspectives on Adaptivity and Learning*, pages 347–403. Springer, Berlin/Heidelberg/New York, 2002.

[8] K. Springer and G. L. Murphy. Feature Availability in Conceptual Combination. *Psychological Science*, 3:111–117, 1992.

[9] R. Sternberg. *Metaphors of Mind: Conceptions of the Nature of Intelligence*. Cambridge University Press, Cambridge, 1990.

[10] H. Strohner, L. Sichelschmidt, I. Duwe, and K. Kessler. Discourse Focus and Conceptual Relations in Resolving Referential Ambiguity. *Journal of Psycholinguistic Research*, 29:497–516, 2000.

[11] H. Strohner and G. Stoet. Cognitive Compositionality: An Activation and Evaluation Hypothesis. In M. K. Hiraga, C. Sinha, and S. Wilcox, editors, *Cultural, Rsychological and Typological Issues in Cognitive Linguistics*, pages 195–208. John Benjamins, Amsterdam, 1999.

System Theoretical Research on Language and Communication: The Extended Experimental-Simulative Method

Hans-Jürgen Eikmeyer, Walther Kindt, and Hans Strohner

Bielefeld University
{HansJuergen.Eikmeyer,Walther.Kindt,Hans.Strohner}@uni-bielefeld.de

1 Introduction

The following contribution presents experiences made with a system theoretical methodology within the frame of the Collaborative Research Center *Situated Artificial Communicators* (CRC 360) at Bielefeld University. Starting point for this methodology is, on the one hand, the belief that theoretically and empirically backed research on the complex subject of natural language communication needs a systematic and interdisciplinary integration of methods. On the other hand, this kind of integration is possible only on the basis of a system theoretical conception of linguistics, which combines the predominating structural analytical approach with a procedural analytical approach. The experts from CRC called this a change in paradigms.

The system theoretical idea of a model, which is the basis of the Bielefeld CRC, conceptualizes communication as the interaction of dynamic systems in a given situation using linguistic utterances. The standard empirical setting used in CRC consists of interactions about a construction which are examined and modeled. In these interactions a constructor has to put together the model of an airplane using parts from *Baufix* while relying solely on the verbal instructions of the instructor. In this way we really have a specific system theoretical constellation: the verbal output of the instructor functions as input for the constructor and is processed in dependency on the external situation and the mental state. At the same time the constructor reacts to the input with his own output by making a construction and/or linguistic utterances which are, if necessary, perceived and processed as input by the instructor.

As an example of the explanation of the system theoretical methodology tried out here, the procedure of the partial project *Strategies for the Securing of Understanding* in the CRC was chosen. The subject of examination of this project is the verbal instructions in which the instructor and/or constructor undertake specific linguistic or mental activities in order to arrive at successful communication (coordination of meaning) for the respective construction

H.-J. Eikmeyer, W. Kindt and H. Strohner: *System Theoretical Research on Language and Communication: The Extended Experimental-Simulative Method*, StudFuzz **209**, 401–417 (2007)
www.springerlink.com   © Springer-Verlag Berlin Heidelberg 2007

using the verbal instruction. The goal of this project is, for one, the formulation of hypotheses about regularities in the use and linguistic realization of communicational strategies; and secondly, the experimental checking of these hypotheses and their validation in simulation or to make them usable for the construction of artificial communicators.

Which procedure seems most effective in which sequential steps for reaching this goal? In contrast to the usual experimental methodology in psycholinguistics we conducted intensive structure analytical research on the basis of an existing theory of understanding [3, 4] in order to find ecologically valid hypotheses. Consequently, we talk about an *Extended Experimental-Simulative Method*. The first step includes the gathering of an extensive corpus on communication in which the phenomena to be looked at can be observed using manifestly linguistic types of utterances or in which reliable interpretative inferences can be made in regard to this phenomenon on the basis of certain linguistic activities. In the project *Strategies for the Securing of Understanding* we used the option of searching for manifestly verbal strategies for the coordination of meaning. A structure analytical reason for the finding of hypotheses takes place regardless of the specific question according to this principle: it can be ascertained that in the given corpus the output of one interlocutor with the property E1 very often is followed by a second interlocutor's output, which has the property E2. In order to gain a first hypothesis which is formulated as precisely as possible an attempt should be made to find out under what contextual constraints E1 is followed by E2.

Moving beyond the customary approach in conversation or discourse analysis, the structure analytical finding of hypotheses today has the possibility of checking and, if needed, modifying the hypotheses found in the first step using a machine evaluation of corpora. A machine evaluation of corpora needs to be available to annotation as a prerequisite. For this it would be convenient if the respective contextual conditions and features could be identified using formal linguistic indicators.

The use of the comparatively costly experimental methods for checking the hypotheses makes sense only if both structure analytical steps of the examination are used optimally for the formulation of the hypotheses. The specific facts and conditions in an experimental procedure are known from psycholinguistics and need no further explanation here. It should be said that neurolinguistic methods are used increasingly in the Bielefeld CRC. The familiarity of these procedures is due to the fact that in the end transitions between states and reactions in human input-output systems are conditioned neurophysiologically.

It should not be assumed that a single experiment suffices to confirm or falsify the hypothesis to be tested. Very often it cannot be explained sufficiently to what extent the experimentally varying contextual conditions are responsible for the dependency between E1 und E2. Furthermore, there are indications that there are other still unconsidered relevant factors. In such

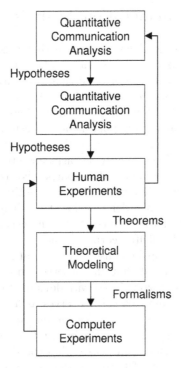

Fig. 1. The flow-chart of the extended experimental-simulative method.

cases it can be useful to modify the hypotheses accordingly and check them again corpus-linguistically before conducting the next experiments.

The experimentally confirmed hypotheses are used in the next step, possibly also using the already available theories on the system interaction to be examined, in order to construct a theoretical model for the system behavior in question. Generally it can be assumed that not all relevant influential factors from the corpus or the experiments conducted are included or controlled. Therefore, very often certain theoretically and/or empirically founded intervening variables are added to the model building in order to arrive at an explanation of wider scope. This allows for a generalization of the validity of the model drafted in contrast to the setting within the corpus and the experiments as well as to certain contexts with differing situational conditions. Of special interest, but also problematic, are the assumptions about intervening variables in respect to the inner states of the systems and their changes.

The final step in the systematic development of a theory is the simulation. It checks whether the laws and assumed conditional constellations underlying the model constructed are sufficient for an explicit and formal modeling of the empirically established system behavior or whether the hypotheses have to be made more precise and explicit.

2 Qualitative Communication Analysis

Research in conversation analysis has shown that successful communication between people relies to a great extent on the interactive coordination of meaning. In this respect it is necessary that the participants in communication know the interactive communicative strategies used in order for successful communication to be modeled. Usually, linguistic discourse research deals with corpora consisting of spontaneous speech. Such corpora have the disadvantage that the communication in these corpora is inhomogeneous in regard to several points: The underlying tasks for interaction and the related communicative expectancies vary and the parts taken over by the communication partners can therefore be defined very differently, meaning that different topics are addressed, etc. In order to allow for a higher comparability and a better generalization of the results of the analysis, it would be useful to work with experimentally elicited corpora. The aspects to be looked at in the communicative behavior can be considered natural in these corpora as well. Therefore, it can be assumed that the observed interaction for the *Baufix*-construction in the CRC-corpus in no way deviates from a "natural" way of behaving. At the same time this corpus serves as a good data basis for the question posed in the project *Strategies for the Securing of Understanding* because instruction dialogues call for a higher level of understanding. This again leads to numerous side sequences with a coordination of meaning. The twenty-two dialogues in the CRC-corpus were transcribed using a relatively easy transcription system and then analyzed for communication problems and their solution using a classification system which was developed in the project. Within the framework of this system it was checked among other things to see whether the participants consider the respective communication problem to be a difficulty in formulating or understanding, and which grammatical form of a sentence is used to solve the problem (e.g. a suggestion in the form of a statement or question), and whether several alternatives for solving the problem are offered, etc. This type of analysis inevitably shows circumstances that occur more frequently or, in the best case, even certain dependencies which form the starting point for the formulation of hypotheses. The constructor in the CRC-corpus, for example, very often uses an inference introduced by the conclusive conjunction *so* (*also* in German) resorting to what he has understood as a strategy and in this way testing the instructor and the degree of success in understanding. Also striking was the fact that questions for clarification from the constructor were usually formulated as alternative questions (using the conjunction *or*) or in form of *wh*-questions (using the question pronoun *which*). This condition suggested checking which type of question depends on what factors. Relatively soon it became clear that the number of referential objects for a suitable interpretation of an utterance is a relevant independent variable.

3 Quantitative Communications Analysis: Computer Assisted Analysis of Linguistic Corpora

The computer assisted analysis of linguistic data uses an annotated corpus of linguistic data as its search space. The core of this data is a transcription of the observed verbal behavior or speech. This information is enriched by additional information which characterizes selections of the verbal behavior in a purpose dependent way. An unlimited number of annotations can be attached to a stretch of speech. The analysis combines search facilities for both the core data and the annotations. In regard to the first, a full text search with regular expressions is used in our system. It is more powerful than a string search since a regular expression describes not only a single string but a set of strings. In regard to the second, annotations assign a finite number of properties to a selection of verbal behavior. The kind of information covered in an annotation is the central property and other properties are assign by attribute-value-pairs, where the attribute subspecifies the aspect of the information given by its value.

In the framework of this CRC 360 we looked at how interlocutors can ensure that they understand one another. They can do so either in a prospective way when they (try to) make sure that no problem arises, or they can do so in a retrospective way when they already have a problem. In construction dialogues where an instructor tells a constructor how to manipulate certain objects, a basic problem is to make sure that both talk about the same objects. In case of an object identification problem due to a lack of information on the side of one of the interlocutors this can be tried to be solved with a request for clarification. Several strategies can be applied in such a case, but, based on eclectic analyses, we were able to formulate an initial set of hypotheses. For their formulation we use the following terminology:

> Depending on the situation, a *specific* or an *unspecific* request for clarification can be formulated. Specific requests for clarification can be a *proposal* (a possible partial object description adding more information to the information already available) or a *list* of two proposals connected with *or*. Unspecific requests for clarification do not name a certain possibility. They are *open questions*, preferably in form of wh-questions.

Hypothesis 1 *The relevant situational parameter is the number of possible reference objects: in case of exactly two objects an or-question is used while a wh-question is used for more than two possibilities.*

This set of hypotheses then claims that the constructor's request for clarification in regard to an underspecified object specification is along the lines of C1, C2, or C3:

I: *take a long bolt*
C1: *the red one?*
C2: *the red one or the yellow one?*
C3: *what colour?*

In order to validate this hypothesis in the corpus one can, in a first step, apply a full text string search for the word *or* in order to identify all possible specific requests for clarification and annotate those occurrences which really are such cases. A full text regular expression search for wh-words (they are actually "*w*"-words only in German and occur in inflected forms as e.g. "*welcher, welche, welches*") will find all occurrences of these words and allow for the classification of those which are really unspecific requests for clarification. Proposals like C1 can not be searched for and have thus to be annotated the hard way. Simple search thus helps find candidates for expressions of a special kind and annotate them accordingly.

Once annotated, one can then search for annotated selections of speech:

- Show all specific requests for clarification
- Show all proposals
- Show all open questions asking for "size«»/"color«»/"length«»" information

Moreover, Boolean combinations of search patterns of the kinds mentioned can be used.

Table 1. The distribution of specific requests for clarification in relation to the number of possible objects.

Number of Possibilities	All requests	C1 cases				C2 cases		
	Abs.	%	abs	% of		abs	% of	
				All requests	All proposals		All requests	All lists
One	126	28	101	**80**	34	3	2	6
Two	193	42	128	66	43	32	17	**65**
More	129	28	67	52	22	14	11	29
Sum	458		300			49		

Table 1 shows some simple results of the numerical analysis. Eighty percent of all requests for clarification found in situations with only one possible object are proposals. This is a significantly higher percentage than that of all proposals. Lists with *or* are the preferred way of formulating requests for clarification in situations with two possible objects (65%). The percentage with respect to all requests is relatively small (17%), but it is significantly higher ($Chi^2 = 6,99$) than the percentage with respect to all requests. The relatively small number of lists is possible due to the relatively high planning activity needed to produce them. These results are completely in line with the hypotheses formulated above.

The annotation as to form and strategy of a request for clarification is only the starting point for a more detailed analysis, cf. Rittgeroth et al. [8].

4 Human Experiments

As a crucial part of the *Extended Experimental-Simulative Method*, experiments fulfill several important research functions:

- They serve as a *link* between *Quantitative and Qualitative Communication Analysis* and the *Theoretical Modeling* part of the method.
- They clarify the *relationships* between the various variables involved in the research design.
- They give valuable insight into the *mental processes* underlying the observable behavior of the subjects.

For a detailed discussion of these research functions see section 7.

In order to illustrate these functions of the experimental component we will describe an experiment carried out as part of a larger research project (cf. Kindt et al. [5]). Specifically in the present study we focussed on the influence of referential ambiguity and time pressure on question strategies. Our hypotheses based on the theoretical approach of situated understanding (e.g. [3, 4]) included in addition to the effects of a semantic factor (see Hypothesis 1 above) also a pragmatic factor:

Hypothesis 2 *Time pressure has a significant influence on the question strategy: The instruction to react as fast as possible results in more specific descriptions than no such instruction.*

The experimental procedure consists of a game of cards between two persons: an experimental confident and a subject without knowledge about the experiment. The cards show arrays of four objects, which can be combined in groups of two or three objects, e.g. two large hexagonal bolts, one small hexagonal bolt, and one small round bolt (see Figure 1). On each trial the confident names a specific object in the array which the subject has to identify. If the confident says "the small bolt", there are only two alternative target objects. However, if the confident says "the hexagonal bolt", there are three alternative target objects. If the subject is not sure about the object intended, she or he is encouraged to ask a question for clarification. The type of these questions for clarification is the dependent variable in the experiment.

The whole experimental procedure was tape recorded. In addition, the experimenter documented potentially relevant behavior of the subject.

The results were in line with the two hypotheses given above (cf. Table 1). A referential field of two objects resulted in more specific questions (e.g.

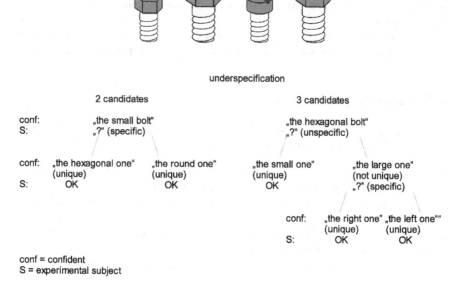

Fig. 2. An experimental situation.

"Do you mean the hexagonal or the round one?") than did a referential field of three objects. In this case, for instance, a preferred question was "Which one do you mean?" Equally, time pressure resulted in more specific questions. These main results of the experiment with human subjects can be transformed into theorems which might serve as an input for the theoretical modeling component of the method.

These theorems are also compatible with the results of *Communication Analysis* of authentic questions for clarification in task-oriented dialogues. Even in cases in which the results obtained from the observation seem to be in conflict with the results from the experiment, a more finely grained analysis was able to show the relevance of the experimentally demonstrated strategies.

Table 2. Main results in absolute and relative numbers of question types.

time pressure	alternatives	specific	unspecific	others	sum
no	2	105 (50,0%)	64 (30,5%)	41 (19,5%)	210 (100%)
	3	98 (42,1%)	93 (39,9%)	42 (18,0%)	233 (100%)
yes	2	119 (63,3%)	48 (25,5%)	21 (11,2%)	188 (100%)
	3	124 (55,1%)	71 (31,6%)	30 (13,3%)	225 (100%)

5 Theoretical Modeling

The theoretical modeling component is the central link between the human and computer experiment components. A theoretical model integrates the experimentally confirmed theorems into a coherent system, which relates the independent variables to the dependent variables. In most cases this is possible only if certain intervening variables between independent and dependent variables are constructed. These hypothetical instances and their functional relations form the creative part of the model and often give reason for critical discussions.

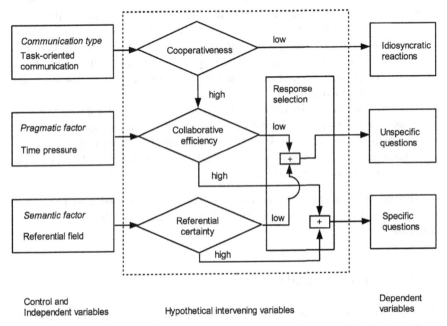

Fig. 3. A simple model of the observed question strategies with its independent, intervening, and dependent variables and its main activation routes.

In order to illustrate the theoretical modeling component we present a simple model of the described scenario for questions for clarification. The model is based on the two theorems which are the main results from the experiment with human subjects. In addition, the fact that there was no tendency towards an interaction between time pressure and referential factors needs consideration. These two significant effects seem to be additive. The model comprises the following four hypothetical intervening variables (cf. Fig. 3):

- *Cooperativeness:* Since a necessary precondition for a successful task-oriented communication is cooperativeness, this intervening variable dominates the whole interaction process. Only if the cooperativeness is high enough, the two independent variables of the experiment can function in a predictable way. Otherwise, the results will be idiosyncratic reactions of the subjects.

- *Collaborative efficiency:* If the subject tries to react cooperatively and there is a moderately high time pressure, the subject should select the most efficient question strategy. Undoubtedly, the most efficient strategy is to ask a question which includes as much information as possible. In this case the cognitive effort may be higher, yet the efficiency of the question will also increase.

- *Referential certainty:* If the referential field consists only of two possible objects, the related knowledge of these two objects and their critical differences should be very good. This referential knowledge is an excellent precondition for planning a specific question, e.g. in the form of "X or Y?« If the referential field consists of three potential objects, the knowledge is more diffuse and the preferred question strategy might be something like "Which one do you mean?"

- *Response selection*: In order to combine the effects of collaborative efficiency and referential knowledge, the model needs an operator for the selection of the response. In figure 3, only the two main important combinations of effects are illustrated. The less important combinations can easily be added to the model.

During the final step of theoretical modeling the described model has to be transformed into a formal system and it has to be implemented as a computer program.

6 Computer Simulation

A model is a reduced and simplified description of a section of reality (cf. Eikmeyer [1]). There is a similar relation between the model and reality. This can be characterized by the fact that the model highlights the essential aspects while it neglects the inessential ones. In addition to this relation the connection between a model and its theory have to be kept in mind. This connection can be better understood through the three levels of description proposed by Marr [6] for information processing models. Such a process takes information as an input and turns it into an output. On the first level a computational theory has to be specified, i.e. a theory which describes what the transition from input to output aims at and why this is suitable. The latter means the specification of the necessary and sufficient conditions of the transformation. These are based on empirical evidence of the process to be described. The second level of description asks for both the representations of the information assigned to the input and the output variables. Moreover, it requires the specification of an algorithm for the transition in the formulation of which the intervening variables play a central role. These two levels can be called the model. Marr's third level, finally, deals with the physical realization of the algorithm. If this realization is done by a computer, it is called a simulation.

The model shown in Fig. 3 is an input-output system, in which the control and independent variables make up the input and the dependent variables make up the output to be specified on the first level. The intervening variables and their connections are used to formulate the algorithm for the transition from input to output. The algorithm is given in an intuitively understandable graphical depiction. A concrete model needs to be more specific, since on the second level representations for both the input and the output have to be specified as well as the details of the algorithm.

Once a model has been specified, it has to be evaluated, i.e. it has to be found out in what respect the model correctly describes reality and in what respect it is false. A computer simulation can be used as a tool for model evaluation. Tests can be repeated almost endlessly, all parameters can be modified and new hypotheses or predictions can be derived.

According to Popper [7] falsifiability is a minimal requirement for scientific models. Marr [6] claims that modeling has to aim at the specification of representations and algorithms. Johnson-Laird [2, p. 52] further adds, that "theories of the mind should be expressed in a form that can be modelled in a computer program".

7 The role of Human experiments in the Extended Experimental-Simulative Method

7.1 The Link Function

Communication Analysis results in a rich description of the communicative processes going on in the intended research field. Usually, this picture gets even more complicated by different results in different case studies. For theoretical modeling the information resulting from *Communication Analysis* is often too complex and too vague. What is needed is an evaluation procedure of the theoretical hypotheses which resulted from the interpretation of the observed behavior sequences. This evaluation procedure is contributed by the experimental method. It includes the following steps.

Hypotheses

Clearly formulated hypotheses are one of the first steps towards theory building. Hypotheses are the result of *Communication Analysis* and a necessary precondition for precise experimentation. Hypotheses are formulated as declarative expressions linking two variable groups of the research topic in the form of 'If group A has property E1, then group B has property E2'. Usually, the variable group A is termed the independent variable and B the dependent variable. In order to get the A-B relationship in a relatively undisturbed way other relevant variables C, which may influence it, have to be thoroughly controlled.

Research Design

On the basis of the selected hypotheses the research design spells out the variable types A, B, and C with respect to the experimental setting. If there is more than one variable in the independent group, a factorial design is given. Since in a factorial design the various interactions between the selected variables have to be considered, the number of variables should be restricted to a manageable size. Due to the risk of measurement interferences, the number of dependent variables also should be as low as possible. One of the most difficult experimental tasks consists in controlling the variable group C. If relevant variables are not included in the control, the design may end in an ecologically invalid situation which is not related to the authentic situation observed in *Communication Analysis*.

Interpretation of Results

The interpretation of the experimental results yield answers to the question whether the results contradict the hypotheses or not. In cases in which the hypotheses have to be rejected some interpretation work has to be done. What are the possible reasons for the failure? They may be found in the underlying hypotheses or in the experimental design. A crucial part of the experimental method is to give some tentative answers and, even more important, to give some hints about how these problems can be resolved. The overall criterion for these suggestions is their compatibility not only with the experimental results but also with the observations during *Communication Analysis*.

7.2 The "Causal" Analysis Function

Theoretical models consist of a logically consistent network of propositions. As already mentioned, experimental hypotheses are formulated to fulfill these requirements. The basis of experimental hypotheses is the conditional relation between the independent variables A and the dependent variables B with respect to the controlled variables C. Since the knowledge of conditional relations is a crucial precondition for intervening during the practical application of the theory, the confirmation of conditional relations is a central task for scientific research. Once established, the conditional relation or dependency between variables A and B can be interpreted in a more specific way: Often it raises the suspicion of a causal relation or the person modeling bases the assumption of causality on the neutral dependency relation, thus enhancing the theoretical impact. (Loosely speaking, conditional relations may be termed "causal", keeping in mind that the term "cause" is a philosophical term and is not dealt with in experimental research.)

7.3 The Mental Function

Somewhat similar to the discussion of the term "causal" analysis is the explication of the term "mental" analysis. By trying to explain the confirmed relations between the independent and dependent variables, researchers rely on certain intervening variables. However, the important question is how to interpret these hypothesized structures or processes. Are they believed to be real instances of mental life or do they serve only a formal function in order to connect the input and output of the observed organism? Researchers have to be careful not to fall back onto the ideologically based positions of mentalism or behaviorism. One procedure generally agreed upon in the experimental community is to stick closely to the operationally defined independent and dependent variables. Anyhow, experiments serve as a valid heuristic basis for formulating hypotheses and intelligent and well founded speculations on mental structures and processes.

8 Management of Linguistic Corpora

8.1 Aim and Functionality

The aim was to implement a software system for handling corpora of communicative interactions. From a linguistic point of view this requires the three following basic functionalities:

- *Transcription:* This means the possibility to represent an interaction between two or more persons in a written form. Interactions include both verbal and non-verbal means and that a representation of the language signal produced by the interlocutors is the core of the gathered data. It has to be ensured that arbitrary information can be added to the core data. The visualization of the data uses a score view with a number of voices for the interlocutor's language signal. Such a view easily codes the relation between an interlocutor and what she is saying and, moreover, depicts overlapping speech.
- *Annotation:* This means the possibility of adding meta information to the core data. This type of information is in no way limited with respect to what it is and how many perspectives on the core data it is representing. However, the meta information has to be formally structured to make it treatable by machines. We chose an attribute-value-based approach, i.e. all meta information is represented by attribute-value-pairs, where the attributes to be used have to be specified.
- *Analysis:* All information of the corpus, the language signal and all annotations, has to be accessible for an automatic mechanism which analyzes the data according to the user's requirements. These include full-text search

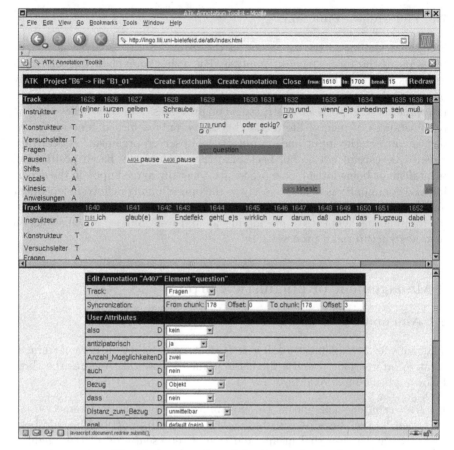

Fig. 4. A screenshot of the annotation tool.

as well as search for attributes, values, or attribute-value-pairs of annotations and combinations of both. All such queries are formulated in a special language.

From a purely practical point of view the two following functionalities were added:

- *Accessibility:* All data has to be easily accessible for a group of several researchers and the system has to be accessible from different places and from any software platform. A web-based approach is optimal for these means.

- *Re-usability:* All information of the annotated corpora has to be exportable in a suitable standardized output format in order to offer an interface to other systems. For practical reasons and based on the current state of the art in text technology we chose XML as our export language.

8.2 Data Model and Technical Aspects

A theoretical data model has to guarantee the formal integrity of the data. We needed a data model for a corpus which is a set of interactions. An interaction is represented by a directed acyclic graph. These graphs contain two types of nodes: those representing a single contribution of a speaker during the interaction – called *text chunks* – and those representing *annotations*.

Text chunks contain an ordered sequence of minimal elements. According to the language transcribed these may be words, morphemes, phonemes, or anything else that might be suitable. All text chunks are related by their respective positions: they start somewhere in another chunk, mostly at its end, but possibly somewhere in the middle. An edge in the graph has this position as the value of its start-attribute. Overlapping speech is thus easily represented.

Annotations are coded similarly: they have two edges pointing to a text chunk and the start- and end-attribute of the respective edges code the relative position of the annotation with respect to the text. Both types of node have a set of attribute-value-pairs attached to them. The set of admissible attributes has to be defined by the user.

Technically, the system uses a client-server architecture. User interaction happens via a web browser communicating with a dedicated web server using the HTTP-protocol. The CGI-interface starts PERL-programs which themselves communicate with a data base (for the persistent storage of the data) and a Prolog-engine (for data analysis). The theoretical data model was mapped onto a relational data base scheme covering both the graph model and the attribute-value based information. In addition to algorithms for data handling, a transformation interface was implemented, which shows the language data in a score view with HTML. A query language was designed for data analysis.

9 Conclusion

The *Extended Experimental-Simulative Method* can be looked at from several perspectives. We will start with a justification for it from a narrower linguistic perspective and then turn to the broader context of cognitive science in general.

The methodology proposed above tries to clarify which significance the methods used for structure- and process-analysis in the different branches of linguistics have for an integrated system-theoretic development of models. These methods are – in contrast to common appreciation and practice – not to be regarded as concurrent but as complementary. Our project was able to show that the analysis of the structure of communication has to combine communication analysis (qualitative and quantitative) with grammar-

and semantic-theoretical methods. At the same time, the relevance of postulated structures in language and communication can be backed up by process-analytic studies of the systems involved. On the other hand, psycholinguistic experiments might have little impact if they are not based on a differentiated language- and communication-analytical fundament.

The future development of cognitive sciences depends not only on progress in theory construction but also on methodological innovation. We need new methodological concepts and procedures which will contribute to a better integration of the cognitive subdisciplines. This paper presents one possible strategy of relating some of the disciplines to each other. Specifically, we propose that communication analysis, experimental research, and computer simulation cooperate in order to build up a new integrated method named *Extended Experimental-Simulative Method*.

The core of this new approach consists of a close relation between human and machine experiments. The obligatory link between these two types of experiments results from precise theoretical models, which can be formalized in adequate computer programs. If the classic experimental-simulative method is applied to complex discourse, it must be extended. The methods of communication analysis have to be included in order to relate the human experiments to ecologically valid discourse. Since authentic corpora are hard to analyze due to their complexity, computer assisted analysis of these corpora is added to the method.

The *Extended Experimental-Simulative Method* is not only able to contribute to the methodological integration of cognitive sciences but can also form a basis for theoretical progress. If all authors in cognitive sciences could agree that good theories should be transformed into formal models, which are to be confirmed in human and computer experiments, then efficient criteria for theory testing would be available. In our opinion these criteria will be better met by system theoretically derived models than by other types of theory.

References

[1] H.-J. Eikmeyer. Simulative Methoden. In T. Herrmann and J. Grabowski, editors, *Sprachproduktion*, pages 51–70. Hogrefe, Göttingen, 2003.

[2] P. N. Johnson-Laird. *The Computer and the Mind*. Fontana, London, 1988.

[3] W. Kindt. Konzeptuelle Grundlagen einer Theorie der Verständigungsprobleme. In R. Fiehler, editor, *Verständigungsprobleme und gestörte Kommunikation*, pages 17–43. Westdeutscher Verlag, Opladen, 2002.

[4] W. Kindt. Koordinations-, Konstruktions- und Regulierungsprozesse bei der Bedeutungskonstitution: Neue Ergebnisse der Dynamischen Semantik. In A. Deppermann and T. Spranz-Fogazy, editors, *be-deuten. Wie Bedeutung im Gespräch entsteht*, pages 34–58. Stauffenburg, Tübingen, 2002.

[5] W. Kindt, H. Strohner, and K. Jang. Rückfragestrategien bei referentieller Ambiguität: Ein Beispiel von Bedeutungskoordination im Diskurs. In I. Pohl, editor, *Prozesse der Bedeutungskonstruktion*, pages 357–374. Peter Lang, Frankfurt a. M., 2002.

[6] D. Marr. *Vision: A Computational Investigation into the Human Representation and Processing of Visual Information*. Freeman, New York, 1982.

[7] K. Popper. *Logik der Forschung*. Mohr, Tübingen, 4 edition, 1987.

[8] Y. Rittgeroth, S. Birkemeier, K. Poncin, and W. Kindt. Rückfragen als Strategie zur Verständigungssicherung. Technical Report 2001/5 SFB 360: Situierte Künstliche Kommunikatoren, available as http://www.sfb360. uni-bielefeld.de/reports/2001/2001-5.html, Universität Bielefeld, Fakultät für Linguistik und Literaturwissenschaft, May 2001.

Part VII

Visual Systems Modeling

The Dimensionality of Text and Picture and the Cross-Cultural Organization of Semiotic Complexes

Wolfgang Wildgen

Bremen University
wildgen@uni-bremen.de

1 Two Basic Moods of Representation

The distinction between picture and text involves a set of basic semiotic challenges. First, pictures are linked in their production to the motoricity of hands, in their receipt to the eye and the visual cortex. Language in its basic form, spoken language, is linked in its production to the motoricity of the human vocal apparatus (from the vocal cords to the lips) and in its perception by the ear and the auditory cortex. The dynamics of these four subsystems and moreover the coordination of the pairs of subsystems in production and reception define the base line of any comparison of picture and text. The fact that written texts map the characteristics of spoken texts onto the dynamics of hands and eyes (to abbreviate the more complete description above) points to the fact that transitions between the two basic modalities have been achieved in the last millennia. If we take abstract signs of the Palaeolithic as point of departure (cf. Wildgen [32, 34]), this (cultural) evolution has been running the last 30,000 years. An even deeper evolutionary opposition opposes manual/facial sign languages and spoken language. The origin of human language after the proto-language of Homo erectus was basically a dominance shift from a slower and less rich system, at least partially based on visual/motor articulations, to a much quicker and richer systems of phonetic/auditory articulation (cf. Wildgen [33]). We have no direct knowledge about the sign language of Homo erectus, but we may guess the characteristics of such a manually based language, if we consider modern signed languages. Due to the use of the manual/visual mode, they show, in spite of being constructed in parallel to existing phonetic languages, characteristic deviations (cf. Emmorey [5], and Lidell [10]). The most characteristic differences concern the diversity of parameters and the relevance of gradient subsystems. As [24] summarizes, spoken language has as major parameter the recombinant system based on phonetic quality. A set of further parameters of vocal dynamics (loudness, pitch, timbre, etc.) rather contribute emotional, social-semiotic and paralin-

W. Wildgen: *The Dimensionality of Text and Picture and the Cross-Cultural Organization of Semiotic Complexes*, StudFuzz **209**, 421–442 (2007)
www.springerlink.com

guistic contents. In contrast to spoken language, the manual/visual systems of American Sign Language (ASL) uses for its classifier systems seven groups of mostly gradient parameters (with at least 30 variables). This example shows that the (spoken) text is heavily dependent on discrete, recombinant dynamics, whereas utterances in a sign language exploit a whole set of gradient systems.

The "language of pictures" is again different from visual/manual sign languages because the manual activities are not restricted to directly observed motions of the hands but rather to the effects of the instrumental use of hands applied to colors, textures via certain media, like a canvas or others. Therefore, the comparison between texts and pictures concerns at least two levels that must be separated:

- the evolutionary basic distinction between a mode of manual/visual and a mode of phonetic/auditory communication,
- the application of manually based techniques to materials (with color or luminosity differences) and media (canvas, paper, glass, etc.) applying specific instruments (crayon, paint-brush, chisel etc.) is basic for products of the visual arts. This applies also to writing. As a consequence, pictures refer to a highly developed culture of materials, techniques and media (for writing cf. Wildgen [32, chapter 5]).

Our analysis will not consider the details of artistic techniques and media, but will concentrate on the issue of the (basic) dimensionality in the semiotic product: a phonetic or written text versus a two dimensional picture.

If de Saussure's axiom that language is basically a linear structure is correct, and the dominant temporal organization of spoken language as well as the linear arrangement of all systems of writing (if their function is not primarily decorative) argue in favor of this basic assumption, then it is fundamentally different from the symbolic organization of pictures (e.g. paintings, photos) and even more different from sculptures and architecture. The dimensionality of any organization of signs is a basic determinant of its structure; it imposes other degrees of freedom and asks for other restrictions. Nevertheless, one has the intuition that sentences/texts and pictures have many features in common, they can cooperate, interact as in comics, illustrations, emblems, etc., they may be translated into one another and, what is more crucial, they respond to a cognitive system with perception, motor-control, memory and imagination which is presupposed as more or less identical in the speaker and hearer and independent from the specific semiotic modality she/he chooses. A model called *Semiotic Cognitive Information Processing* (SCIP) has been put forward by Rieger:

"[It] will allow for a pictorial representation of the semantic space structure as computed from the text corpus describing the real world situations" [18, p. 390].

In the case of mixed text/picture sign-structures different inputs from an external reality – called "exo-reality" in Rieger [18] – have to be mapped into a unitary "semantic space structure". Similar problems of integration starting from sensorial inputs with different spatial, temporal and dynamic characteristics arise with all *symbolic forms*, e.g. language, art, myth (religion), techniques, ethical rules (laws), economic rules of exchange (monetary systems) – cf. Cassirer [3], Sandkühler & Pätzold [21] and Wildgen [32, chapter 10]. In the case of text/picture the questions, therefore, are: What (spatial/temporal/dynamic) organization do text and picture have in common (we shall later expand the comparison to sculpture and architecture)? Can we conceive a semiotic framework in which both the similarities and the differences between text (sentence) and picture are mapped?

2 How Valid is Saussure's Axiom about the Linearity of Language? A First Confrontation with Pictures

The linearity of language is much more controversial than Saussure's axiom makes us believe. The parallelism of phonetic features and in general of features that may describe the phonological, morphological, syntactic, semantic information found in language has become common opinion in information based models [17] and in "Head Driven Phrase Structure Grammar" (HPSG). In these models vectors of features are combined into matrices and thus a multidimensional descriptive space is constructed. The tree-graphs of constituent and valence analysis also have an analytic second dimension given by the descriptive labels attached to constituents or valence bound entities. Even if these supplementary dimensions are the consequence of descriptive devices and therefore artificial, there exists a phenomenological multi-dimensionality in intonation and paralinguistic information which points to parallel information channels and interaction (structural binding) between these channels. One could argue that parallel lines of coding are still linear and the interaction may be restricted to points of coordination. Another complication which has to be considered is the direction of the process of speaking/reading. The simplest model was that of a linear automaton which from left to right reads one element, replaces it or not and goes on. The discussion in Chomsky [4] on the format of generative (production) grammars already showed the restrictions of a linear (unidirectional) automaton. Not only do discontinuous constituents occur, the action of the automaton must consider information given in specific places in the sequence already passed; in the case of a "garden path", it must even go back to a specific place and redo the analysis. Thus, the linearity of language must accept moves back and forth and the range of these moves depends on the information given in single places (constituents). Even future places not yet reached by the analysis may be relevant. As a consequence the linearity of language has a dominant direction (on the time axis) but it also has a memory of *relevant* places in the past and reacts to structural places not yet reached (but asked for, necessary to come). In figure (1) I try to give a

a) Unidirectional process

b) Dependence on past or future steps of the process

c) Garden path and reanalysis

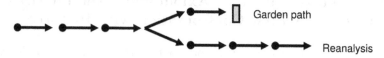

Fig. 1. Major deviations of the unidirectional linearity of language.

schematic (not a precise) picture of the kind of linearity involved in language processing (mainly in language production).

Another counterargument to linearity in language processing could be the duality of *syntagmatic* and *paradigmatic* relations [8, p. 59]. The syntagmatic relations correspond roughly to the contents of figure (1). The paradigmatic relations open a field of choices or variations (cf. Wildgen [26] for a systematic analysis of semantic and pragmatic variation). At any moment, and before any move on the line is made, the speaker may consider a set of possibilities allowed by the prior choices. As the sequence does not strictly determine this choice, the set of alternatives at any moment opens a second dimension (of freedom) governed by morpho-syntactic rules and semantic pragmatic choice criteria. The choice made is responsible for the style, the literary quality, and the rhetorical effect of the uttered sequence. This is a real challenge to Saussure's axiom of linearity, although the control and selection of variation and choice points to the dominant linearity of language (forward and backward fitting) and thus tells us that a model of language processing is predominantly linear. The difference between syntagmatic and paradigmatic relations refers to the distinction between on-line production (working memory) and choice out of a subset of permanent "pieces" (long term memory). The scope of this subset depends on the available lexicon (which is static) and the current process stage (which is dynamic). The stylistic choices open a multi-linear field of associations and form-meaning correspondences which is definitely not subjected to the strict linearity of linguistic production (consider the difficult choice made by a poet). I consider therefore stylistic (and esthetic) variation

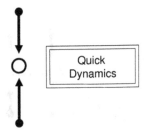

Fig. 2. A point like attractor.

to go beyond the linearity of language and to restrict the validity of Saussure's axiom.

In the case of pictures (e.g., drawings or paintings) it is immediately obvious that the eye which reads and the hand which draws/paints the picture operate basically in two dimensions. Even in the case of drawing where the hand performs linear moves, these have many different directions, i.e., they have many orientations in two dimensional space. In paintings, linear (directed) strokes may be performed (and even be visible as in van Gogh's paintings), but what is dominant is the composition of a surface out of subsurfaces; therefore the composition is not linear and neighborhood of colors and shapes is defined in two dimensions.[1] If a perspective is constructed, a third dimension in space is simulated and hierarchical orders may exist, i.e., many phenomena mentioned in the discussion of linearity in language reappear again if we analyze pictures. Nevertheless, pictures are grounded in a two-dimensional structure. If words, sentences, texts have characteristic linear boundaries, pictures have two-dimensional boundaries. Thus, the shape of the frame: be it rectangular, quadratic, circular, oval, etc., and all the dynamics inherent in a picture are influenced by the fact that they end at this border-line or start from it.

We may generalize this basic insight: If the dimensionality of a phenomenon is equal or smaller than one, $d \leq 1$, then we either have a point ($d = 0$) or a line ($d = 1$).[2] The dynamics are locally either stable, i.e. they are controlled by a point-like attractor or they perform a transition by passing a border or fold line, as shown in figure (2).

The first dynamics (of stability or state) are called quick (instantaneous) dynamics, the second are called slow (unfolding) dynamics. The singularities of the slow dynamics are called bifurcations. The fold is the first and simplest bifurcation pattern; it creates or abolishes a state of stability and thus defines

---

[1] Cf. Wildgen [28, part two], where neighbourhood in a cellular automaton is discussed and applied to describe the dynamics of narrative structure.

[2] If we consider fractal dimensions, e.g. $0 < d < 1$, we may arrive at the Cantor set, i.e., a line broken into line-segments or even points.

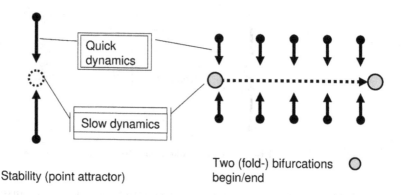

Fig. 3. Two basic types of dynamics (quick and slow) and the fold-archetype: birth/death.

a stable process with limitations in one dimension (co-rank or number of state variables = 1); cf. figure (3).

3 From one Dimensional Language to Two-Dimensional Pictures

We shall first consider the one dimensional space (of language) which is devoid of linguistic signs, i.e., silence, and the two-dimensional space (of pictures) which is blank. What kind of dynamics may we find in this "virgin"-situation? In the case of language one may infer that silence just began (communication stopped) or that silence will finish (communication is just about to start). The void linear space has implicitly a vector pattern of begin/end, as illustrated in the central window of figure (4).

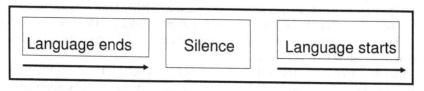

Fig. 4. Virtual dynamics of "silence" in communication.

In the case of a frame without picture, a blank canvas, the situation is more complex. The "linear" space of silence of language has as its correlate a (denumerable infinite) set of regular surfaces (I neglect non- or semi-regular surfaces): the equilateral triangle, the square, the regular polygons with $5, 6, 7, \ldots, n$ corners. For simplicity sake, I shall just consider the square.

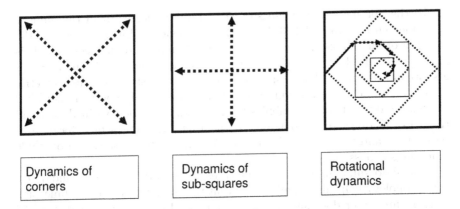

| Dynamics of corners | Dynamics of sub-squares | Rotational dynamics |

Fig. 5. Virtual dynamics of a square (diagonal, horizontal/perpendicular and spiral force-lines).

What are the dynamics of a square frame (without picture)? It is obvious that the corners, the diagonals which link them, the regular grid of squares which may compose it, or may be inscribed or circumscribed define a whole family of implicit paths and thus dynamical potentialities. Figure (5) shows this basic observation and adds typical paths (for further comments on the "pictorial base space" cf. Saint-Martin [20, chapter four]).

The comparison of language and picture shows that the transition from $d = 1$ to $d = 2$ leads to a dramatic increase in the potential dynamics: A frame without picture (pictorial "silence") has a complexity which goes far beyond that of a one-dimensional linguistic "silence". Before we begin to fill the void spaces, we should ask, if this increase of latent structure continues steadily with $d = 3$, $d = 4$, ... The answer which already impressed Plato (or his dialogue partner Timaeus) is that the story does not go on as one would guess. The major reason is that, although we find infinite regular polygons, we only find five regular polyhedrons (the Platonic solids), in 4-space we find six regular hypersolids, in 5-, 6-, 7-space only three [23, p. 91]. This means that the dimensionality does not induce a monotonic increase in the number of basic forms, on the contrary it involves restrictions which reduce this number. In order to complete somewhat the argument (which cannot be followed in detail here) one has to consider, that there is still a steady increase in the number of corners (and therefore of implicit dynamic fields, cf. above):

- two end points in a line segment,
- four corners in a square,
- eight corners in a cube,
- sixteen corners in a four-dimensional cube,
- 32, 64, 128, ... corners if we increase further the dimensionality of the cube (cf. Stewart [23, p. 90f.]).

To the non-monotonic increase (even dramatic decrease) in the number of regular entities ($d = 1, 2, 3, 4, \ldots, n$) corresponds a dramatic clash in the stability of unfoldings (process-types) with different (internal) co-rank. This is the heart of Thom's classification theorem (cf. Wildgen [27, p. 7-18] for a short introduction). The cuspoïds (co-rank 1) correspond to the geometric dimension 2 and have an infinite (denumerable) set of types; in co-rank 2 these are reduced to three umbilics (corresponding to Klein's dihedrons) and three E-unfoldings (corresponding to the Platonic solids which have three basic types and two duals). With co-rank 3, no finite classification of stable unfoldings is possible.

This short summary of basic regularities discovered in geometry and differential topology helps us to understand that the transition from one dimension to two, three, four does have dramatic structural consequences and it would be a silly mistake to believe that one has just to add some more features to the body of results obtained in the case of one-dimensional structures (e.g., language) in order to describe pictures which are basically two-dimensional. Another silly argument would be that if pictorial structures are very different (qualitatively different) from linguistic ones, one should just forget the results of linguistic analysis and begin the analysis of pictures *ex ovo*, as if they had nothing to do with language. In both cases do we have symbolic forms (cf. Cassirer [3] and Wildgen [31]) and basically these symbolic forms use the same perceptual, mnemonic and imagistic resources. The dimensionality is therefore the key to the difference between language and pictures. The common (cognitive) base of both modalities allows for the blending of linguistic and pictorial signs and their contribution to one universal type of human understanding.

4 Implicit Force-Fields and the Organization of Content

The space of silence in language may be filled by a sentence (we simplify the real processes). It inherits the borders of this space such as: beginning/end and is governed by relative probabilities in a linear sequence, i.e., the set of possible first constituents and dependent on it of second constituents, etc. The production grammars put forward since Markov's first proposals (by Harris, Chomsky, and others) elaborated this basic idea (and added the special cases of context sensitivity, transformation, reanalysis, etc.). I will just take this tradition as given and ask how a similar process may look like in the case of picture-production/analysis.

First, we have seen that even the ideal paths in a square (let alone non-ideal or chaotic paths) are multiple. I have mentioned the diagonals, the square grid (vertical and horizontal symmetry lines) and a spiral moving from the outside to the center or vice-versa.[3]

---

[3]In the late sixteenth century Giordano Bruno (1548-1600) made a similar analysis of mnemo-technical systems based on square grids and paths in a structured square (cf. Wildgen [29, p. 140,170]).

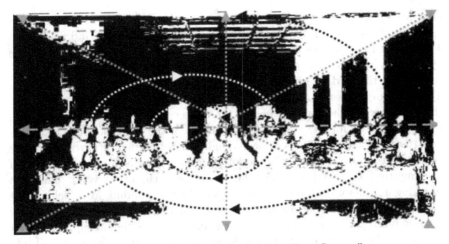

Fig. 6. The force fields in Leonardo's "Last Supper".

In producing a picture (on a void surface) these force-fields are relevant and they depend naturally on the shape of the picture (be it rectangular, square, circular, elliptical, etc.). A strong preference is given to rectangular frames which are near to the ideal (the square) but introduce a basic asymmetry.[4] If we take the painting the "Last Supper" of Leonardo da Vinci (cf. Wildgen [30, 35] and [32, chapter 6]), the prominent table of the supper fills the basic horizontal line and Christ marks the intersection with a vertical line of symmetry. The diagonals correspond to the slightly deformed lines of perspective (see the ceiling and the tapestry at left and right) which produce the illusion of three-dimensionality. Figure (7) reconstructs the basic force-fields.

As this example shows, all three force fields we analyzed in the case of a void frame are used to organize specific contents (surfaces, figures, persons in space) in Leonardo's fresco. The head (ear) of Jesus is at the center of all force fields. The sub-centers of the groups of apostles lie in the intersections between the horizontal axis and the symmetric spiral which end at Jesus' head (ear). The rectangle of the whole fresco breaks the symmetry of the (ideal) square.[5] The perspective generates a subdivision of the background space into three equal zones. In the central zone are situated: Jesus, John (at the right of Jesus), and Thomas, James Major (at the left of Jesus); Judas is already outside of this field although he has the second position at the right of Jesus. Peter and Philip are at the intersections of these fields. Geometrically we

---

[4]An asymmetric ideal is defined by the "golden proportion" based on the irrational number $\frac{1}{2}(1 + \sqrt{5}) = 1.61803399\ldots$.

[5]If the sides a, b of the rectangle fulfil the golden proportion, it can be subdivided into smaller rectangles fulfilling the same proportion ad infinitum; under this aspect this rectangle is also "ideal". The fresco fills the whole breadth of the dining room in Santa Maria delle Grazie in Milan. The proportion is roughly 1 : 2 and thus not in the golden proportion (roughly 1 : 1.662).

have a blending of two orders: the symmetrical subdivision of the group of apostles into $6 + 6$ and $(3 + 3) + (3 + 3)$ and the three background fields with Jesus and three apostles in the middle and four apostles at the right and the left (Judas has his arms on the table and thus sits in a plane nearer to the spectator); this order is basically $4 + 4 + 4$ (+ Judas in a frontal position). The table organizes the spatial distribution of the persons, which are all in the lower part of the frame (which is therefore vertically in a disequilibrium); the same is true for the trunks, heads, hands of the persons above the table and the feet below; there is a clear dominance of the body parts above the table. Thus, the geometrical rigor of the force lines is broken by a set of asymmetries. The information of the picture is at the first level of analysis a breaking and deformation of symmetries and corresponding force fields. This fits the idea of information as the history of deformation demonstratively focused on in [9]; i.e. the regularity of the base space is the presupposition for the generation of "information" on deformations, asymmetries inscribed into the (pre-informative) space.

Our analysis only considered the fundamental restructuring of a space void of content but structured by force-fields dependent on a frame. As soon as specific contents, i.e. a person or a configuration of persons, objects (e.g., flowers, fruit, dead animals in a "nature morte") or abstract configurations are introduced, these contents "graft" local spaces and dynamics upon the dynamically organized pictorial frame. Thus Jesus and his twelve apostles implant their own configuration into the painting. The new dynamical relations may be gravitational (the apostles sit or stand at the table), map events and actions (giving, holding something), or symbolic acts (gestures and/or glances). This means that the two-dimensional space contains several sub-spaces which introduce their own structure and dynamics into the picture. They may be coordinated by the overall structure but they still create conflicts, oppositions, deformations in the already deformed base space. The basic content complexes organized in Leonardo's painting are:

1. The table in the fore-ground.
2. The perspective of the dining room, the windows, the landscape visible through the window, the subdivision of the background into three equal sub-fields.
3. The arrangement of 12 apostles (grouped by 4×3) on both sides of Jesus.
4. The gestures (body poses) and glances of Jesus and his apostles superimpose a further dynamical structure.

The blending of these different content complexes constitutes the central message of the painting. At the same time it creates a pattern of structural layers which is not basically different from what we know about linguistic structures:

- The basic (linear) dynamics (start/end) of the sentence define a starting field where we often find a subject, a middle field where verbal constituents and valence governed noun phrases/pronouns are found and (sometimes) a closing field. The specific filling depends on the type of language and on the pragmatics of the utterance in question.

- The verb in the center of a valence pattern introduces a local space which partially controls the linear dynamics. The dynamical patterns of verb valences have been described in dynamic semantics [27, 28].
- In the periphery of valence patterns the hierarchically nested nominal, verbal and adjectival phrases complete the picture and adverbial modifiers or inflectional markers further specify the time/mode/aspect (TMA) of the central event/action reported.

In the case of classical paintings, which transport a narrative content and may be "translated" into a text or illustrate a given text, the basic organization of the painting adapts the patterns found in language to the conditions of a two-dimensional representation and its inherent dynamics (which are different from a linear pattern, although they can embed such patterns).

If we continue this line of thought to sculpture and architecture, new types of restrictions are added, which may overwhelm the patterns found in sentences and pictures. Thus the sculpture as a free-standing physical object is submitted to the gravitational force field. (The objects represented in the painting should not contradict our gravitational imagination but gravitation does not effect them directly.) Thus we may wonder, if Mary in Leonardo's painting of St. Anne may fall from St. Anne's lap, but in a corresponding sculpture gravitational forces may really destroy this unstable configuration. Therefore sculptors like Henry Moore formulated as the central aim of their art that sculptures must "stand" or "lie" naturally. In an architectural design the physical, technical restrictions become dominant, because a building must be statically and functionally "consistent". The domain of artistic freedom left for a semiotic message is therefore heavily restricted by static and functional considerations.

In the line of our framework, it would be a challenge to analyze abstract paintings where the "contents" seem to be absent. I presume that colors and shapes still have enough "content" that it remains a relevant problem how to fit them into the force-field of a void frame and even a void frame may attract specific contents and thus be "content-filled" by the sympathetic viewer. I shall turn to a last, more general question related to dimensionality: the compression or "flattening" of space and the role of dissipative systems.

5 The Reduction of Semiotic Complexity and the Coding of "Lost" Information

Semiosis is itself a dramatic selection and a reorganization which maximizes order and recurrence if compared to the non-semiotic world we may guess to exist behind all the diverse manners of semiosis (cf. the plurality of "symbolic forms" described by Cassirer).[6] As we have shown, similar real world situations (in a three-dimensional spatial frame) can be flattened to two dimensions in the picture and one dimension in language and music. The temporal continuum is broken into discontinuous segments not only in linguistic temporal categories, but also in comics, and to a further extent in the pictures and pixels which make up a film or a video tape. The sensory brain, memory, imagination add a huge diversity of features, qualities, characteristics to the spatial and temporal base-space, and thus "blow" it up to a number of feature dimensions which may have the magnitude of the set of adjectives or adjective pairs in a language (cf. Osgood [13] for the construction of a Semantic Differential). The phenomenon of "space-diffraction" asks for a strategy of compression. If the dimensionality of space-time (R^4) is reduced to three, four, two, one dimension we call this strategy "flattening". The operation underlying Osgood's Semantic Differential exploits the fact that the many dimensions of a quality-space exemplified in the lexicon of adjectives are statistically interdependent; they may be clustered, such that a few abstract dimensions are found (by factor analysis).[7]

From a more theoretical perspective one may consider complex dynamical systems to be either dissipative or chaotic. Compression means that either a few dominant "slaving" factors must be found or an algorithm must be formulated which generates the chaotic system (and has simple rules of generation). The first strategy is described in Nicolis & Prigogine [12], the second in Peitgen et al. [14].

Stewart [23, p. 94] compares the behavior of a dynamical system in n-space with a fluid. Motion in n-space is like a "whole bunch of initial points, moving along these curves". If the system is without fluctuation or friction, if it is "Hamiltonian", "then the fluid is incompressible" (ibidem). This means that the n-space cannot be reduced to $n - x$ dimensions. But natural systems have fluctuation, friction, statistical interdependence, etc., i.e., they are dissipative or far from thermodynamic equilibrium in the sense of Nicolis and Prigogine [12]. In this state of non-equilibrium, the probabilities of different states, the forces of different factors are dramatically divergent. It could be shown for many physical and non-physical systems that they are slaved by very few strong forces, which Haken [6] calls the "order parameters". The compression and reorganization controlled by these strong forces is called "self-organization" as the new order seems to emerge by itself (in reality the order parameters reduce all other forces to irrelevance and thus select a specific pattern which was invisible [but existent] before the effect of self-organization

---

[6]Cf. Wildgen [31] and [32, chapter 9]. The major types are: language, myth, science, art, techniques, ethics.

[7]I do not discuss more specific temporal, aspectual and modal information contained in sentences and texts (cf. Brandt [2, chapter II (Analyses du temps)]). The richness of these fields may constitute a basic difference in comparison with pictures, sculptures etc., although the medium of film recovers many of the specificities found in language with the means of a visual code.

occurred).[8] The "fluid" in n-space may thus be compressed to 2-space if there are two control factors which reduce all other factors to noise (for Haken the laser is the prototype machine which compresses all frequencies of light to one wave-length with high energy).

The chaos-theoretical strategy has as its prototype the multiple copying machine with transformations (scaling, rotation, mirroring, and other affine-linear transformations; cf. Peitgen et al. [14, chapter 5]). If the transformational and copying process is iterated, a fractal picture is created. All this complexity, however, is generated by a very small set of basic operations (and their iteration). This sketch of general ideas of "compression" in dissipative and chaotic dynamics is intended as background for the following suggestions.

In the case of language, the categorization of vowels is a straightforward application of self-organization. The two basic formants are the phonetic order parameters; phonological oppositions, e.g., in a language with only three vowels $[i][u][a]$, may select three prototypes and define corresponding frontier-lines (cf. Petitot-Cocorda [15, chapter 3]). This allows for a quick decision, say, for $[a]$; even if the articulators have many (≈ 30) degrees of freedom and the specific mix of frequencies changes with the speakers, the situations and the position of the vowel in a word. For shape recognition similar diffusion models have been proposed (cf. Petitot-Cocorda [16, p. 293-296]). In the case of pictures Mandelbrot [11] has demonstrated the fractal geometry of natural forms and his followers have created a fractal "art".

After these first illustrations of the mechanisms underlying dimensional "compression", I will come back to the comparison of picture/sculpture/architecture and language:

1. An architecture (3-space) is represented in an illusionist painting (trompe l'œil), but one can neither enter the room nor move before it without destroying the illusion.
2. A sculpture is represented in a mural painting; one part may be sculpted, the other painted. In a proper position against the wall, it may be difficult to grasp the difference between 3-space and 2-space.
3. A text describes a landscape, a building, a person either directly or as represented in a painting.
4. A sentence contains an action scenario (in 3-space + time) in its valence structure, e.g., "Eve gives Adam an apple in the garden Eden". The action in 3-space + time is flattened to a sentence with verb and case assignments/linear order.

In all examples, the basic 3-space with time is flattened to a 2-space (without time), or even to a 1-space (a sequential pattern). Is the other information lost in the compression or may it be recovered? I just enumerate some answers:

---

[8]For a programmatic linguistic application of these ideas cf. Wildgen & Mottron [36]. In 1983 Prof. Haken was the key-note speaker at the DGfS-congress in Bielefeld with Ballmer, Eikmeyer and Wildgen as co-referents. This was probably the first joint venture between linguistics and synergetics.

- The technique of perspective (rediscovered in the Renaissance) codes artificially for the third dimension; gestures, glances, frozen actions code for the temporal dimension (cf. the analysis of dynamics in Leonardo's "Last Supper" in [32]).
- The technique of valence patterns (control of NPs), case assignment, etc., codes for the spatial parameters and allows their flattening into a sequence of verb (V) + subject (S) + object (O), etc. (in different orders dependent on the type of langue: SVO, SOV, VSO, etc.).
- All the non-spatial or non-temporal dimensions are coded for by attributes/shape modifications/colors in a painting or by lexical differentiations in a sentence.

The compression thus leads necessarily to a system of coding levels which must be such that the most relevant (not all) features may be recovered. Common knowledge or context finally help to fully translate pictures into language and language into pictures.

6 Symbolic Creativity and Cross-Cultural Dynamics

In order to simplify cross-cultural comparison I shall stick to the example "Last Supper" or "Group of persons at a dinner-table". The second topic seems to be rather universal, but if we take "table" and "persons sitting at the table" it becomes clear that Plato's symposium, one of the archetypes of this situation, had the persons rather lie than sit, and the table, if present, was not prominent, and even in Leonardo's painting half of the apostles are either standing or raising. The table and the seats may be very low as in Japan or food may just be displaced on a carpet. Women may be present (one could argue, if in the real biblical events women were participating or not and who cooked and served the plates). Such variations are probably culturally relevant. For instance, the earlier prototype of a meal in Christianity was rather a meal given to poor members of the community in the house of a more wealthy person (agape), the separation of Judas from the other apostles was parallel to the medieval struggle against heretics, who were excluded from the communion (excommunicated); after the Reform wine became a more prominent element on the table, and in Italy in the 16th century the paintings show rather scenes of opulent and prestigious meals. In both cases the meaning component of sacrifice became rather secondary.

In the 20th century some novels and films have totally desecrated the "Last Supper". Thus Marco Ferreri in the film "La Grand Bouffe" (1973) or Peter Greenaway in "The Thief, the Cook, his Wife and her Lover" (cf. the analysis of Roelens in [19]) transform the biblical dinner into an extreme meal, where in the first case the participants kill themselves by eating, in the second case the meal ends with scenes of cannibalism. These short remarks show that the topic itself, the object of picture or text, is variable across cultures. The interpretation, the reading of these sign-structures will be even more relative. The only things properly conserved are:

1. The activity of eating with the mouth helped by hands and eventually instruments.
2. The size (approximate) of the human group that comes together for a common meal.
3. The central position of a leader.

The (cultural) variability concerns the room, table, chairs, dishes, the kind of food, and the composition of the group (its social structure). The linguistic variability (e.g., in translations of the biblical text) concerns the lexicon of items referring to this scene like verbs, nouns, adjectives, and their grammatical, morphological, semantic properties. At the lexical and syntactic level the field of verbs and the associated case-frames may be important structures that diverge culturally. In the case of the iconographic tradition related to the biblical episode of the "Last Supper", cultural differences are marked by the outfit and the physiognomy of the persons (Jesus, his apostles, bystanders), by the room and its furniture. The fundamental difference concerns the need for precision and concretion that is less urgent in a text than in a (classical) painting. In the following I shall analyze two different pathways of cultural variation:

1. The path of pictorial abstraction. I will show that it does not coincide with the kind of abstraction, which is fundamental for language.
2. The path of intertextual deformation, mainly in the direction of satire or parody.

7 Abstraction in Paintings

Andy Warhol has systematically assessed the topic of the "Last Supper" in a series of works in 1986. The different paintings [25] apply the technique of collage introduced by Max Ernst and a super realistic style known as Pop art. In this tradition he simplifies Leonardo's figures to contours, lines and monochrome surfaces. The collage assembles different parts of the painting, rotates and blends them. The dish on the table is remade in a different perspective. This variant of Leonardo's painting shows clearly the phenomenon of citation and of selective rearrangement. Nevertheless, it is not a satire and could be understood as a new proposal in the classical religious iconography (Warhol was a practicing Orthodox Catholic).

Another type of abstract presentation of the multi-person topic (as in the "Last Supper") was chosen by Paul Klee. In a painting of 1923 titled "Die Sternverbundenen" (persons linked by stars) a number of human bodies (4, 5 or more) are organized in a rectangular plane together with geometric surfaces ("stars"). Possibly one (partial) body in the center is nearer to the

Fig. 7. Andy Warhol, "The Last Supper", 1998 [25, p. 81].

viewer (larger) than the others, thus a perspective is still given. In general, the characterization of the persons is reduced to a minimum (head, trunk, four limbs). This corresponds roughly to a simple lexicon of human body parts; its central relation is called part/whole relationship (partinomy). The spatial disposition uses a lower zone with a baseline (where the persons stand or sit) and a higher zone, where the "stars" are distributed. The order of disposition is almost regular (one could superpose a grid with two rows and five (six) cells on each row. Thus, from the point of view of a two-dimensional composition (with a third dimension alluded to) this painting still respects the mode of pictorial organization valid for Leonardo. A complex narrative content like that in the "Last Supper" can, however, no more be represented in a painting in Klee's style.

The third example comes from the classical "abstract" painter Wassily Kandinsky. His first, totally abstract painting was probably "Komposition VII" (1913) or a water color which prepared it. The painting I have chosen is called "Rotes Oval" (Red Oval), and was made in 1920. The reason why I chose to comment on it is that it has a (deformed) rectangle (like a table) in its center and further objects on it, at it, around it (cf. Jesus and his apostles sitting at the table); it thus shows a formal correspondence to Leonardo's "Last Supper". Instead of persons, dishes, bread one can only distinguish color-surfaces. The most prominent one, the oval, has a clear geometrical contour and a vivid color (in a constant hue 'red'). This center could fit the role Jesus plays in Leonardo's painting. Rather compact color-surfaces are around it (8-10 different surfaces). A diagonal is marked from bottom-left to the right upper corner; this may remind us of Leonardo's perspective (its diagonal parts).

Fig. 8. Paul Klee, "Sternverbundene", 1923 [7, p. 135].

One could take many examples of modern paintings which allow for a rather formal comparison with Leonardo's painting. It is clear that neither Klee nor Kandinsky cite any content of Leonardo (whereas Warhol does), but there are basic laws of figural composition that are still in vigor in all these paintings. In more diverse cultures, some of these principles may not be observed and, as in the case of Chomsky's Universal Grammar (U.G.), one may ask what the common human base for pictorial expression is (what a U.P. = Universal Picture) is? As there are paintings dated to 40,000 B.C. (cf. Wildgen [33] and [32, chapter 6]) but no linguistic remnants of that age (not before writing was invented, i.e., 3,000 B.C.), it is much easier to answer the question for paintings (U.P.) than for language (U.G.). In any case, cultural variation must be seen in comparison with cross-cultural communalities or even with universal features.

8 Satirical Deformations

Any topic which has gained importance and prestige in a culture may be the object of a satire, of a comical deformation. This is true for biblical motives (perhaps more restrained by the control of the Vatican until the last century) but pervasive since the Renaissance for topics in antique mythology, literature,

Fig. 9. Wassily Kandinsky, "Rotes Oval", 1920 [1, p. 123].

philosophy, art, and even natural science (until the 17th century). The citation may have the character of a parody (pastiche) as the "The Last Supper" by Gradimir Smudja [22, p. 9], where Leonardo (as Jesus) sits on a table with different European painters at his left and right. The spatial frame, the table, the grouping of persons correspond to those in Leonard's "Last Supper". The principle at work is one of replacement: Take a given famous painting (here Leonardo's "Last Supper") and replace Jesus by Leonardo, the super-painter, and his apostles by painters who occupy a similar role in the history of art, but on a lower level. The satirical (even blasphemous) content consists in the comparison of Jesus and his apostles with a famous painter and the next level of painters in a hierarchy of fame. A more dramatic conflict between an artist who cites Leonardo's "Last Supper" and the religious authorities occurred in the case of Buñuel. In his film "Viridiana" a group of beggars organizes a

Fig. 10. A central picture in Buñuel's film "Viridiana".

meal (at the costs of the nun Viridiana who is absent) and poses for a photo, which cites Leonardo's "Last Supper".

The Vatican and, due to its authority, the Spanish government, banished the film for this citation. The author, Buñuel, however, contested the blasphemous interpretation. In fact he had only cited Leonardo and not directly the Bible. As the iconography of Leonardo's "Last Supper" shows, numerous painters in the 16th century had already transformed the topic into a banquet (Jacobo Bassano, Veronese and Tintoretto [35]). In general one can observe three basic lines of development:

1. In the case of a biblical narrative, the text is dominant (it has originated from God), the painting is understood as an illustration. Insofar as the biblical texts were not in everyone's hand, the paintings became a rather independent medium of communication, however controlled by the church. This control was politically still relevant in the 1960s, when Buñuel's film was banished (in Italy and Spain).
2. The topic "Supper" becomes independent from its textual source, and is considered for its own sake and may be related to a basic human experience. In this way, the pictorial tradition becomes an independent province of the fine arts. This may result in a kind of globalization ("mondialisation"). Maria with the child Jesus can now be understood by a non-Christian society as a representation of the intimate relation between mother and child; the "Last Supper" can be understood all around the world as a representation of a common meal (although there is a bias for societies, where males dominate).

3. The formal features of a specific iconographic tradition may survive as culturally relevant structures even in the case where the painting is devoid of meaning (in a narrative sense). Thus, even abstract paintings can continue a given iconographic tradition. An interesting question is if a textual tradition can make the same move towards abstraction that modern abstract art did. Dada-literature and concrete poesy followed this route, but they were much less successful. Thus one may assume that there is a fundamental difference between language and pictures related to the manners of abstraction. Language is from its origin on relatively abstract, because it has developed its referential function rather late, whereas pictures have always (even in the Palaeolithic period) possessed a range of phenomenological diversity between very vivid, realistic pictures and highly schematic sign-structures (cf. Wildgen [32, 34] for further discussions of this topic).

9 Conclusion

It is theoretically and empirically rewarding to analyze texts (sentences) and pictures as two different ways of solving the same semiotic problem: How can world-information be compressed into a basically low-dimensional representation? What is the subsidiary system of coding levels which allows for the reconstruction of an imaginistic 3-space (+ time) and thus for the understanding of the picture or the text (sentence)? At the heart of these questions lies the phenomenon of self-organization or order selection. Some aspects of the problem may be clarified if the generative mechanisms of chaos are studies with reference to pictures and texts (sentences) – others ask for a statistical model of semiotic self-organization. As a general result, the organization of content based complexes, i.e. of meaning, depends on a proper understanding of the dimensionality which dominates a given symbolic form (language, figurative art) and on the discovery of the basic coding strategies which are able to compensate the information loss due to dimensional compression.

The transformation of a textual topic in the history of modern art (Kandinsky, Klee, Warhol) showed that the specific type of bi-dimensional semantics of paintings based on lines and color-surfaces has been put into the foreground in modern art, whereas modern literature, although it partially took the same path, did not follow the trend towards abstraction in the same fashion. Thus, if we compare text and picture in the time of Leonardo and today we observe that on the one side linguistic texts are always less concrete, less spatially specified than pictures. On the other side, pictures allow for a radical type of abstraction, which is not (easily) accessible to texts. Concerning cultural diversity, one must distinguish rather superficial (conventional) differences. They show up in the lexicon and less in the grammar of different languages and in the décor and details of paintings. The question, if a U.G. (U.P.) of pictures and languages exists was formulated but not answered.

References

[1] U. Becks-Malormy. *Wassily Kandinsky. 1866-1944. Aufbruch zur Abstraktion.* Benedikt, Köln, 1993.

[2] P. A. Brandt. *Dynamique du sens. Études de sémiotique modale.* Aarhus University Press, Aarhus, 1994.

[3] E. Cassirer. *Philosophie der symbolischen Formen,* volume 1-3. Wissenschaftliche Buchgesellschaft, Darmstadt, 1923-1929.

[4] N. Chomsky. *Syntactic Structures.* Mouton, The Hague, 1957.

[5] K. Emmorey. *Language, Cognition and the Brain: Insights from Sign Language Research.* Erlbaum, Mahwah (NY), 2003.

[6] H. Haken. *Synergetik. Eine Einführung.* Springer, Berlin/New York, 2 edition, 1983.

[7] W. Herzogenrath, A. Buschhoff, and A. Vorwinkel. *Paul Klee – Lehrer am Bauhaus (Ausstellungskatalog, Kunsthalle Bremen, 30.11.2003-29.02.2004).* Hauschild, Bremen, 2003.

[8] R. Jakobson. *On Language. Edited by L. R. Waugh and M. Mouville-Burston.* Harvard University Press, Cambridge, Massachusetts, 1990.

[9] M. Leyton. *Symmetry, Causality, Mind.* MIT-Press, Cambridge, Massachusetts, 1992.

[10] S. Lidell. Sources of Meaning in ASL Classifier Predicates. In K. Emmorey, editor, *Perspectives on Classifier Constructions in Sign Language.* Erlbaum, Mahwah (NY), 2003.

[11] B. B. Mandelbrot. *Die fraktale Geometrie der Natur.* Birkhäuser, Basel, 1977/1992.

[12] G. Nicolis and I. Prigogine. *Exploring Complexity. An Introduction.* Freeman, New York, 1989.

[13] C. E. Osgood. *Focus on Meaning.* Mouton, The Hague, 1976.

[14] H.-O. Peitgen, H. Jürgens, and D. Saupe. *Bausteine des Chaos Fraktale.* Springer, Berlin/New York, 1992.

[15] J. Petitot-Cocorda. *Les catastrophes de la parole. De Roman Jakobson à René Thom.* Maloine, Paris, 1985.

[16] J. Petitot-Cocorda. *Physique du Sens. De la théorie des singularités aux structures sémio-narratives.* Editions du CNRS, Paris, 1992.

[17] C. Pollard and I. A. Sag. *An Information Based Syntax and Semantics,* volume 1: Fundamentals. CSLI, Stanford, 1987.

[18] B. B. Rieger. Semiotic Cognitive Information Processing: Learning to Understand Discourse. A Systemic Model of Meaning Constitution. In R. Kühn, R. Menzel, U. Ratsch, M. N. Richter, and I.-O. Stamatescu, editors, *Adaptivity and Learning. An Interdisciplinary Debate,* pages 347–403. Springer, Berlin/New York, 2003.

[19] N. Roelens. Cènes, Banquets et Festins. In S. Caliandro and A. Beyaert, editors, *Espaces perçus, territoires imagés en art,* pages 99–119. L'Harmattan, Paris, 2004.

442   Wolfgang Wildgen

[20] F. Saint-Martin. *Semiotics of Visual Language.* Indiana U. P., Bloomington, 1990.

[21] H.-J. Sandkühler and D. Pätzold, editors. *Kultur und Symbol. Die Philosophie Ernst Cassirers.* Metzler, Stuttgart, 2003.

[22] J. Spahr. *Paradies & Pastiches.* Christoph Meriam, Basel, 1991.

[23] I. Stewart. *Does God Play Dice? The Mathematics of Chaos.* Penguin Books, London, 1989.

[24] L. Talmy. Recombinance in the Evolution of Language. *Chicago Journal of Linguistics,* 39, 2005 (forthcoming).

[25] A. Warhol. *Catalogue Published for the Exhibition "Andy Warhol. The Last Supper", Munich 27 May – 27 September 1998.* Crantz, Ostfildern-Ruit, 1998.

[26] W. Wildgen. *Differentielle Linguistik, Entwurf eines Modells zur Beschreibung und Messung semantischer und pragmatischer Variation.* Niemeyer, Tübingen, 1977.

[27] W. Wildgen. *Catastrophe Theoretic Semantics. An Elaboration and Application of René Thom's Theory.* Benjamins, Amsterdam, 1982.

[28] W. Wildgen. *Process, Image, and Meaning. A Realistic Model of the Meanings of Sentences and Narrative Texts.* Benjamins, Amsterdam, 1994.

[29] W. Wildgen. *Das kosmische Gedächtnis. Kosmologie, Semiotik und Gedächtnistheorie im Werke von Giordano Bruno (1548-1600).* Peter Lang, Frankfurt a. M., 1998.

[30] W. Wildgen. Die Darstellung von Hand (Gestik) und Auge (Blick) in einigen Werken von Leonardo da Vinci. In *Proceedings of the Annual Meeting of the German Association of Semiotics,* Kassel, August 2002.

[31] W. Wildgen. Die Sprache – Cassirers Auseinandersetzung mit der zeitgenössischen Sprachwissenschaft und Sprachtheorie. In H. J. Sandkühler and D. Pätzold, editors, *Kultur und Symbol. Die Philosophie Ernst Cassirers,* pages 171–201. Metzler, Stuttgart, 2003.

[32] W. Wildgen. *The Evolution of Human Languages. Scenarios, Principles, and Cultural Dynamics.* Benjamins, Amsterdam, 2004.

[33] W. Wildgen. The Paleolithic Origins of Art, its Dynamic and Topological Aspects, and the Transition to Writing. In M. Bax, B. van Heusden, and W. Wildgen, editors, *Semiotic Evolution and the Dynamics of Culture,* pages 117–153. Peter Lang, Bern, 2004.

[34] W. Wildgen. Éléments Narratifs et Argumentatifs de l' "Ultime Cène" dans la Tradition Picturale du XIIe au XXe Siècle. In S. Caliandro and A. Beyaert, editors, *Espaces perçus, territoires imagés en art.* Harmadan, Paris, in press.

[35] W. Wildgen. *Time, Motion, Force, and the Semantics of Natural Languages.* Antwerp Papers of Linguistics. Universiteit Antwerpen, in press.

[36] W. Wildgen and L. Mottron. *Dynamische Sprachtheorie. Sprachbeschreibung und Spracherklärung nach den Prinzipien der Selbstorganisation und der Morphogenese.* Brockmeyer, Bochum, 1987.

Appendix

List of Contributors

Prof. Dr. Gabriel Altmann
Ruhr-Universität Bochum
Sprachwissenschaftliches Institut
(emer.)
Stüttinghauser Ringstr. 44
D-58515 Lüdenscheid, Germany
RAM-Verlag@t-online.de

Dr. Harald Atmanspacher
Institut für Grenzgebiete der
Psychologie und Psychohygiene e.V.
Department of Theory and Data
Analysis
Wilhelmstr. 3a,
D-79098 Freiburg, Germany
haa@igpp.de

Stefan Bordag
Universität Leipzig
Institut für Informatik
PO Box 920
04109 Leipzig, Germany
sbordag@informatik.
uni-leipzig.de

Dr. Hans-Jürgen Eikmeyer
Universität Bielefeld
Fakultät für Linguistik und Litera-
turwissenschaft
PO Box 10 01 31
D-33501 Bielefeld, Germany
hansjuergen.eikmeyer@
uni-bielefeld.de

Prof. Dr. Peter Gritzmann
Munich University of Technology
Centre for Mathematical Sciences
Gabesbergerstr. 43,
D-80333 Munich, Germany
gritzman@ma.tum.de

Prof. Dr. Gerhard Heyer
Universität Leipzig
Institut für Informatik
PO Box 920
D-04009 Leipzig, Germany
heyer@informatik.uni-leipzig.
de

Prof. Dr. Janusz Kacprzyk
Systems Research Institute
Deputy Director for Research
Polish Academy of Sciences
ul. Newelska 6
01-447 Warsaw, Poland
kacprzyk@ibspan.waw.pl

Prof. Dr. Walther Kindt
Universität Bielefeld
Fakultät für Linguistik und Litera-
turwissenschaft
PO Box 10 01 31
D-33501 Bielefeld, Germany
Walther.Kindt@uni-bielefeld.de

446   List of Contributors

Prof. George J. Klir
Binghamton University,
Department of Systems Science &
Industrial Engineering,
PO Box 6000
Binghamton, NY 13902, USA
gklir@binghamton.edu

Prof. Dr. Reinhard Köhler
Linguistische Datenverarbeitung
Universität Trier
Universitätsring 15
D-54286 Trier, Germany
koehler@uni-trier.de

Prof. Dr. Winfried Lenders
Institut für
Kommunikationsforschung und
Phonetik der Universität Bonn
Poppelsdorfer Allee 47 D-53115
Bonn, Germany
Lenders@uni-bonn.de

Prof. Dr. Edda Leopold
Hochschule für Angewandte
Wissenschaften Hamburg
Fachbereich Medienwissenschaften
Stiftstraße 69
D-20099 Hamburg
leopold@mt.haw-hamburg.de

Dr. Alexander Mehler
Juniorprofessor for Text Technology
Universität Bielefeld
Fakultät für Linguistik und Literaturwissenschaft
PO Box 10 01 31
D-33501 Bielefeld, Germany
Alexander.Mehler@
uni-bielefeld.de

Dr. Alfonso Medina-Urrea
Universidad Nacional Autónoma de
México
Instituto de Ingeniería (UNAM)

Ciudad Universitaria, Del. Coyoacán,
Apartado Postal 70-472, CP 04510,
Distrito Federal, México
AMedinaU@iingen.unam.mx

Prof. Dr. Dieter Metzing
Universität Bielefeld
Fakultät für Linguistik und Literaturwissenschaft
PO Box 10 01 31
D-33501 Bielefeld, Germany
Dieter.Metzing@uni-bielefeld.de

Dr. Leonid Perlovsky
Technical Advisor
Air Force Research Laboratory
Sensors Directorate / SNHE
80 Scott Dr.
Hanscom AFB, MA 01731, USA
Leonid.Perlovsky@hanscom.af.mil

Jens Pönninghaus
Universität Bielefeld
Fakultät für Linguistik und Literaturwissenschaft
PO Box 10 01 31
D-33501 Bielefeld, Germany
Jens.Poenninghaus@
uni-bielefeld.de

Prof. Dr. Gert Rickheit
Universität Bielefeld
Fakultät für Linguistik und Literaturwissenschaft
PO Box 10 01 31
D-33501 Bielefeld, Germany
Gert.Rickheit@uni-bielefeld.de

Prof. Dr. Jürgen Rolshoven
Informationsverarbeitung
Sprachliche Informationsverarbeitung
Albertus-Magnus-Platz
D-50923 Köln, Germany
rols@spinfo.uni-koeln.de

Kari Sentz
Los Alamos National Laboratory
P.O. Box 1663
Los Alamos, NM 87545, USA
ksentz@lanl.gov

Prof. Dr. Hans Strohner
Universität Bielefeld
Fakultät für Linguistik und Litera-
turwissenschaft
PO Box 10 01 31
D-33501 Bielefeld, Germany
Hans.Strohner@uni-bielefeld.de

Prof. Dr. Michael Stubbs
Universität Trier
Anglistik
Universitätsring 15
D-54286 Trier, Germany
stubbs@uni-trier.de

Prof. Dr. Wolfgang Wildgen
Fachbereich 10 Sprach- und
Literaturwissenschaften
Universität Bremen

PO Box 33 04 40
D-28334 Bremen, Germany
wildgen@uni-bremen.de

Prof. Dr. Lotfi Zadeh
Computer Science Division
Director, Berkeley Initiative in Soft
Computing (BISC)
University of California
Berkeley, CA 94720-1776, USA
zadeh@cs.berkeley.edu

Dr. Slawomir Zadrożny
Warsaw School of Information
Technology (WSISiZ)
ul. Newelska 6
01–447 Warsaw, Poland
zadrozny@wsisiz.edu.pl

Prof. Dr. Arne Ziegler
Karl-Franzens Universität Graz
Geisteswissenschaftliche Fakultät
Mozartgasse 8/II
A-8010 Graz, Austria
arne.ziegler@uni-graz.at

Index of Names

Abnous, R., 308
Ahuja, R. K., 108
Aigner, M., 110
Albus, J. S., 364, 365
Alexandrov, M., 280
Allan, K., 34, 44
Altmann, G., 16, 22, 192, 203, 206,
    210–212, 214, 215, 225, 445
Alvarado García, M., 278, 289, 291, 297,
    298
Andreev, N., 277
Aristotle, 5, 54, 305, 308, 367, 378
Atkins, B. T. S., 250
Atlan, H., 85, 87
Atmanspacher, H., 22, 79, 81–88, 445

Baayen, R. H., 126
Bacon, F., 233, 238–240, 251
Baeza-Yates, R., 4, 343, 345, 346
Baillet, A., 80
Balatoni, J., 83
Ball, J., 241, 242
Ballard, B. W., 34, 37
Ballmer, T., 433
Banarjee, S., 182
Bar Hillel, Y., 82
Barnbrook, G., 241–243, 247
Barwise, J., 17, 19, 34, 88, 145
Batali, J., 15
Bates, J. E., 84
Baxendale, P., 5
Bayes, T., 128
Beech, D., 263

Beesley, K. R., 267
Behaghel, O., 198
Belew, R. K., 343
Bellman, R. E., 365
Bennett, J., 5
Berry, L., 12
Berry, M. W., 123
Berry-Rogghe, G. L. M., 214
Bezdek, J. C., 62
Biber, D., 323, 325, 326
Biemann, C., 171
Biermann, A. W., 34, 37
Bird, S., 257, 258, 263
Birkemeier, S., 407
Black, M., 1
Boas, F., 277
Bonnesen, T., 99
Bookstein, A., 349
Boole, G., 340–342, 345, 346, 348, 353,
    354, 359, 406
Bordag, S., 12, 15, 19, 22, 171, 445
Bordogna, G., 340, 348, 350
Borgwardt, K. H., 108
Bosc, P., 340, 350
Bossaert, W., 277, 280–283
Bow, C., 258, 263
Brückner, T., 323
Brandenberg, R., 108
Brandt, P. A., 432
Brants, T., 181
Bray, T., 255, 264
Brill, E., 181
Brinker, K., 324, 325

Subject Index